Re-envisioning Remote Sensing Applications

Re-envisioning Remote Sensing Applications

Perspectives from Developing Countries

Edited by

Ripudaman Singh

CRC Press
Taylor & Francis Group
Boca Raton London New York

CRC Press is an imprint of the
Taylor & Francis Group, an **informa** business

First edition published 2021
by CRC Press
6000 Broken Sound Parkway NW, Suite 300, Boca Raton, FL 33487-2742

and by CRC Press
2 Park Square, Milton Park, Abingdon, Oxon, OX14 4RN

ISBN: [978-0-367-50239-3] (hbk)
ISBN: [978-1-003-04921-0] (ebk)

Typeset in Times
by Deanta Global Publishing Services, Chennai, India

Contents

SECTION II Hydrology, Microclimates and Climate Change Impacts

SECTION III Land Use/Land Cover Analysis Applications

Preface

The present volume, *Re-envisioning Remote Sensing Applications: Perspectives from Developing Countries*, is primarily concerned with providing up-to-date papers in all major remote-sensing application areas, and also to envisage future research agenda, showing the prospects of upcoming remote sensing practices. In all, eighteen chapters, covering almost all major fields of remote-sensing applications, as well as newly emerging uses, are presented in one place. More than fifty researchers from various developing countries have contributed to these chapters which are categorized into four broad overlapping areas of applications, namely *Agricultural, Soil and Land Degradation Studies; Hydrology, Microclimates and Climate Change Impacts; Land-Use/Land-Cover Analysis Applications;* and *Resource Analysis and Bibliometric Studies.* All contributed chapters, with applications of remote sensing in these broad areas, have been presented through case studies focusing on different parts of India, and analyzing varied applications of remote-sensing techniques, describing research from other developing countries, such as Bangladesh, Ethiopia, Nepal, Nigeria, and South Africa. Of the eighteen chapters, six chapters are penned by twenty-one international researchers from all these case study countries, as well as others, such as Canada and China.

Current developments in *geospatial technologies*, and their fusion with *information and communications technology (ICT)*, has been transforming the nature and analysis of geospatial data. Apart from the geographers, who use maps as their basic tool for interpretation of spatial analysis, *geographic information systems (GIS)* and *global positioning systems (GPS)* enabled the development of diverse maps, which are now readily available to everyone. Google Earth and other such spatial data providers are providing maps built into every smartphone as a basic tool to be used by anyone, to streamline the movements from one place to another in an easy and time-saving mode. Nowadays, all major subjects, irrespective of their spatial connotations, are using one or other variants of these easily obtainable maps. Diverse remote-sensing applications, on the other hand, have further stimulated the use of digital geospatial maps in wide-ranging subjects and domains. In addition to geographers, applications of remote sensing by *agriculture scientists, archaeologists, anthropologists, botanists, civil engineers, computer scientists, forestry researchers, geologists, hydrologists, meteorologists, soil scientists*, and other such *scientists/technologists* are increasing day-by-day, and in the current age of *Big Data, Machine Learning, Internet of Things (IoT)* and *Artificial Intelligence (AI)*, the remote-sensing applications are ever expanding. So, with advances and new innovations in *satellites, sensors*, and *ICT*, remote sensing has witnessed a huge transformation, with relevance to almost all domains of knowledge. Presently, it is in the phase of evolving interactions with *cloud computing, data sciences*, and *fusion of various technologies*, and these collaborations are bound to be strengthening and growing in varied remote sensing applications in the future.

Re-envisioning Remote Sensing Applications envisages these ever-increasing innovations in arenas of *geospatial technologies* and their uses. The ease of access to geospatial data has cumulative effects on its applicability in applied fields, such as *agricultural monitoring, archaeological excavations, disaster management, natural resource monitoring, forestry, global climate change, mineral and hydrological studies, smart city and urban & regional planning*, and other such diverse domains. The interface of remote sensing with advanced technologies, such as *IoT, smart digital* and *computing devices, smart mobile apps, tracking sensors* and others in the *learning environment* can be useful in accessing *smart learning.* Availability of *open source geospatial data* is leading to the processes and applications of remote sensing to *deep learning*, where its applications with programming languages such as *R, Python* and *AI* are classifying remote sensing data for varied applications. Extensive uses of *machine-learning algorithms* for remote-sensing data are generating actionable intelligence and surveillance monitoring nowadays. Various *civilian uses of GPS*, through automobiles and smartphones, are going to impact *smart remote-sensing applications* in the near future, and will expand with the progress in *smart city models* and *monitoring.* Advances in *drone, unmanned aerial vehicle (UAV) technologies*, and remote-sensing sensors are providing a range of geospatial information with high resolution, which are improving day by day. Remote sensing is bound to be re-creating itself with advanced technologies, and moving beyond human and Earth's horizons. Remote sensing of *micro spaces* as well as *extra-terrestrial spheres*, are predicted to increase RS applications at the micro levels and penetrating into extra-planetary explorations. The phase of '*post-remote sensing*' is envisioned where it would be moving beyond its traditional boundaries and reaching new horizons.

Ever-increasing need for monitoring and surveillance, particularly in the *post-COVID-19* scenario, is also bound to encourage novel remote-sensing applications, with more penetration of remote-sensing equipment through the digitally connected smart cities into the communities, localities, and individual buildings. Such remotely sensed data, which could be monitored from anywhere and at any time, is going to change its applications in the new environments. It is envisioned that remote-sensing applications will be providing Earth observations to support response and recovery in critical uses. These *re-envisioned remote-sensing applications* will certainly provide a new spectrum of various applications of remotely sensed data and geospatial techniques in the future. It is hoped that the present volume will provide scholars in the field of remote sensing, as well as those aspiring to explore the unknown, most of the resources and methodologies required in advancing the cause of research and developing new ideas for re-envisioning novel remote-sensing applications.

Ripudaman Singh

Acknowledgments

I extend my profound acknowledgments, first and foremost, to the Indian Institute of Remote Sensing and Indian Society of Remote Sensing, Dehradun, for providing me with the opportunity to initiate and accomplish this publication endeavor. I am grateful to the Taylor & Francis Group in general and CRC Press in particular (especially *Dr. Gagandeep Singh* and *Mr. Lakshay Gaba*) for guiding me, from time to time, to achieve the completion of this assignment. My sincere gratitude is due to my teachers, *Principal (retd.) Dr. Darshan Singh Manku Ji*, who stimulated the geographer in me and inspired me to pursue higher studies in geography; *Professor Emeritus Gopal Krishan Ji*, who always encouraged me to look beyond the horizon and to explore the unknown; *Professor Surya Kant Ji*, who taught me Remote Sensing courses during my studies and was instrumental in making my bonds with IIRS stronger and more enduring through the EDUSAT programs. Thanks are due to *Shri P.L.N. Raju*, Director NESAC, for always helping me with IIRS queries. I offer my gratitude to all the contributors, reviewers, and well-wishers, for their timely support, advice, and encouragement. Lastly, I am indebted to my inspiration, *my father* (*Late*) *Professor Baldev Singh Vig*, who introduced me to the field of Education and Scholarship, and my family, particularly my better half, *Dr. Manpreet Kaur* for her utmost support and for giving me ample time to accomplish this task. It is hoped that the scholars in the field of remote sensing and all those aspiring to explore the unknown, will find this book full of resources of value in furthering their cause of research and for re-envisioning novel remote sensing applications.

Acknowledgments

Editor

Dr. Ripudaman Singh, is Professor of Geography at Lovely Professional University. After completing his doctoral research from Panjab University, Chandigarh, India, with UGC-JRF/SRF fellowships, he has been a faculty member in various institutions, including Post Graduate Government College, Sector-11, Chandigarh (2005–2010); Central University of Punjab, Bathinda (2010–2011); Lyallpur Khalsa College, Jalandhar (2011–2012), and Lovely Professional University, Phagwara (since 2012). He has also served Government College, Hoshiarpur (2004) on an honorary basis and Panjab University, Chandigarh as Guest Faculty (2005–2006). He was the recipient of the 'Young Geographers Award' from the National Association of Geographers, India (NAGI) in 2005.

Dr. Singh has special interests in Remote Sensing and GIS, Development Geography, Regional Development and Planning, Geography of India, Political Geography, Applied Geography and Disaster Management. Since 2012, Dr. Singh has been serving as the Coordinator of IIRS-EDUSAT Programs at Lovely Professional University (LPU). He has published more than fifty research papers and articles in reputed international/national journals and has three books to his credit. His other two books are in press. Apart from this edited volume, he is also editing a book titled *Twenty-First Century India: Land, People, Economy*, being published by Springer. Along with various seminars, Dr. Singh has convened an International Conference at LPU and has edited two special issues of Scopus-listed journals, publishing the papers from that conference.

Dr. Singh specializes in the field of regional development. He has been working on regional disparities in India and analyzing how these disparities have evolved and how their nature, extent and intensity have varied over time and space. His PhD thesis on 'Trends in Regional Disparities in India since Independence: A Geographical Analysis' has been highly acclaimed and published as *Regional Disparities in Pre- and Post-Reform India* (Saptrishi, Chandigarh, 2016), and currently he is in receipt of an ICSSR-IMPRESS major research grant on Revisiting Regional Disparities in India 2020. Remote Sensing Applications in various fields of Geography, including Urban & Regional Planning, are among his core research areas.

Editor

Contributors

Pawan Agrawal
Dr. APJ Abdul Kalam Government College
Dadra and Nagar Haveli
Dadra and Nagar Haveli, India

F.O. Akinluyi
Department of Remote Sensing and GIS
Federal University of Technology
Akure, Nigeria

Abdullah-Al-Faisal
Department of Urban and Regional Planning. Rajshahi University of Engineering and Technology, Rajshahi, and
Climate change and Disaster Management Division, Center for Environmental and Geographic Information Services (CEGIS), Dhaka, Bangladesh

Abdulla-Al Kafy
Department of Urban and Regional Planning
Rajshahi University of Engineering and Technology (RUET)
Rajshahi, Bangladesh
and
ICLEI South Asia
Rajshahi City Corporation
Rajshahi, Bangladesh

Abdullah Al Rakib
Department of Urban and Regional Planning
Rajshahi University of Engineering and Technology (RUET)
Rajshahi, Bangladesh

Md. Arshadul Islam
Department of Urban and Regional Planning
Rajshahi University of Engineering and Technology (RUET)
Rajshahi, Bangladesh

Tahera Jahan Ashrafi
Department of Urban and Regional Planning
Chittagong University of Engineering and Technology
Chittagong, Bangladesh

Amanpreet Kaur Benipal
CSIR-CSIO
Chandigarh, India

Anil Bhardwaj
Punjab Agricultural University
Ludhiana, India

Bimal K. Bhattacharya
Agriculture and Land Ecosystem Division (BPSA Group; EOAPSA Area)
Space Applications Centre (ISRO)
Ahmedabad, India

A.G. Chachadi
Earth Sciences Department
Goa University
Goa, India

Lewlynn de Mello
The Energy and Resources Institute
St Cruz
Goa, India

O. Y. Ekundayo
Department of Geography and Environmental Science
University of Fort Hare
Alice, South Africa

and

Department of Meteorology and Climate Science
Federal University of Technology
Akure, Nigeria

Farhana Firdaus
School of Instrumentation Science and Opto-Electronics Engineering
Beihang University
Beijing, China

Dodge Getachew
Department of Geography and Environmental Studies
Debre Berhan University
Debre Berhan, Ethiopia

Tapan Ghosh
Punjab Remote Sensing Centre
Ludhiana, India

Tirthankar Ghosh
Amity Institute of Geo-Informatics and Remote Sensing (AIGIRS)
Amity University
Noida, India

Md. Hasnan Sakin Sarker
Department of Urban and Regional Planning
Rajshahi University of Engineering and Technology (RUET)
Rajshahi, Bangladesh

Md. Hasib Hasan Khan
Department of Urban and Regional Planning
Rajshahi University of Engineering and Technology (RUET)
Rajshahi, Bangladesh

Nur Hussain
Department of Geography and Earth Sciences
McMaster University
Ontario, Canada

Muhaiminul Islam
Department of Urban and Regional Planning
Rajshahi University of Engineering and Technology
Rajshahi, Bangladesh

A.M. Kalumba
Department of Geography and Environmental Science
University of Fort Hare
Alice, South Africa

Sanjay Kataria
Bennett University (Times of India Group)
Greater Noida
Uttar Pradesh, India

Jatinder Kaur
Department of Climate Change and Agricultural Meteorology
Punjab Agricultural University
Ludhiana, India

Ravinder Kaur
Department of Geography
Panjab University
Chandigarh, India

P. K. Kingra
Department of Climate Change and Agricultural Meteorology
Punjab Agricultural University
Ludhiana, India

Mahender Kotha
Earth Science Department
Goa University
Goa, India

Deepak Kumar
Amity Institute of Geo-Informatics and Remote Sensing (AIGIRS)
Amity University
Noida, India

Parmod Kumar
Remote Sensing and GIS Expert
Mahalanobis National Crop Forecast Centre
New Delhi, India

Neha Munjal
Lovely Professional University (LPU)
Phagwara, India

Md. Nazmul Huda Naim
Department of Urban and Regional Planning Chittagong University of Engineering & Technology (CUET)
Chattogram, Bangladesh
and
Bangladesh Rural Advancement Committee (BRAC)
Cox's Bazar, Bangladesh

Rahul Nigam
Agriculture and Land Ecosystem Division (BPSA Group; EOAPSA Area)
Space Application Centre (ISRO)
Ahmedabad, India

E. C. Okogbue
Department of Meteorology and Climate Science
Federal University of Technology
Akure, Nigeria

Israel R. Orimoloye
Centre for Environmental Management
University of the Free State
Bloemfontein, South Africa
and
Disaster Management Training and Education Centre for Africa
University of the Free State
Bloemfontein, South Africa

R. K. Pal
Punjab Agricultural University Regional Research Station
Bathinda, India

Ashwini Pai Panandiker
Earth Science Department
Goa University
Goa, India
and
The Energy and Resources Institute
St Cruz
Goa, India

Brijendra Pateriya
Punjab Remote Sensing Centre
Ludhiana, India

Abhishek Pathania
Punjab Engineering College
Chandigarh, India

Muhammad Rizwan
Department of Geography and Earth Sciences
McMaster University
Ontario, Canada

Sattyam
Remote Sensing and GIS Expert
Mahalanobis National Crop Forecast Centre
New Delhi, India

Harkanwaljot Singh Sekhon
Department of Climate Change and Agricultural Meteorology
Punjab Agricultural University
Ludhiana, India

Raj Setia
Punjab Remote Sensing Centre
Ludhiana, India

Md. Soumik Sikdar
Department of Urban and Regional Planning
Chittagong University of Engineering and Technology
Chittagong, Bangladesh

Ravinder Singh
Department of Geography
Panjab University
Chandigarh, India

Ripudaman Singh
Department of Geography
Lovely Professional University
Phagwara, India

Shiv Singh
Bennett University
Greater Noida, India

Som Pal Singh
Department of Climate Change and Agricultural Meteorology
Punjab Agricultural University
Ludhiana, India

Surender Singh
Lovely Professional University
Phagwara, India

Anil Sood
Punjab Remote Sensing Centre
Ludhiana, India

Sagar Taneja
Punjab Remote Sensing Centre
Ludhiana, India

Pawan Thapa
Department of Geomatics Engineering
Kathmandu University
Kathmandu, Nepal

Md. Yeamin Ali
Dan Church Aid
University of Rajshahi,
Rajshahi, Bangladesh

Prologue
Re-envisioning Remote-Sensing Applications

Ripudaman Singh

Since the dawn of humanity, it has piqued man's inquisitiveness to look beyond the horizon and to explore the unknown. This urge has been instrumental in the growth and expansion of human knowledge and the development of new technologies. In every age, there have been advances in technology, which augmented cultural and civilizational progress. Of the various revolutions that changed and aided humans to leap forward and advance their technologies further in accruing the gathered knowledge into new ideas and insights, the major ones included discovering fire; crafting the wheel; making the magnetic compass; developing the printing press; inventing the electric light; and creating the computer. With their respective discoveries and inventions, all these innovations revolutionized the progression of humanity to climb higher and higher (Table P.1).

First among all these technological innovations was the discovery of fire, which was substantial in illuminating the dark caves, providing warmth during the ice ages, protecting humans from wild animals and cooking raw plant and animal tissues into delicious, readily digestible food. It was one of the first and most important tools of humanity, discovered around 1.5 Mya (million years ago), that was essential in moving toward the evolution of cultures. Fire has supported all the other developments of modern technologies, such as ceramics, to metal working, to the nuclear industry and so on (Gowlett, 2016). Fire was the first and foremost discovery, which differentiated humans from other animals and from which man moved towards civilizational attainments. It is also argued that, distinctively, human language, mind and society emerged during a revolution, and that '*human revolution*' was accomplished by our ancestors somewhere between 70,000 to 200,000 years ago (Mellars and Stringer, 1989). All this resulted in the association of people and the spread of early human cultures over the globe. With the end of the last Ice Age (some 12,000 years ago), man moved out of the caves and started settling along the river channels and domesticating animals and cultivating the food grains. Eventually, it led to the '*agricultural revolution*' around 8,000 years before common era (BCE), when foragers became farmers (Barker, 2006). Thereafter, the rise of river valley civilizations was sustained by wheel crafting, which was useful not only in creating earthen potteries for storing the grains, but also in creating the wheel-driven cow carts, which were helpful in transporting the goods, materials and people from one place to another.

TABLE P.1
Technological Discoveries/Inventions and Revolutions

Technological Discovery/ Invention	Achieved	Preceded by or Leading to a Revolution	Changes Arisen
Fire (1.5 million years ago)	Illumination in dark, warmth from cold, safety from animals, cooking food	Human Revolution	Spread of human cultures
Wheel (5000 years ago)	Domestication of animals, crops, transportation	Agricultural Revolution	River valley civilizations
Magnetic compass (11–12th century)	Navigation and extended trades to Columbian exchange	Geographical Revolution	Age of discoveries and discovery of New World
Printing press (15th century)	Spread of ideas and information through printed material	Scientific Revolution and Industrial Revolution	Reformation and scientific thinking, industrialization and urbanization.
Electric light (19th century)	Illumination in homes, controlling heat, manufacturing industries, electric trains, photography	Modern Revolution	Modern way of life and modern society. Metropolitanization
Computers (20th century)	Fast computing, robotics, satellites, Artificial Intelligence etc.	Digital and Information Revolution	Computerization, Information and communications technology, globalization

These frequent movements of people also required better understanding of directions and locations. This led to the growth and production of geographical knowledge. Slowly, the well-traveled trails evolved into caravan routes, with the most prominent one being the 'Silk Road' linking the eastern and western civilizations, for trade and movement of people and ideas (Hansen, 2012). Making magnetic compasses furthered the movements of people in the right directions and the extension of navigation to open seas. The magnetic compass was instrumental in the discovery of new lands and riches by the medieval Europeans, leading to exploration, new discoveries and innovations (Gurney, 2004). In a way, it was a true 'geographical revolution' that brought the age of discoveries, new explorations and many new technological innovations to follow. The mid-fifteenth century witnessed another revolution, with Gutenberg's printing press; with this invention, the creation of knowledge and books were no longer restricted to the elite. The spread of ideas and information through printing became an agent of change (Eisenstein, 1980). Europe started changing from medieval to the modern age. It further brought the age of reformation, the age of discoveries and of scientific inventions (More, 2000). Accumulating knowledge and ideas begot new ideas and inventions ushering in the 'scientific revolution' (Cohen, 1994). This scientific revolution merged with elements of enlightenment, generating an industrial erudition, which ultimately gave rise to the new technologies of the 'industrial revolution' (Mokyr, 2008, 2012).

As fire was instrumental in revolutionizing humanity in the first place, the invention of electric light re-revolutionized modern civilization in the nineteenth century. Electric light was able to produce visible light from electric current through Humphry Davy's invention of incandescent light in 1802. Earlier, Alessandro Volta had invented the electric battery in 1800. Thomas Edison became successful by patenting the electric light bulb in 1879 (Guarnieri, 2015). Electric light illuminated homes, rotated fans to dispel heat, electric furnaces manufactured goods day and night in factories, electric trains moved faster and faster, so that everything started getting done faster, furthering the advance of automation. The 'modern revolution' led to the transformations of modern life, which couldn't be imagined without the climax of the industrial revolution, reached through the usage of fossil fuel energy resources and the invention of electric light. The speed of these transitions was a distinctive feature of the modern revolution, where the pace of change has also increased; indeed, this change has been so decisive that it forces us to approach the modern revolution differently from all earlier revolutions (Christian, 2004). Apart from electric lighting, one of the major and foremost uses of light was made in the invention of photography, when Nicéphore Niépce (1827) became successful at capturing the first preserved photograph of a real-world scene, 'The view from the window at Le Gras' (Baatz, 1997). Photographs taken by pigeons and *via* air balloons became the first remote-sensing experiments.

Last of all the six major revolutions, 'creation of computers' has totally transformed modern life. The evolution of computers from simple computing calculations to the age of artificial intelligence has certainly led to the evolution of the ‘digital revolution’, which could be attributed to the 1970s to 1990s, when enterprise computing and communication technologies became deployed within businesses and corporations, and post-1990s computerization transitioning from operations to the strategy segment (Andriol, 2005). Computers have transcended human thought, even coupling it with newer and newer inventions and discoveries. Computerization has led to cumulative developments and progression in every sphere, with an ‘infosphere’ being created with the ‘information revolution’ (Floridi, 2014). Among all the above-mentioned technological discoveries and inventions, the last three (printing press, electric light and computers) have completely changed and transformed the world through industrial, modern and digital/information revolutions. Production of knowledge and information has been accumulating in each case and this accumulated knowledge is now amassing beyond the boundaries of common human thought.

With the invention of the camera and photography, real-world objects could be captured through photographs and the information in these images could be analyzed and utilized from planning and defense perspectives. Even before the invention of camera and photographs, studies related to colors of light, its reflectance and images of objects obtained through their interactions with light beams were underway a century before. Table P.2 depicts some of the key milestones related to the advances leading to the evolution and growth of remote sensing. These milestones had a great bearing on the growth of remote sensing techniques in the later periods. By the mid-eighteenth century, Johann Heinrich Lambert was one of the first scientist in this field, and, in his famous work ‘*Perspectiva Liber*’ (1759), highlighted principles related to photogrammetry. These principles were useful in the interpretation of aerial photographs and later with respect to satellite imageries. Remote sensing is also an aligned technology, like photogrammetry, which derives information from such images. By the turn of the nineteenth century, in 1800, infrared radiation had been discovered by Sir William Herschel, while measuring the temperatures of different parts of sunlight passing through a prism, where he found that light at wavelengths beyond the red spectrum were invisible infrared radiation. Further details about the infrared spectrum were later revealed by J.B.L. Foucault in 1847, and by 1873, entire electromagnetic spectrum had been explained by J.C. Maxwell. A century later, thermal infrared remote sensing emerged as an important field of remote sensing (Colwell, 1983). Meanwhile, photography evolved in the 1820s in France through the efforts of Nicéphore Niépce in 1827 and became established as a profession by 1839. Aerial photography started with Gaspard-Felix Tournachon in 1858, who came up with the idea of mapmaking and surveying through aerial photographs as early as 1855 (PAPA, 2020).

The American Civil War (1861–65) provided a platform for American balloonists to take more and more aerial photographs to estimate their opponents’ strengths. This initiated the military and surveillance usages of initial remote sensing through balloons. Aerial photography, through the use of kites, was developed by English meteorologist E.D. Archibald in 1882, when he used a string of kites with a camera attached to the last one (PAPA, 2020). As happened during ancient and medieval times, carrier pigeons were the most expedient messengers

TABLE P.2
Key Milestones in Remote-Sensing Advances

1759 *Perspectiva Liber* by J.H. Lambert (Photogrammetry)	**1957** First satellite, *Sputnik*, launched by the USSR
1800 Infrared radiation discovered by Sir William Herschel	**1958** First American satellite, *Explorer 1*, launched
1827 First photograph taken by Nicéphore Niépce	**1959** First photograph of Earth from space
1839 Photography begins as a profession	**1960** *TIROS* satellite launched (remote-sensing use)
1847 Infrared spectrum of sunlight by J.B.L. Foucault	**1964** Multiband photography for Earth resources (NASA)
1858 Aerial photograph by Gaspard-Felix Tournachon	**1972** *Earth Resources Technology Satellite* by US
1859 Aerial photography from balloons starts	**1973** *Skylab* US space station launched
1873 Electromagnetic spectrum by J.C. Maxwell	**1982** *Thematic Mapper* with 30 m resolution and seven bands
1882 Aerial photography from kites (E.D. Archibald)	**1986** *Hyperspectral sensors* developed
1902 Airplane designed by the Wright brothers	**1986** *SPOT 1* satellite launched by France
1903 Pigeon carrier camera designed (J. Neubronner)	**1988** Indian remote-sensing satellite *IRS-1A* launched
1909 Photography starts from airplanes	**1990** *RADARSAT* projects in Canada
1916 Aerial reconnaissance in World War I	**1991** *European Remote-Sensing Satellite* (*ERS-1*)
1934 American Society of Photogrammetry established	**1999** *Landsat 7* with Enhanced Thematic Mapper Plus
1935 Radar developed in Germany	**1999** *IKONOS* with 1-m resolution launched
1940 Applications of non-visible Part of EMS	**2001** High-resolution commercial satellite *QuickBird*
1944 Manual of photogrammetry by American Society	**2006** *Cloud Computing EC2* by Amazon
1947 Spectral reflectance of natural material by *Krinov*	**2008** *GeoEye 1* (0.41- to 1.65-m resolution) launched
1950 Military research and development in Cold War	**2009** *Bhuvan*: Gateway to geospatial world launched by India
1950s Evelyn Pruitt coined the term "remote sensing"	**2019** *Google Earth Engine* for satellite images data
1956 R.N. Colwell used infrared for plant disease detection	**2020** *EagleView* processed its 100 millionth image

for delivering information and messages before the development of the modern postal service, and Julius Neubronner was successful at creating miniature pigeon carrier cameras in 1903 (PDR, 2020). These miniature cameras were designed so that individual exposures were taken at regular intervals, linked to the folding of the pigeon's feathers.

In the meantime, the Wright brothers designed the first successful airplane in 1902, and such airplanes were able to capture aerial photographs by 1909. World War I (1914–18) provided an appropriate testing ground for aerial photography through aerial reconnaissance. The post-war phase witnessed the establishment of The American Society of Photogrammetry in 1934, and radar technology was developed before World War II by 1935 in the US, Germany and other European countries. During the second world war, military surveillance gained prominence and, by 1940, applications in non-visible parts of the electromagnetic spectrum started. Krinov (1947) characterized the spectral reflectance of natural materials. With this, identification of different spectral signatures advanced the remote-sensing applications further.

For two centuries, from the mid-eighteenth (1759) to the mid-twentieth century (1950s), much progress was made, from the principles of photogrammetry (by Lamberts) to the beginning of photography to aerial photography, leading to the '*remote sensing*' word being coined by Evelyn Pruitt in the 1950s, to describe the processes involved in the identification, observation and measurement of radiation from the objects on the Earth's surface. These observations using various sensors, like cameras, scanners, radar, Lidar, etc. through different platforms (viz. balloons, kites, aircrafts, satellites, drones, unmanned aerial vehicles or UAVs), from a considerable height from the Earth's surface (ranging from a few hundred meters to thousands of kilometers). These observations are recorded on a suitable medium, such as photographic film, videotape, magnetic or digital formats; as the reflectance of each and every object is different, so there are different spectral signatures for each and every object over the Earth's surface. Remote sensing is all about acquiring the information about any object or phenomenon over the Earth's surface, without making any physical contact with the object.

Remote sensing as a subject became established in the 1960s. The Cold War era (1950–1990) witnessed the competition between the United States and the USSR, when both superpowers were trying to dominate through technological advancements over each other. *Sputnik* and *Explorer* were sent to space by the Soviets and the Americans in 1957 and 1958, respectively. TIROS (Television and Infra-Red-Observation Satellite) was the first US meteorological satellite (for remote-sensing purposes) launched in 1960 for global weather observations. With the improvements in satellite technologies in the 1960s and 1970s multiband photography was started in 1964 by NASA and the *Skylab* space station was launched in 1973 by the United States. The 1980s witnessed the improvements in resolution (30 m) and increased bands (seven) with thematic mapper and hyperspectral sensors. From the mid-1980s to 1990–91, France (*SPOT 1* in 1986), India (*INSAT IRS 1A* in 1988), Canada (*RADARSAT* in 1990) and the European Union (*ERS 1* in 1991) launched their indigenous remote-sensing satellites, which augmented regional advancements in remote-sensing technologies; in particular, India has emerged globally as a leading producer of remote-sensing data. Moreover, it has developed indigenous space technology, which is amongst the best and most cost-effective remote-sensing technologies (Foust, 2006).

With the turn of the new millennium, again there were major advances in remote sensing as the Enhanced Thematic Mapper Plus (ETM+) was launched through *Landsat 7* in 1999. In addition, the commercial remote-sensing satellites *IKONOS* (1999) and *QuickBird* (2001) were helpful in achieving greater precision in remote-sensing data with resolution of less than 1-m. If we compare the spatial resolution of satellite imageries, four decades ago, Thematic Mapper (1982) had a spatial resolution of 30 m, and now *GeoEye* (2008) has a resolution of less than 0.5 m, with an increased number of bands giving greater precision and accuracy in spatial data. With the introduction and widespread adoption of *cloud computing* (2006 onwards), the availability of computer system resources and online data sharing, the storage and usage of spatial data have increased many fold, and this has been further increased with the launch of *Google Earth Engine* in 2019. As geospatial data can be collected and analyzed through various sensors and methodologies, in addition to the traditional ones (Global Positioning System [GPS]), surveying, photogrammetry, remote sensing), radar/Lidar scanning, mobile mapping, geo-locations/geo-tagging and volunteered geographic information (VGI) etc., the amount and volume of archival (geo)spatial data of petabyte-scale have become freely available from international agencies such as the *Copernicus Program* of the European Union, the USGS, NOAA of America and *Bhuvan* (2009) Geoportal of ISRO, India. Supplementary accessibility of spatial data, through the flood of big data in all fields of investigations, is also posing a challenge to efficient and scalable processing of remotely sensed data in the context of its various applications (Sun et al., 2019). This requires a re-look at traditional remote sensing and its fusion with the big data, data sciences and other futuristic artificial intelligence (AI) methodologies.

The idea for compiling this edited volume originated from the fact that the rapid changes in technologies were also prompting their improved and enlarged applications in various domains, with remote sensing not being an exception to this change. *Re-envisioning Remote-Sensing Applications: Perspectives from Developing Countries* aims at providing varied application themes of remote sensing, which are not so traditional, but are new with respect to the upcoming technologies. In addition, the new areas of research, which are being brought forward for the creation of new knowledge, are applying remote-sensing techniques for spatial data analysis and usage. Broadly speaking, remote-sensing applications can be grouped into six major groups, such as agricultural, soil and land degradation studies; hydrology, microclimates and climate change impacts; land use/land Cover analysis; natural resource management; urbanization and urban problem analysis; and applications in disaster management, models and planning (Figure P.1). Considering the limited size of the present volume, only the first four broad applications are being presented here, while others will be presented in another volume of such remote-sensing applications. Apart from the introductory prologue and the concluding epilogue, the present book is organized into four sections of overlapping areas of research, having eighteen chapters on varied remote- sensing applications:

Prologue: Re-envisioning Remote-Sensing Applications
I *Agricultural, Soil and Land Degradation Studies*
II *Hydrology, Microclimates and Climate Change Impacts*
III *Land Use/Land Cover Analysis Applications*
IV *Resource Analysis and Bibliometric Studies*
Epilogue: Future Research Agenda

The Prologue on *Re-envisioning Remote-Sensing Applications* provides the background ideas for this book and its overall structure, and presents brief overviews of individual chapters. As described earlier, it was the technological advances in each age, that have augmented the cultural and civilizational progress. The new ideas, spawning new innovations and advanced technologies, have led to the culmination of important revolutions in history, including the *human revolution, agricultural revolution, geographical revolution, scientific/industrial revolution, modern revolution and digital/information revolution.* The evolution of improvements in computers have exceeded human thinking in modern times, with the accumulation of knowledge and information to innovate new technologies in the past two and half centuries having advanced remote sensing to higher and newer trajectories. The present volume is a collection of 18 different chapters, grouped into four sections with respect to their remote-sensing applications, which are detailed below.

Section I of the book, entitled *Agricultural, Soil and Land Degradation Studies*, comprises four chapters related to crop yield estimations, agrometeorological applications and degradational impacts of urban expansion. Chapter 1, "Wheat Yield Prediction through Spectral Indices, Using Agro-Meteorological Data" classified the images and pixels related to the wheat crop for generating the wheat mask for generating the profiles for spectral indices. Thus, yield predicted from the spectral index models was found to be closer to the actual yield, where the percentage relative deviation was found to be maximal in the case of Normalized Difference Vegetation Index (NDVI), as compared with the minimal value for Infrared Percentage Vegetation Index (IPVI), indicating the suitability of IPVI for providing the best yield prediction for wheat in the study area. Chapter 2, "Impact of Variation in Agro-Meteorological Parameters on Wheat Production, and Comparison of Yield Predicted by Agro-Meteorological Indices with Remote Sensing," for the same study area computed six meteorological indices. The yield predicted from the agro-meteorological index models were found to be closer to the actual yields for both the drought and wet years. Minimum relative deviation occurred in the cases of Accumulated Growing Degree-days (AGDD) and Photo Thermal Units (PTU) for the drought and wet years, respectively, and maximum relative deviation was found for Helio Thermal Units (HTU) in both years. Agro-meteorological multi-regression model prediction of wheat yield for the wet year was closer to the actual yield as than for the dry year. Predicted results of wheat yield by spectral vegetation indices of remote sensing data were closer to the actual yield than those predicted by agro-meteorological indices. Chapter 3, "Geostationary Meteorological Satellites for Agrometeorological Applications in India," found that retrieved meteorological and value-added agrometeorological products and indicators from the Indian geostationary satellite (*INSAT, K1*) approved to be valuable tools for assessing crop growth and condition in near-real time, which helps planners to make decision on the regional to national scale. For the future, high-resolution and multi-spectral data from ISRO, geostationary missions such as INSAT 3D-S and GISAT, will promise more and more accurate agrometeorological parameters to guide agriculture decision-making systems more accurately. The last chapter of this section, Chapter 4, "Prediction of Urban Expansion and Identification of its Impacts on the Degradation of Agricultural Land: A Machine Learning-Based Remote-Sensing Approach in Rajshahi, Bangladesh," applies a cellular automata (CA) model to depict LULC analysis, suggesting a significant increase in the built-up area (+11.71%) and a corresponding decrease in agricultural land (–12.40%) in the study area. The significant loss of agricultural land destabilizes biodiversity, unbalances the ecosystem services, and increases the food insecurity.

Section II, *Hydrology, Microclimates and Climate Change Impacts,* comprises four chapters on remote-sensing applications, with case studies on Sukhna Lake, southwestern Punjab, the Sundarbans region and Bangladesh. Chapter 5, "Spatio-Temporal Variations of Water Surface Temperature and its Effect on the Microclimate of Sukhna Lake in Chandigarh (India)," depicts water surface temperature as an important proxy measure of climate variability. The study

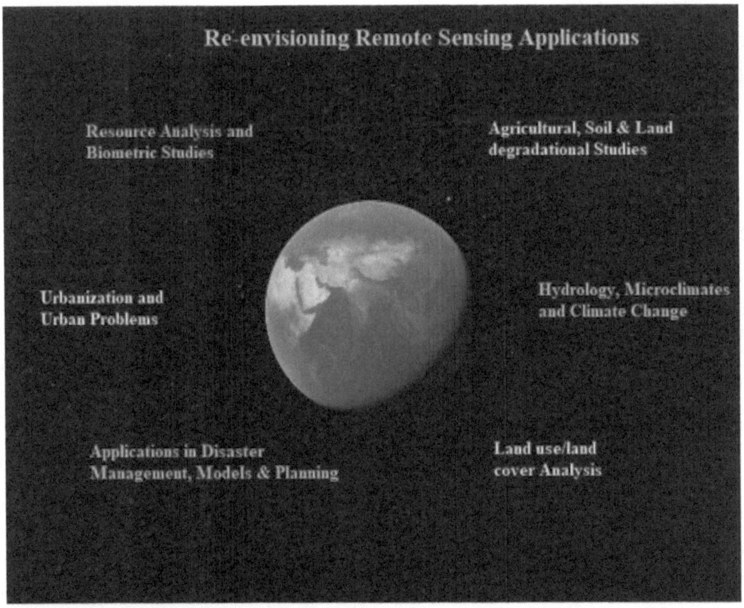

FIGURE P.1 Re-envisioning Remote Sensing Applications.

analyzed water surface temperature of the Sukhna Lake between 2001 and 2018, using Landsat satellite data (Landsat 5, 7 and 8) and its effect on the microclimate of the lake. Geospatial analysis of the lake water surface found a temperature increase by 1.98 °C from 2001 to 2011 and by 3.48 °C from 2011 to 2018. An increase in water surface temperature is mainly attributed to urbanization and locally rising air temperatures. The microclimate effect of lake water was found to be significant up to 360 m, cooling the environment by 0.32, 0.17 and 0.14 °C in the years 2001, 2011 and 2018, respectively. These findings suggest that water surface temperature has a significant effect on the water budget (mainly *via* evaporation), water storage, climate change and the lake ecosystem at the micro level. Chapter 6, "Assessing the Impact of Climate Variables on Soil Salinity, Using Remote Sensing and GIS," analyzed the salt-affected areas in the parts of south-western Punjab, which were classified using Landsat satellite data during the pre-monsoon, monsoon and post-monsoon seasons for the years 2003, 2008 and 2013. Normalized Difference Salinity Index (NDSI) was used for automatic extraction of salinity from the satellite images. In general, it was found that salinity was higher during the pre-monsoon seasons and the area of the salt-affected region was highest during the pre-monsoon season of 2003. The overall accuracy of salt-affected areas was 85%, with a Kappa coefficient of 0.35. The values of NDSI, rainfall and temperature corresponding to ground truth points were extracted from the raster images and the relationships among these variables indicated that NDSI was significantly negatively correlated with rainfall but significantly positively correlated with temperature. It also suggested that integration of salinity with climate variables and with other factors, like slope and depth to water level etc., in GIS may help in decision-support system developments, which would be useful for policy makers and scientists in planning the strategies to manage the salt-affected areas and soil resources. Chapter 7, "Monitoring Climate Change-Induced Variations in the Indian Sundarbans Region, Using Geospatial Sciences," finds the Indian Sundarbans to be one of the most vulnerable ecosystems facing climate change and global warming challenges. Significant loss of land between the years 2000 and 2020 in some of the southernmost islands of the region namely, Sagar, Ghoramara, Bulchery, Jambudwip, etc., was observed. Chapter 8 of this section, "Remote Sensing of Photosynthesis, Vegetation Productivity and Climate Variability in Bangladesh," focused on remote-sensing-based seasonal vegetation productivity and photosynthetic activity, employing satellite retrieved solar-induced chlorophyll fluorescence (SIF) observations in Bangladesh. This study used GOSIF satellite and SIF to explore photosynthetic function, and to show variability in photosynthetic activity over Bangladesh from 2005 to 2018, with a parallel comparison of satellite-retrieved vegetation greenness, using additional satellite measurement of Gross Primary Production (GPP), Vegetation Indices (Vis), and Land Surface Temperature (LST) variability. Moderate Resolution Imaging Spectroradiometer (MODIS)-based LST measurement was also used for assessing climatic response. The findings highlighted the dependency between vegetation productivity and photosynthetic activity, which might be a better approach for future phenological research.

Section III, *Land Use/Land Cover Analysis Applications*, comprises six chapters dealing with two case studies from India, i.e., the Chandigarh periphery and Goa; and one study from each of Nepal, Bangladesh, Nigeria and Ethiopia. Chapter 9, "Rural Land Transformation in Chandigarh Periphery: A Spatio-Temporal Analysis," highlights land transformation in Chandigarh's periphery, using measurements obtained from satellite images, classified according to land use, from the past forty years. Land use and land cover changes and trends of urban built-up area expansion, its pattern of change, growth and urbanization, and the impact of Chandigarh on its surrounding periphery is thoroughly analyzed, which is a pre-requisite for better urban and regional planning. Chapter 10, "Examining the Temporal Change in Land Cover/Land Use in Five Watersheds in Goa, India, using a Geospatial Approach," discusses how hydrologic response of watersheds to physical (land use) and climatic (rainfall and air temperature) change is a critical element of water resource planning and management. Considering this, the watershed areas of five rivers (the Kalna, Sal, Talpona, Galjibag, and Khandepar) across the state of Goa were analyzed from 1977 to 2018 through the use of remote-sensing data from *LISS III*, *Landsat 1–5*, and *Landsat 8*. Analysis of the most recent data, from 2009 to 2018, shows a decrease in the forest cover in all five watershed areas, except in the Galjibag watershed. These results can provide useful insights in predicting future land-use scenarios to carry out hydrological modeling exercises on these five watershed areas and on the prediction of water flows. Chapter 11, "Impacts of Land Use and Land Cover Change on Land Degradation: A Case Study in Dolakha District, Nepal," using ArcGIS with *Landsat* images and Open Data Nepal, ICIMOD land use and cover data produced the LULC map of Dolakha district. The study found that barren land had significantly increased in the study area by 7%, 27% and 22% during the years 2001, 2010 and 2018, respectively, leading to land degradation. This assessment provides the land status and guidelines for proper land management in the Himalayan country. Chapter 12, "A Remote-Sensing-Based Approach to Identifying the Influence of Land Use/Land Cover Change on the Urban Thermal Environment: A Case Study in Chattogram City, Bangladesh," finds urbanization leading to massive land use/land cover changes and significantly contributing to the increase in urban land surface temperature (LST). It identified the variation in LULC, and its influences on the urban thermal environment such as LST, urban heat island (UHI), and urban thermal field variance index (UTFVI) for the city of Chattogram, Bangladesh, through multi-temporal Landsat Thematic Mapper/Operational Land Imager (TM/OLI) satellite images for the years 1999, 2009 and 2019. Using the support vector machine algorithm, the LULC was classified, and the accuracy of the categorized maps was found to be greater than 85%. The spectral radiance model was used to extract urban thermal information from satellite images. The analysis of LULC estimates suggests a significant net increase in urban built-up areas (+3.51%) and a reduction in vegetation cover (−6.81%). The mean LST distribution shows that built-up areas were recorded in a high-temperature zone, followed by the lowest temperature in vegetated cover and water bodies and a trend of increasing surface heat stress in the study area. Chapter 13, "Modeling Land Use/Land Cover Change Using

Multi-Layer Perceptron and Markov Chain Analysis: A Study on Bahir Dar City, Ethiopia," depicted that, for sustainable urban and regional development, land use/land cover (LULC) information of urban centers has become vital. The LULC change of Bahir Dar city, using remote-sensing satellite images for 1999, 2009 and 2019, and ancillary data, were used. The model validation (crosstab and validation modules), using the reference map of 2019, provided overall Kappa coefficients of 0.8391 and 0.8293 for scenarios I and II, respectively, which allowed prediction of LULC for 2029. The change analysis result revealed that the study area has been experiencing rapid land transformation since 1999, especially in being transformed from agricultural land to built-up areas. Analysis revealed a more pronounced conversion prediction for the upcoming decade, especially in the urban periphery. So, a coordinated effort is required from all stakeholders for sustainable urban development. Chapter 14, "Geoinformatics Approach in Desertification Evaluation Using Vegetation Cover Changes in the Sudano-Sahelian Region of Nigeria from 2000 to 2010," used Normalized Difference Vegetation Index (NDVI) to monitor changes in vegetation cover related to desertification in the Sudano-Sahelian region of Nigeria using MODIS, specifically the MOD13A3 monthly Vegetation Index (Terra images) data at 1000 m pixel size for the years 2000–2010. All through the study years, the classes depicting Classes E and F vegetation cover decreased, while there was an increase in classes that represented scanty vegetation, such as desert and bare-land, steppe, except for the semi-desert class, which decreased, giving way to a nearly corresponding increase in the steppe class, hence confirming the occurrence of desertification in the study region during the study period.

Section IV, *Resource Analysis and Bibliometric Studie*s, comprises four chapters in all. Chapter 15, "The Role of Geospatial Technology in Crop Growth Monitoring and Yield Estimation," highlighted that assessing crop growth during the growing period and predicting crop productivity is gaining importance for estimation of seasonal production prior to crop harvest. In developing countries, particularly those with greater agricultural predominance, reliable and timely estimates of crop production as a result of different types of abiotic stresses/climatic aberrations, namely heat stress, drought or pest infestation etc., are particularly important. Advance estimates of such adverse effects can be of great help for strategic planning and measures to avoid any food shortages in the country. A good review of the related literature, using remote-sensing techniques for crop estimations, is made. Chapter 16 primarily focuses on A Review of Remote-Sensing Applications for Crop Residue Burning Assessments. Considering the major problems being posed by crop residue burning, apart from the environmental and health issues, residue burning also hampers economic growth and development activities, particularly in the developing countries. Crop residue burning produces large amounts of pollutants and greenhouse gases, that have adverse impacts on human health, air quality and climate on the global as well as regional scales. Applications of remote-sensing technology in the assessment of crop residue burning is an upcoming research topic. The literature is reviewed through the sub-themes of monitoring crop residue burning, biomass estimation and estimation of gases emitted due to residue burning. Chapter 17, "Indian Water

Resources Research, Using Remote Sensing as Reflected by the Web of Science during 2009–2018: A Bibliometric Study," presents a bibliometric analysis of Indian research on water resources cited in the Web of Science over the period in question. Findings show that more than 100 countries were involved in collaborative research on water resources, using remote-sensing applications. Chapter 18, "Research Practice on Remote Sensing: A Bibliometric Study of Indian Scholars," presents the bibliometric analysis of remote-sensing research by Indian scholars, with publications from the well-known database, Web of Science for the 10-year period, 2009–2019. In total, there were 3282 documents retrieved, from 50 sources and contributed by 5553 authors. The majority of the papers were multi-authored, with a collaboration index of 1.75. V.K. Dadhwal of the Indian Institute of Space Science and Technology, Kerala, was found to be the most prominent author (in terms of the number of citations) amongst all the reviewed researchers.

Lastly, towards the end of the book, *Epilogue: Future Research Agenda* projects the agenda for future research in remote-sensing applications. It is envisioned that, in the coming years, there would be more and more remote-sensing applications with cloud computing, big data and Internet of Things. Open source remote sensing and the use of programming languages, such as R, Python and Artificial Intelligence, would grow to new, higher levels. Smart remote sensing, with respect to better interfacing with GIS and usage with smart phones is also anticipated. Fusion remote sensing with data integrations would be a promising means of obtaining high-quality spatio-temporal remote-sensing products and analysis. Four-dimensional remote sensing, as well as extra-terrestrial applications of remote sensing are also envisaged. The current COVID-19 pandemic will also have its impacts on remote-sensing applications. More remote-sensing applications on pandemic effects will be developed. Global lockdowns, due to COVID-19, have improved environmental conditions and air pollution has been drastically reduced all over the world. Comparisons and change detections during the lockdowns and earlier satellite data would be the major focus for upcoming research. The future research applications of remote sensing would reach a stage of '*post remote sensing*', where traditional remote-sensing applications will be merging into the innovative arenas of extra-terrestrial as well as ultrasmart-phone-based analysis and applications. It would be focusing on other Earths (other planets) and side-by-side spatial data would be readily getting analyzed and processed by next-generation ultrasmart phones. It is anticipated that the rapid changes in remote-sensing technologies would be triggering improved and expanded applications in various domains. *Re-envisioning Remote-Sensing Applications: Perspectives from Developing Countries* is presented here to collectively showcase the varied applications of remote sensing and to envisage their interaction with upcoming technologies.

REFERENCES

Andriol, S. J. (2005). *The 2nd Digital Revolution*. Cybertech Publishing: Hershey.
Baatz, W. (1997). *Photography: An Illustrated Historical Overview*. Barron's: New York.

Barker, G. (2006). *The Agricultural Revolution in Prehistory: Why did Foragers Become Farmers?* Oxford University Press: Oxford.
Christian, D. (2004). *Maps of Time: An Introduction to Big History.* University of California Press: Berkeley.
Cohen, H. F. (1994). *The Scientific Revolution: A Historiographical Inquiry.* University of Chicago Press: Chicago.
Colwell, R. N. ed. (1983). *Manual of Remote Sensing, Volume I: Theory, Instruments and Techniques, Second Edition.* American Society of Photogrammetry and Remote Sensing ASPRS: Falls Church.
Eisenstein, E. L. (1980). *The Printing Press as an Agent of Change.* Cambridge University Press: Cambridge.
Floridi, L. (2014). *The Fourth Revolution: How the Infosphere is Reshaping Human Reality.* Oxford University Press: Oxford.
Foust, J. (2006). The Other Rising Asian Space Power. *The Space Review: Essays and Commentaries About the Final Frontier in Association with Spacenews.* Retrieved from: https://www.thespacereview.com/article/768/1
Gowlett, J. A. J. (2016). The Discovery of Fire by Humans: A Long and Convoluted Process. *Philosophical Transactions B*, 371: 20150164. http://dx.doi.org/10.1098/rstb.2015.0164
Guarnieri, M. (2015). Switching the Light: From Chemical to Electrical. *IEEE Industrial Electronics Magazine*, 9(3): 44–47. doi:10.1109/MIE.2015.2454038
Gurney, A. (2004). *A Story of Exploration and Innovation.* WW Norton: New York.
Hansen, V. (2012). *The Silk Road: A New History.* Oxford University Press: New York.
Mellars, P. and Stringer, C. eds. (1989). *The Human Revolution.* Edinburgh University Press: Edinburgh.
More, C. (2000). *Understanding the Industrial Revolution.* Routledge: London.
Mokyr, J. (2008). The Institutional Origins of the Industrial Revolution. In Helpman, E. ed. *Institutions and Economic Performance.* Harvard University Press: Cambridge, MA, pp. 64–119.
Mokyr, J. (2012). *The Enlightened Economy: An Economy History of Britain, 1700–1850.* Yale University Press: Yale.
PAPA (2020). History of Aerial Photography: Dark Room in a Hot-Air Balloon. *Professional Aerial Photographers Association. International.* Retrieved from https://www.professionalaerialphotographers.com/content.aspx?page_id=22&club_id=808138&module_id=158950&sl=1562403579
PDR (2020). Dr. Julius Neubronner's Miniature Pigeon Camera. *The Public Domain Review.* Retrieved from https://publicdomainreview.org/collection/dr-julius-neubronner-s-miniature-pigeon-camera
Sun, J. et al. (2019). An Efficient and Scalable Framework for Processing Remotely Sensed Big Data in Cloud Computing Environment. *IEEE Transactions on Geoscience and Remote Sensing*, 57(7): 4294–4308.

Section I

Agricultural, Soil and Land Degradation Studies

1 Wheat Yield Prediction through Spectral Indices, Using Agro-Meteorological Data

Amanpreet Kaur Benipal, Anil Sood and Surender Singh

CONTENTS

1.1 INTRODUCTION

India is the second-largest wheat producer in the world after China, followed by the United States of America, France and the Russian Federation. India's share of global wheat production and exports was recorded at 11.78 percent and around 0.40 percent in the year 2015–16, respectively. Geospatial technology involves tools that are increasingly used for predictive analyses, such as the prediction of natural resources, like wheat yield. Remote-sensing techniques have demonstrated their potential in providing information on the character and distribution of various natural resources. Possible areas of application related to agriculture are crop acreage and production forecasting, management of land and water resources, crop condition assessment and estimation of damage caused by floods, drought, disease epidemics, etc. (Sahai and Dadhwal 1990, Premalatha and Nageshwararao 1994).

The aim of the present study is to analyze the variation in agro-meteorological parameters during low and high rainfall *Rabi* seasons and their impact on wheat production using remote sensing and geographic information systems (GIS) and to compare the wheat production results obtained, using remote sensing and GIS with historical data.

1.2 STUDY AREA

The study was conducted in the Shaheed Bhagat Singh Nagar (formerly known as Nawanshehar) district of Punjab, India, located between the latitudes of 30° 58' 00" to 31° 14' 00" North and longitudes of 75° 47' 00" to 76° 31' 00" East. The district adjoins Hoshiarpur in the north, Jalandhar and Kapurthala in the west, Rupnagar in the east and south-east and, in the south, the Ludhiana district of Punjab. It is bounded by the River Sutlej in the south. The district comprises of five blocks, namely Aur, Banga, Balachaur, Nawan Shehar and Saroya, and it has four towns and 473 inhabited villages. Balachaur is the largest block and Saroya is the smallest one (Figure 1.1). The Saroya and Balachaur blocks are rainfed areas, with the rest of the blocks having access to irrigation through tube wells.

1.3 MATERIALS AND METHODS

1.3.1 Identification of Wet and Drought Years

Data on rainfall during the wheat-growing season (*Rabi* season – October to March) for the years 2005–06 to 2011–12 was tabulated (Table 1.1) and the years with minimum *Rabi* season rainfall (46 mm during 2007–08) and maximum *Rabi* season rainfall (239.5 mm during 2010–11) were selected as drought and wet years, respectively, to study the impact of variation in agro-meteorological parameters on wheat production.

Considerable variation in mean monthly maximum and minimum temperatures was observed. The month of May is generally the hottest and January is the coldest month. The mean maximum monthly temperature of the area is 40°C in

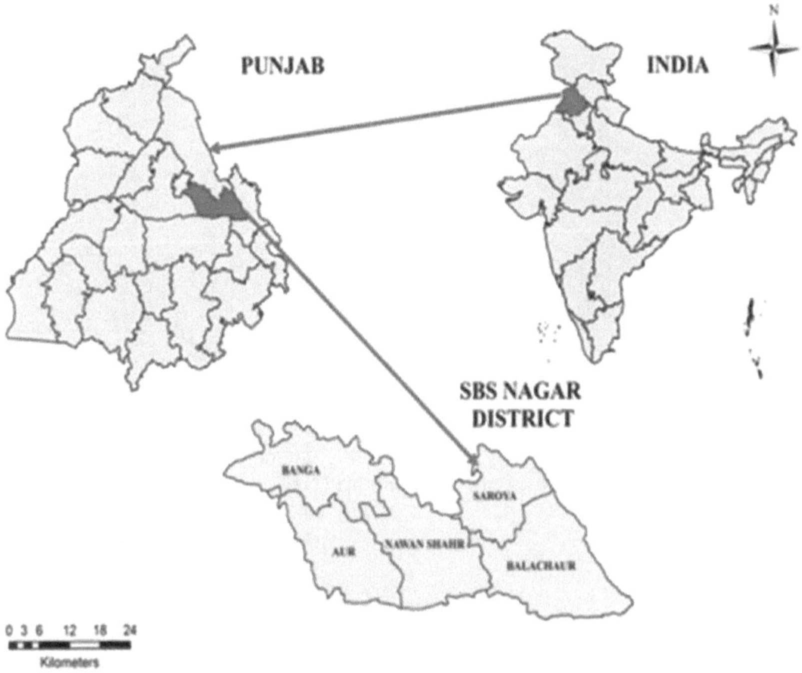

FIGURE 1.1 Location of study area.

TABLE 1.1
Rainfall Recorded at PAU Research Station for *Kandi* Area, Ballowal Saunkhri During *Rabi* Season (October to March)

Year	*Rabi* Season Rainfall (mm)
2005–06	96.4
2006–07	143.2
2007–08	46
2008–09	91.9
2009–10	54.4
2010–11	239.5
2011–12	95.4

summer (May) and mean minimum temperature is 5°C in winter (January). Mean monthly temperature remains above 20°C for about eight months (Figure 1.2). The area is away from the tropical zone and therefore qualifies as a hyperthermic soil temperature regime. The average annual rainfall of the district is 604.8 mm. There is large-scale year-to-year fluctuation in rainfall amounts. About 80%

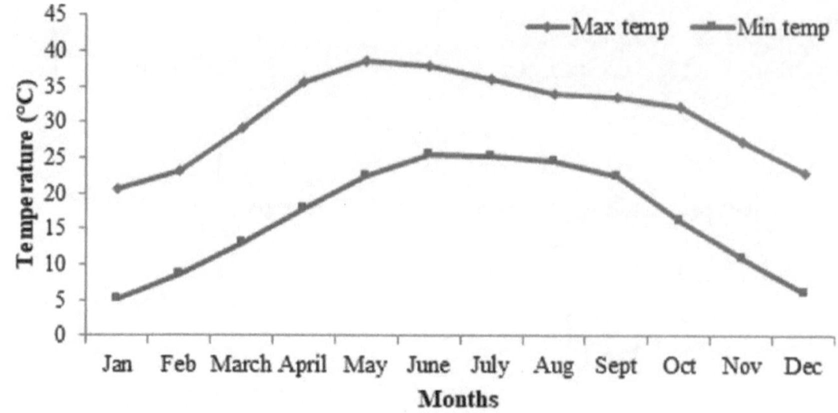

FIGURE 1.2 Average maximum and minimum temperature recorded at the PAU Research Station for *Kandi* area, Ballowal Saunkhri (2005–2012).

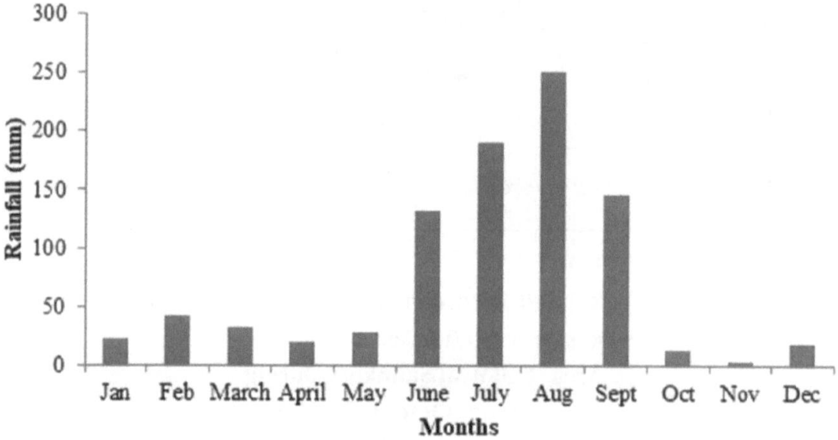

FIGURE 1.3 Average rainfall recorded at the PAU Research Station for *Kandi* area, Ballowal Saunkhri (2005–2012).

of annual rainfall occurs during the monsoon season (Figure 1.3), with a large amount of rainwater being lost as surface runoff, causing flooding and large-scale soil erosion.

1.3.2 Data/Statistics Used

1.3.2.1 Wheat Yield Statistics

Historical district-level wheat yield statistics, provided in the Statistical Abstract of Punjab from 1991 to 2012 (Anon. 1991–2012), were compiled and used in this study. Block-level yield data for the years 2007–08 and 2010–11 was also obtained

TABLE 1.2
Wheat Area, Yield and Production in SBS Nagar District from 1991 to 2012

Year	Area ('00 ha)	Prod ('00 t)	Yield (kg/ha)
1991–1992	830	2770	3336
1992–1993	835	2835	3393
1993–1994	840	2900	3450
1994–1995	840	2870	3412
1995–1996	600	2260	3866
1996–1997	580	2370	4080
1997–1998	570	2000	3567
1998–1999	620	2650	4282
1999–2000	670	3080	4597
2000–2001	700	3130	4463
2001–2002	710	3240	4566
2002–2003	750	3110	4149
2003–2004	750	3110	4149
2004–2005	720	3030	4203
2005–2006	720	3030	4203
2006–2007	720	3000	4163
2007–2008	720	3280	4561
2008–2009	760	2970	3913
2009–2010	740	3160	4271
2010–2011	740	3480	4708

from the office of the Chief Agriculture Officer, SBS Nagar district of Punjab and used in yield modeling. District-level wheat yield, acreage and production are presented in Table 1.2.

1.3.2.2 Satellite Data

Single-date cloud-free IRS LISS-III data at the maximum vegetation crop growth stage of wheat (February) and multi-date AWiFS digital data (November to March), acquired for the years 2007–08 and 2010–11, were analyzed to determine area under wheat and spectral indices, respectively. The FCC (False Color Composites) of the LISS-III scene, covering the district, is presented in Figure 1.4 (Table 1.3).

1.3.2.3 Software Used

ERDAS IMAGINE 9.1 DIP software was used for image processing and classification. District boundary extraction and final output maps were generated by using ArcGIS 10 software. Statistical analysis was carried out, using Microsoft Office Excel 2010.

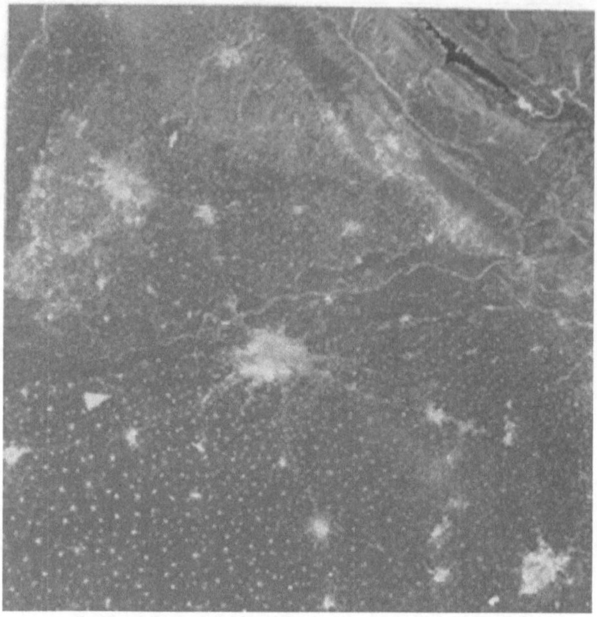

FIGURE 1.4 FCC of LISS-III image covering the study area.

TABLE 1.3
Specifications of LISS-III and AWiFS Sensors

Specifications	AWiFS	LISS-III
Spectral band (μm)	0.52–0.59	0.52–0.59
	0.62–0.68	0.62–0.68
	0.77–0.86	0.77–0.86
	1.55–1.77	1.55–1.77
Spatial resolution (m)	56	23.5
Swath width (km)	140	740
Repeativity (d)	5	24
Quantization (bit)	10	8

1.3.3 Remote-Sensing Techniques for Crop Acreage and Yield Estimations

1.3.3.1 Image Processing, Classification and Digital Data Analysis

The single-date data from wheat (13 February for year 2007–08 and 9 February for year 2010–11) were classified following use of the unsupervised classification algorithm, using district boundary as the mask. The major steps followed in acreage estimation are shown in Figure 1.5.

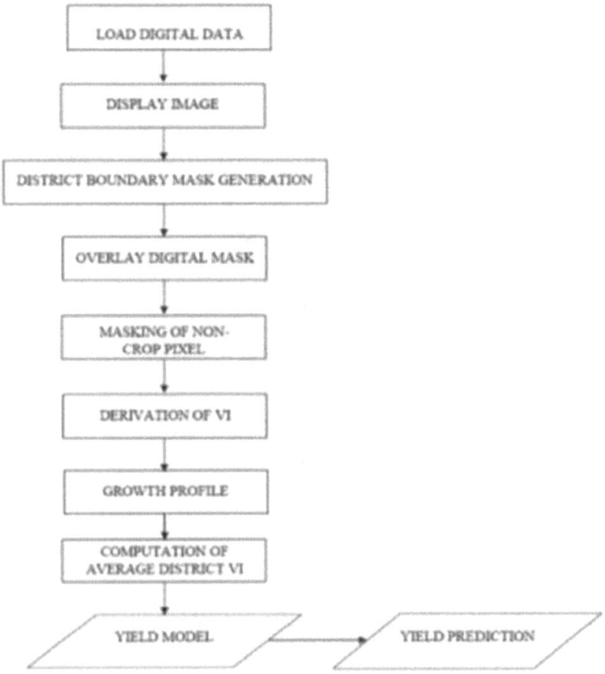

FIGURE 1.5 Methodology for unsupervised image classification, estimation of wheat acreage and wheat production.

1.3.3.2 Digital Data Loading and Extraction

The digital data from computer discs (CDs) were downloaded to the computer from the ERDAS IMAGINE 9.1 digital image processing (DIP) software. The four bands of the digital data were displayed on the display terminal and, using linear enhancement techniques, FCC was generated for identification of various features. Data corresponding to the study area were extracted for further analysis.

1.3.3.3 Geo-Referencing

Satellite data were registered from already-registered IRS LISS-III and AWiFS images, using the geo-referencing facility of ERDAS IMAGINE 9.1 DIP software. Large numbers of Ground Control Points (GCPs), well distributed over the entire image, were identified on the raw image and on the corresponding already-registered image. Then, image-to-image registration was performed, using second-order polynomial models. Root-mean-square-error for image-to-image geo-referencing was less than 0.5 pixel for both IRS LISS-III and AWiFS for individual pixels and <1 pixel for the overall image. These geo-referenced images were used to extract the study area, using the district boundary mask. The geo-referenced FCC image of LISS-III is shown in Figure 1.6.

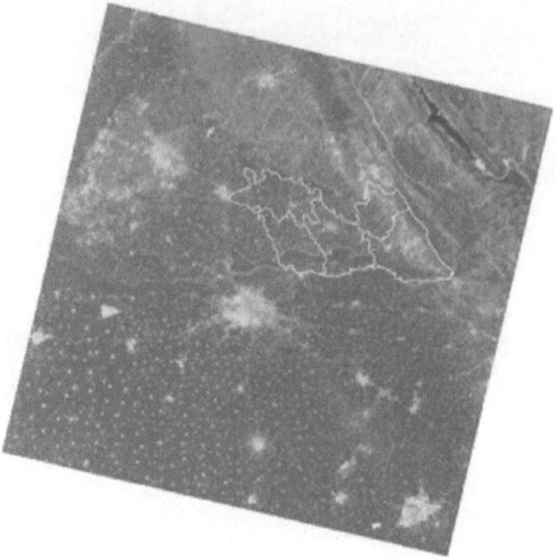

FIGURE 1.6 Geo-referenced FCC image of IRS LISS-III with SBS Nagar district block boundaries.

1.3.3.4 District Boundary Mask Generation

The district boundary mask for the study area was extracted from the Punjab districts shape file in ArcGIS 10 software. This vector layer of district boundary was imported to ERDAS IMAGINE 9.1 DIP software and then the satellite image was subset, using district boundary. Wheat acreage estimation was made by analyzing the pixels inside the boundary.

1.3.3.5 Unsupervised Classification

Unsupervised classification (isoclustering) was performed on the subset images of LISS-III data, and the pixels belonging to the wheat crop were identified, a separate class of wheat crop images generated and the wheat acreage estimated from remote-sensing data. Final output maps were prepared in a GIS environment (ArcGIS 10 software).

1.3.3.6 Computation of Spectral indices

After classification of the images, gray-scale images of indices were generated, in which Digital Number (DN) values varied from low to high. Normalized Difference Vegetation Index (NDVI), RVI (Ratio Vegetation Index), Infrared Percentage Vegetation Index (IPVI) and Difference Vegetation Index (DVI) profiles were generated from AWiFS data. Equations used for the computation of vegetation indices are shown in Table 1.4.

TABLE 1.4
Computation of Different Vegetation Indices Used in the Study

Sr. No.	Index Name	Equation	Reference
1.	Ratio Vegetation Index, RVI	NIR / Red	Jordon (1969)
2.	Normalized Difference Vegetation Index, NDVI	(IRR–Red) / (NIR+Red)	Rouse et al. (1974)
3.	Infrared Percentage Vegetation Index, IPVI	NIR / (NIR + Red)	Crippen (1990)
4.	Difference Vegetation Index, DVI	NIR–Red	Tucker (1979)

1.3.3.7 Wheat Acreage and Yield Estimation

Crop acreage estimation through remote sensing includes the estimation of per cent relative difference (% RD) (Equation (2)) in wheat area by comparing the remote-sensing estimates with the estimates from the Bureau of Economics and Statistics (BES) for the respective years. The pixels under the wheat class were classified using isoclustering unsupervised classification, following the complete enumeration approach and aggregated to achieve the wheat acreage in the district as:

$$A_w = N_{WP} * S_R \tag{1.1}$$

where A_w = area under wheat (ha), N_{WP} = number of wheat pixels identified, S_R = spatial resolution of the sensor used (23.5 m in the case of LISS-III).

1.3.3.8 Wheat Yield Modeling

Spectral indices were used, along with the historical yield data, to develop different yield prediction models. Multiple regression/regression analysis of each of the selected subsets was carried out and regression coefficients were determined. The final yield prediction equation was selected, based on the highest simple or multiple correlation coefficients (r).

1.3.3.9 Spectral Indices Model

Simple correlation analysis was performed between block-level wheat yield in 2007–08 and 2010–11 and vegetation indices derived from AWiFS data of season. Indices showing significant correlations with yield were further used to develop regression equations for prediction of wheat yield over the whole district.

1.3.3.10 Model Performance Evaluation

Wheat yield estimates, obtained by using the spectral indices models, were compared with the historical yield record from the BES. The relative deviation (RD) values were computed as a measure of accuracy, as follows:

$$RD = \frac{(Y_m - Y_h)}{Y_m} \times 100 \quad (1.2)$$

where Y_m = model-estimated wheat yield (kg/ha), Y_h = historical wheat yield according to BES (kg/ha).

1.4 RESULTS

The spectral indices of the crop, such as NDVI, DVI, IPVI and RVI, are indicative of the crop growth status; from these indices, it is possible to measure the crop growth conditions in an area. It has been found that single-date satellite data at the maximum vegetative growth stage of the wheat crop is most appropriate for area estimation. In the present study, wheat acreage and production from the SBS Nagar district in Punjab has been estimated by remote sensing and historical data.

1.4.1 Wheat Acreage and Production for the Years 2007–08 and 2010–11

Acreage estimation for the SBS Nagar district was carried out, following unsupervised classification using the complete enumeration approach by overlaying the district boundary mask onto the geo-referenced IRS Resourcesat-1 (P6) LISS-III satellite data at the peak vegetative growth state of the wheat. False Color Composites (FCC) and classified images of 2007–08 and 2010–11 are shown in Figure 1.7 and Figure 1.8, respectively. Wheat classes were grouped together and shown in yellow color on the classified images, whereas all other non-agricultural features, such as built-up areas, roads, canals, rivers, wastelands, fallow lands and other crops, forests and plantations, etc., were grouped together into one class as a "non-wheat" class and indicated by blue color in the classified images. The estimated wheat acreage of SBS Nagar district, using remote-sensing techniques, in comparison with estimates given by BES for both years, is summarized in Table 1.5. The wheat acreages estimated for year 2007–08 and 2010–11 were 68,244 ha and 71,274 ha, respectively. When acreages estimated through satellite images were compared to those from historical data from the BES for the years 2007–08 and 2010–11, the per cent relative deviation (% RD) was found to be less than 5 per cent in both years (–5.50% for the year 2007–08 and –3.82% for the year 2010–11). The area under wheat had increased in the wet year, 2010–11, by 4.43% relative to the area (68,244 ha) in the drought year, 2007–08. This increase in wheat area may be attributed to the high soil moisture content resulting from high rainfall encouraged farmers to sow a larger acreage. The incremental increase in wheat area was mostly in the rainfed Shivalik foot-hills range (shown in the zoomed window of Figure 1.9), but there was a negligible increase in the lower parts of the district. This may be due to the difference in physiographic conditions of the district, as the lower part

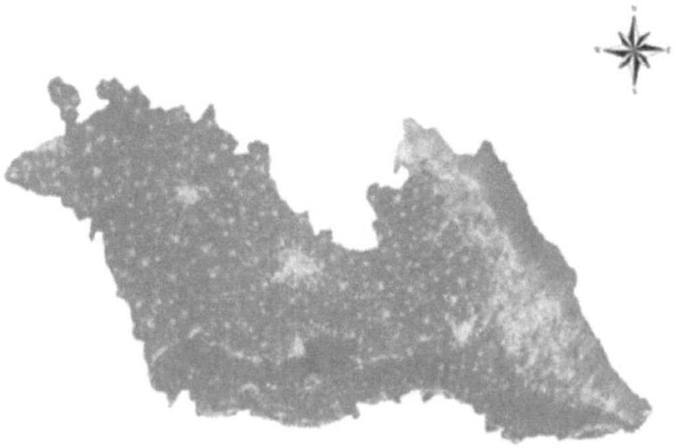

FCC of IRS LISS-III image of SBS Nagar District (2007-08)

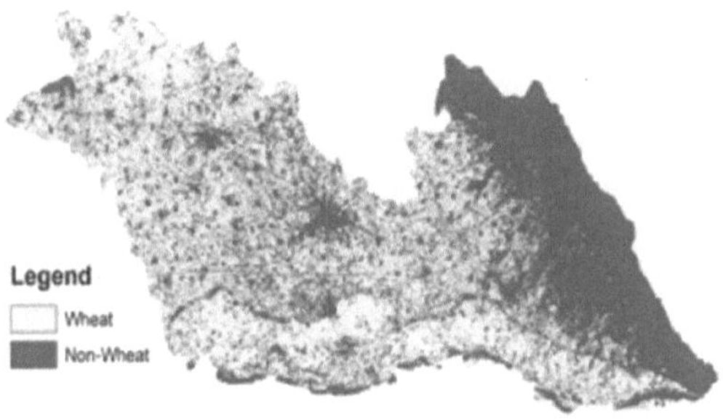

Classified Image of SBS Nagar District for Wheat (2007-08)

FIGURE 1.7 False Color Composites (FCC) and classified image from SBS Nagar District for year 2007–08.

of the district was mainly piedmont plain, alluvial plain and flood plain, and this area has assured irrigation facilities, through either canal or tube wells, whereas the upper part of the district falls in the Shivalik foot-hills, which is mostly rainfed. Wheat production in the drought year (2007–08) was 311,260 tonnes, compared with 335,557 tonnes in the wet year (2010–11), with the production increasing during the wet year by 7.8% relative to the drought year. The wheat yield was 4.561 t/ha and 4.708 t/ha during the drought and wet years, respectively, representing an increase of 3.2% in the wet year.

1.4.2 Spectral Vegetation Indices for Year 2007–08 and 2010–11

Four spectral vegetation indices, namely DVI, IPVI, NDVI and RVI, were computed as per Table 1.5 (Table 3.3 in Chapter 3), from multi-date AWiFS Image data for the years 2007–08 and 2010–11. All the spectral indices are related to

FCC of IRS LISS-III image of SBS Nagar District (2010-11)

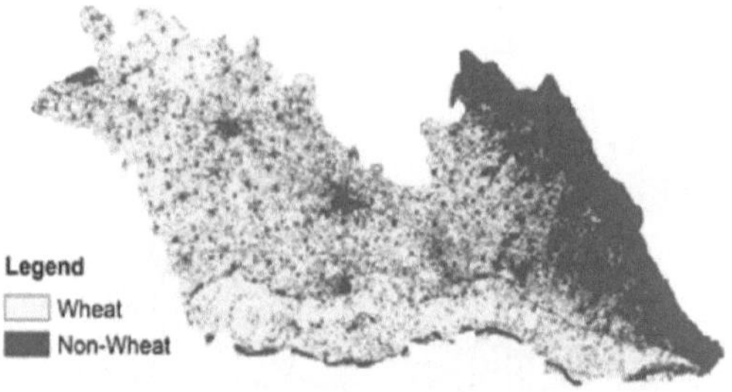

Classified Image of SBS Nagar District for Wheat (2010-11)

FIGURE 1.8 False Color Composites (FCC) and classified image from SBS Nagar District for year 2010–11.

TABLE 1.5
Estimated Wheat Acreage Using Remote Sensing (LISS-III) and Relative Deviation from BES Estimate and Wheat Production for the Drought and Wet Years

	Wheat Acreage ('000 ha)		Relative Deviation	Wheat Production
Year	RS	BES	(%)	(kg)
2007–08	68.244	72	-5.50	311261
2010–11	71.274	74	-3.82	335558

the maximum vegetative growth stage of the crop. Higher values of the indices indicate better crop growth and *vice versa*. The gray-scale images of DVI, IPVI, NDVI and RVI are presented in Figures 1.10a, 1.10b, 1.11a, 1.11b, 1.12a, 1.12b, 1.13a and 1.13b. The temporal profile of these indices are graphically presented in Figure 1.14 for the years 2007–08 and 2010–11.

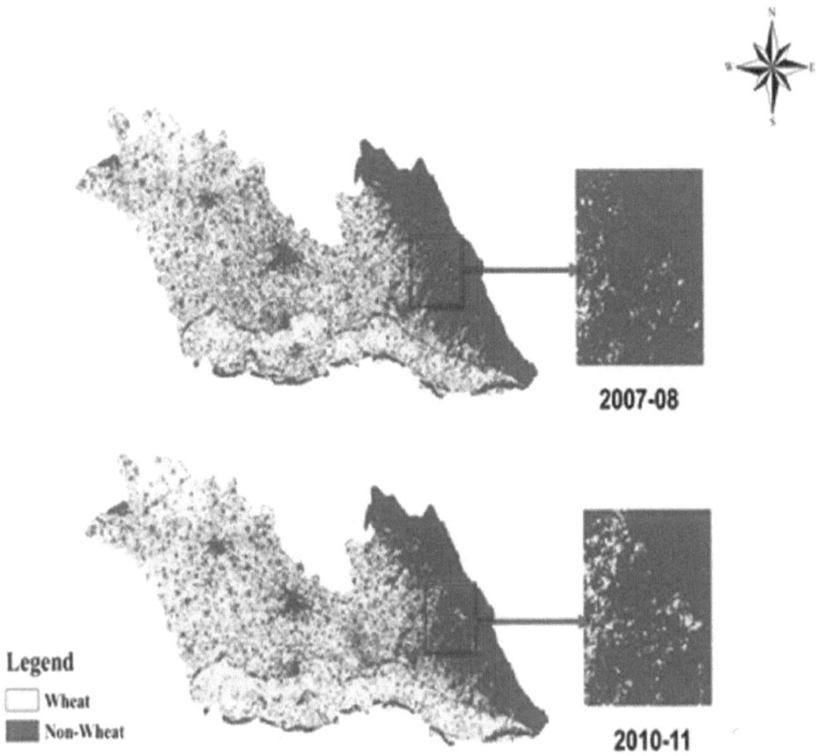

FIGURE 1.9 The increase in wheat area in SBS Nagar District for year 2010–11, compared with 2007–08.

The Difference Vegetative Index (DVI) (Lillesand and Kiefer 1987, Richardson and Everitt 1992) has an infinite range. Results revealed that the value of DVI increased with crop age up to the month of February (184 to 330, Figure 1.10b), i.e., up to the most advanced vegetative growth stage of wheat, and thereafter decreased in the wet year 2010–11, and in the drought year 2007–08, the value increased up to the month of February (211 to 274, Figure 1.10a) and thereafter started decreasing. The Infrared Percentage Vegetation Index (IPVI) is restricted to values between 0 and 1. In Figure 1.11a and 1.11b, the value of IPVI increased from 0.76 to 0.91 (drought year 2007–08) and from 0.81 to 0.93 (wet year 2010–11) with increasing crop age up to the month of February, i.e., up to the highest vegetative growth stage of wheat, and thereafter decreased in both years. The values of the Normalized Difference Vegetation Index (NDVI) (Kriegler et al. 1969, Rouse et al. 1973) vary between −1 and 1. Figures 1.12a and 1.12b, showing that the values of NDVI increased with crop age up to the month of February (0.82 and 0.87 in the case of the drought year 2007–08 and the wet year 2010–11, respectively) up to the highest vegetative growth stage of wheat, and thereafter

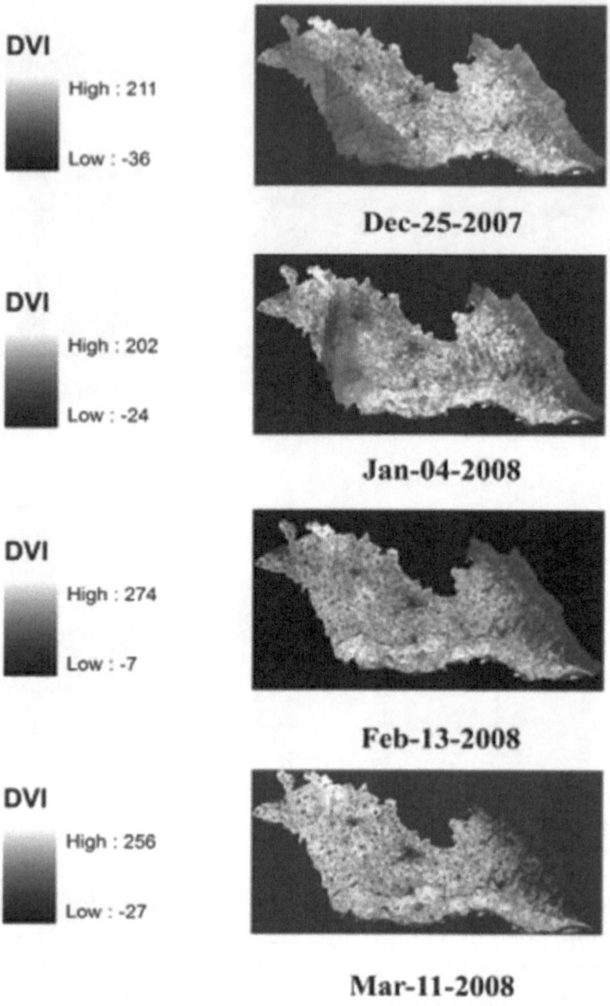

FIGURE 1.10 DVI profile of wheat crop in SBS Nagar District for (a) 2007–08 and (b) 2010–11.

decreased in both years. The Ratio Vegetation Index (RVI) (Jordan 1969) ranges from 0 to infinity. The value of RVI increased from the month of December (3.22 and 4.46 in 2007–08 and 2010–11, respectively, as shown in Figures 1.13a and 1.13b) with crop age up to the month of February (i.e., 10.33 and 15.45 in the drought year 2007–08 and the wet year 2010–11, respectively), decreasing thereafter in both years.

In general, the values of all the indices increased with crop age up to the month of February, i.e., up to the maximum vegetative growth stage of wheat, and

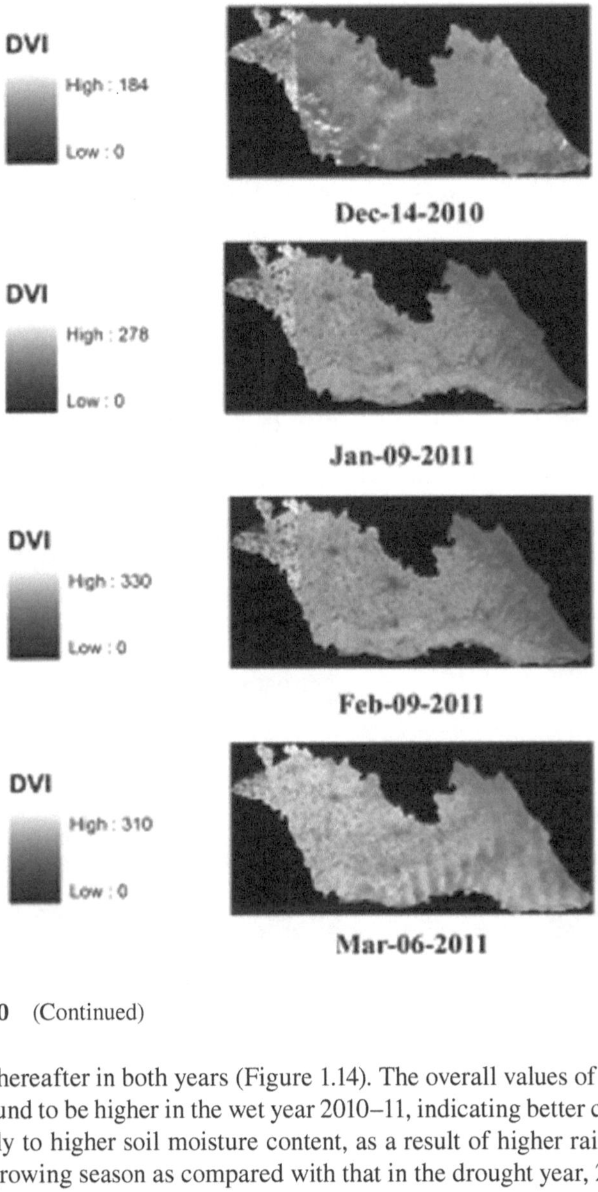

FIGURE 1.10 (Continued)

decreased thereafter in both years (Figure 1.14). The overall values of all the indices were found to be higher in the wet year 2010–11, indicating better crop growth, due probably to higher soil moisture content, as a result of higher rainfall during the wheat growing season as compared with that in the drought year, 2007–08.

1.4.3 Wheat Yield Prediction through Models

1.4.3.1 Spectral Indices Models

Simple linear regression analysis showed significant positive relationships among wheat yield and the spectral vegetation indices, namely NDVI, RVI, IPVI and

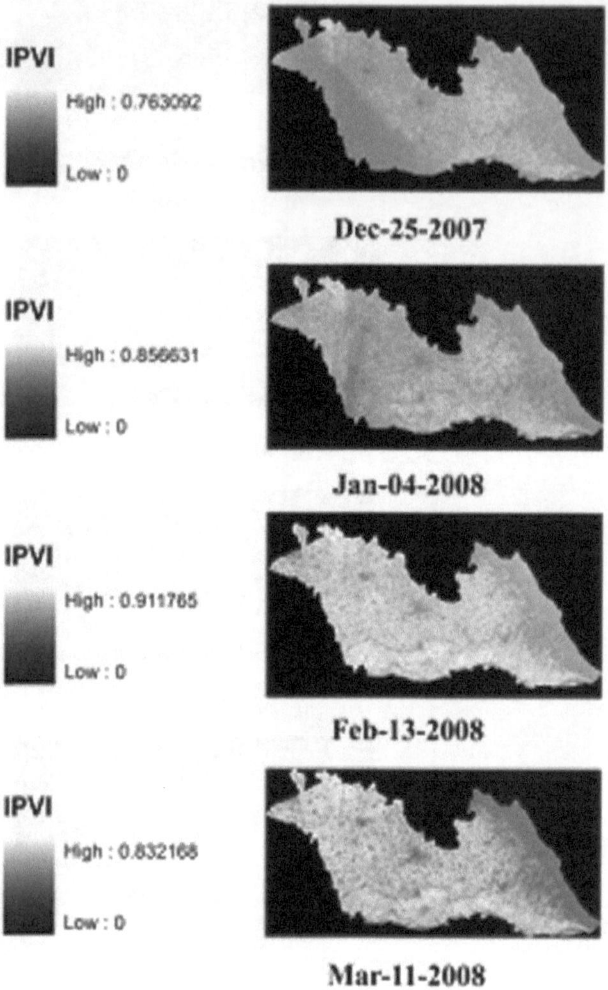

FIGURE 1.11 IPVI profile of wheat crop in SBS Nagar District for (a) 2007–08 and (b) 2010–11.

DVI (Figures 1.15a and 1.15b). The coefficient of determination (r^2) varied from 0.858 for NDVI to 0.884 for RVI in year 2007–08, whereas it ranged from 0.828 for NDVI to 0.872 for RVI in the year 2010–11. Among the vegetation indices, RVI showed the highest r^2 with wheat yield in both years, i.e. 2007–08 ($r^2 = 0.884$) and 2010–11 ($r^2 = 0.872$). Predicted wheat yield results, using different vegetation indices and the relative deviation from the BES estimate, are shown in Table 1.6. The results revealed that the yields predicted by using each of the vegetation indices were closer to the actual yields than were the yields estimated by BES,

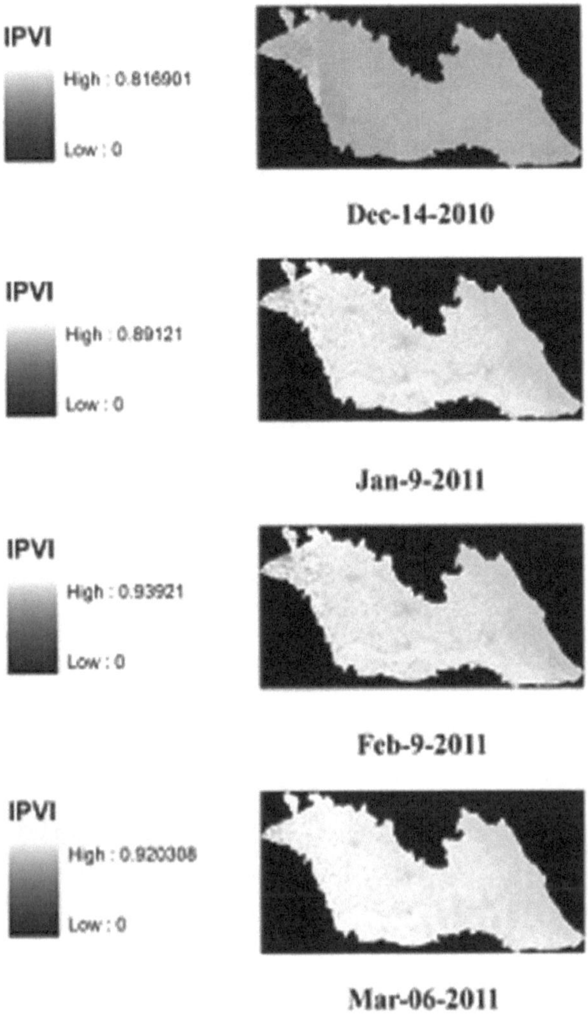

FIGURE 1.11 (Continued)

Punjab, for both years. The per cent relative deviation for both years was less than 2.5. Among the vegetation indices, the per cent RD was maximal in the case of NDVI for both years (−2.31% in 2007–08 and −2.14 in 2010–11), whereas the RVI-predicted yield had the lowest per cent RD for both 2007–08 (RD = −0.90%) and 2010–11 (RD = −0.99%). Similarly, the yield estimated by IPVI was similar with that overestimated by BES for 2007–08 (RD = −0.21%) and 2010–11 (RD = −0.59%). Furthermore, the yield predicted by DVI underestimated the yield predicted by BES in 2007–8 (RD = 1.64%) and in 2010–11 (RD = −2.01%). The

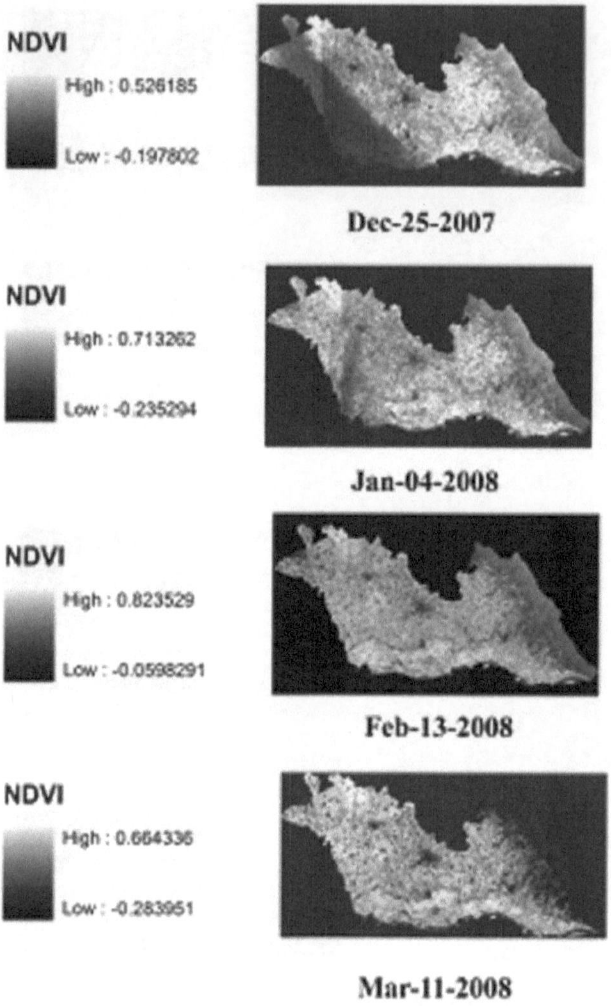

FIGURE 1.12 NDVI profile of wheat crop in SBS Nagar District for (a) 2007–08 and (b) 2010–11.

yields predicted by RVI, IPVI and DVI were closer to the BES estimates during the year 2007–08 than in 2010–11.

1.5 CONCLUSION

Remote sensing is a reliable and faster technology for estimating area under a crop than the traditional methods, which are more time and labor consuming.

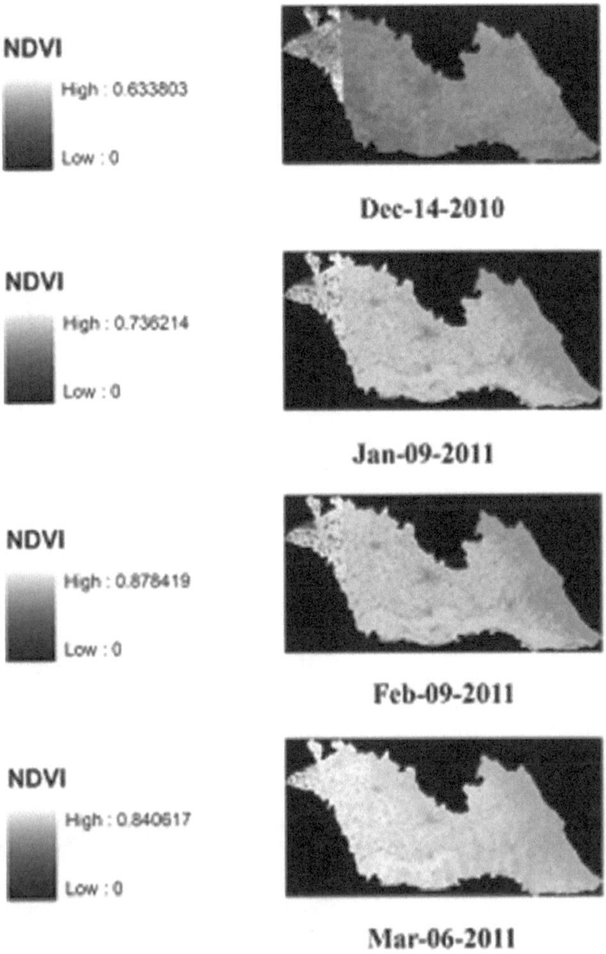

FIGURE 1.12 (Continued)

The wheat area during the drought year (2007–08) was 68,244 ha as compared with 71,274 ha during the wet year (2010–11), as estimated using remote-sensing technology. Wheat production and yield were higher during the wet year (2010–11) than in the drought year (2007–08) by 7.8% and 3.2%, respectively. The yield predicted with the spectral index models were close to the actual yield (RD < 2.5%) for both the drought and wet years. The per cent relative deviation was maximal in the case of NDVI for both years (–2.31% in 2007–08 and –2.14% in 2010–11), whereas IPVI predicted the yield with the lowest per cent relative deviation for both 2007–08 (RD= –0.21%) and 2010–11 (RD=–0.59%).

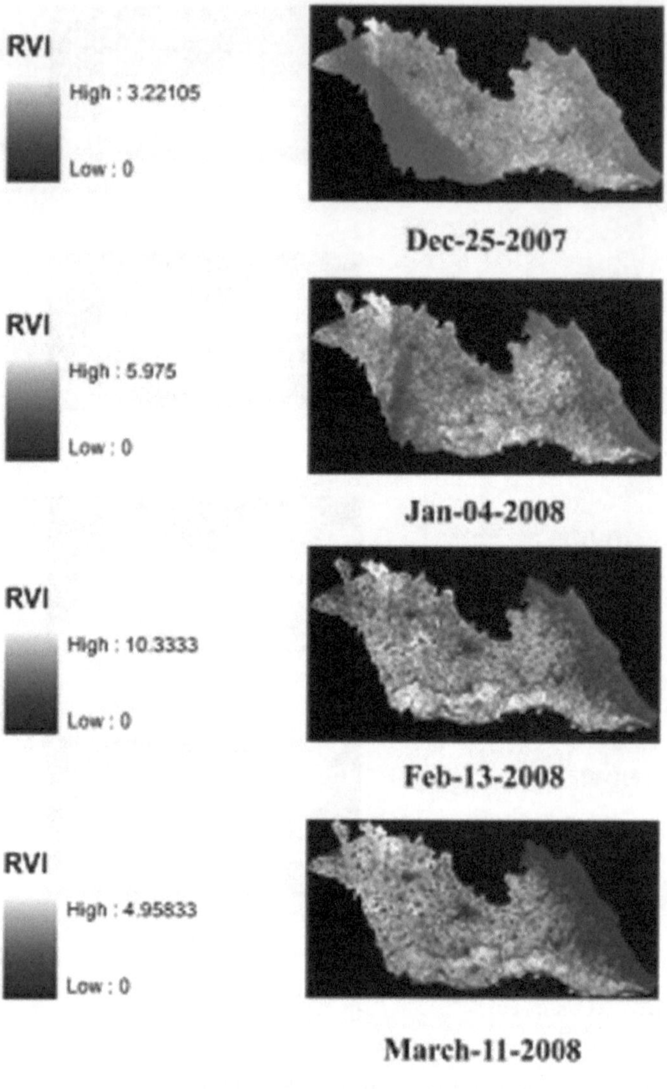

FIGURE 1.13 RVI profile of wheat crop in SBS Nagar District for (a) 2007–08 and (b) 2010–11.

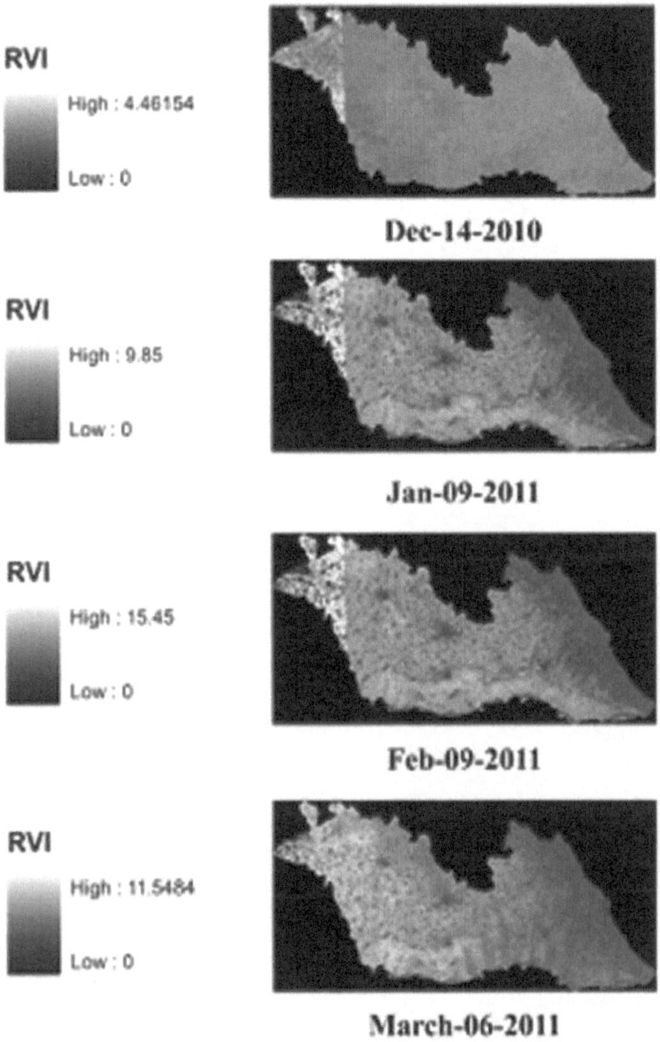

FIGURE 1.13 (Continued)

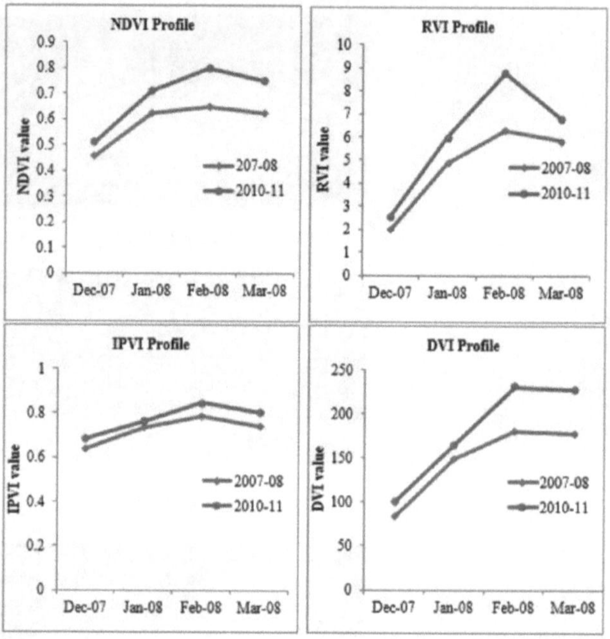

FIGURE 1.14 Growth profile of wheat in SBS Nagar district for different vegetation indices for years 2007–2008 and 2010–11.

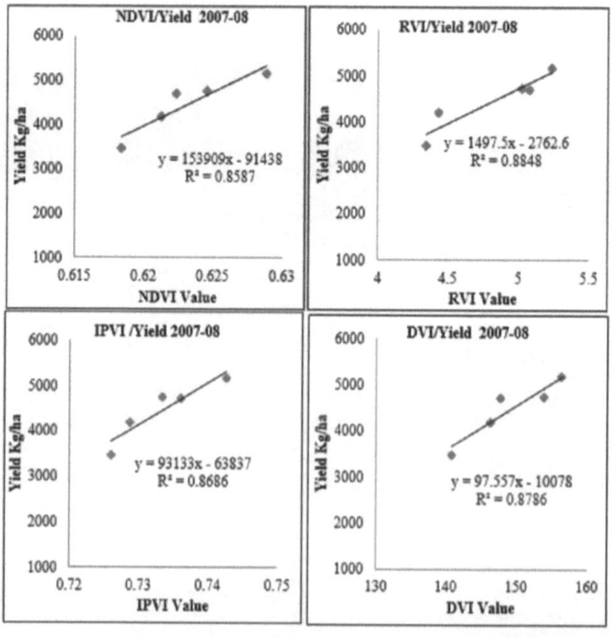

FIGURE 1.15A Relationship between wheat yield and vegetation indices for the drought year 2007–08.

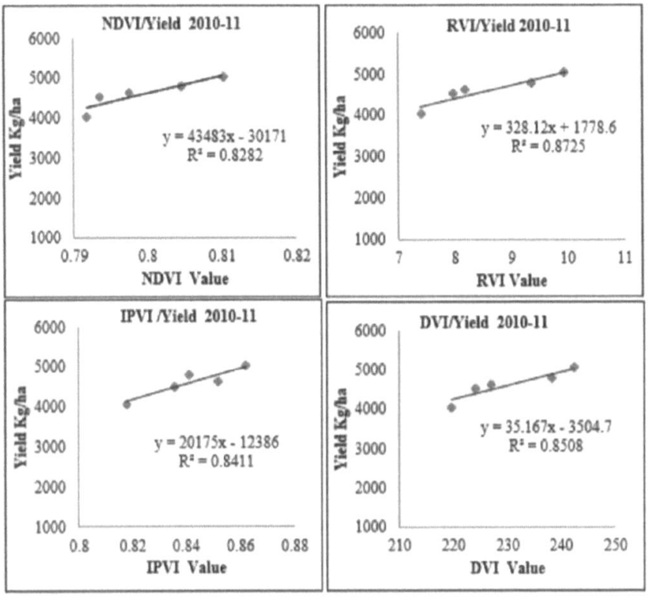

FIGURE 1.15B Relationship between wheat yield and vegetation indices for the wet year 2010–11.

TABLE 1.6
Prediction of Wheat Yield by Different Vegetation Indices and Relative Deviation from the BES Estimate for the Years 2007–08 and 2010–11 of SBS Nagar District

		Prediction by Vegetation Indices (kg/ha)			
Year	**BES (kg/ha)**	**NDVI**	**RVI**	**IPVI**	**DVI**
2007–08	4561	4458 (−2.31)	4520 (−0.90)	4551 (−0.21)	4487 (−1.64)
2010–11	4708	4609 (−2.14)	4662 (−0.99)	4680 (−0.59)	4615 (−2.01)

* Figures in parentheses show the % relative deviation from the BES estimate

REFERENCES

Anonymous (1991–2012) *Statistical Abstracts of Punjab.* The Economic Advisor, Punjab.

Crippen R E (1990) Calculating the vegetation index faster. *Remote Sens Environ* **34**: 71–73.

Jordan C F (1969) Derivation of leaf area index from quality of light on the forest floor. *Ecology* **50**: 663–666.

Kriegler F J, Malila W A, Nalepka R F and Richardson W (1969) Preprocessing transformations and their effects on multispectral recognition. In *Proc of the 6th International Symposium on Remote Sensing of Environment.* University of Michigan, Ann Arbor, MI, pp. 97–131.

Lillesand T M and Kiefer R W (1987) *Remote Sensing and Image Interpretation.* 2nd Ed. John Wiley and Sons, New York, p. 721.

Premlatha M and Rao P P N (1994) Crop acreage estimation using ERS-1 SAR data. *J Indian Soc Remote Sensing* **22**: 139–147.

Richardson A J and Everitt J H (1992) Using spectra vegetation indices to estimate rangeland productivity. *Geocarto Int* **1**: 63–69.

Rouse J W, Haas R H, Deering D W, Schell J A and Harlan J C (1974) *Monitoring the Vernal Advancement and Retrogradation (Green Wave Effect) of Natural Vegetation.* NASA/GSFC Type III Final Report, Greenbelt, MD, p. 371.

Rouse J W, Haas R H, Schell J A and Deering D W (1973) *Monitoring Vegetation Systems in the Great Plains with ERTS.* 3rd ERTS Symposium, Vol. 1, NASA SP-351, pp. 309–317.

Sahai B and Dadhwal V K (1990) Remote sensing in agriculture. In Verma J P and Verma A (eds.) *Technology Bunding and Agrarian Prosperity.* Malhotra Publishing House, New Delhi, pp. 83–98.

Tucker C J (1979) Red and photographic infrared linear combinations for monitoring vegetation. *Remote Sens Environ* **8**: 127–150.

2 Impact of Variation in Agro-Meteorological Parameters on Wheat Production and Comparison of Yield Predicted by Agro-Meteorological Indices with Remote-Sensing Data

Amanpreet Kaur Benipal, Anil Bhardwaj and Surender Singh

CONTENTS

2.1 INTRODUCTION

Despite having only 1.53% of the area of India, Punjab produces about 12% of the cereals produced in the country. Wheat is the main crop in Punjab and contributes 43.4% (2013–2014) share to the centre pool (Anonymous 2014a). Punjab makes up 11.35 % of country's wheat area and has the highest productivity, at 4.28 qt/ha in India (Anonymous 2015). This is due to assured irrigation, use of high-yield varieties, higher-than average agricultural inputs and the adoption of new farming technologies. Seven leading wheat-producing districts in Punjab include Sangrur, Bathinda, Ludhiana, Patiala, Mukstar, Fazillka and Ferozpur, which together contribute 48 per cent of the state's output of wheat (Anonymous 2014b).

Agro-meteorological parameters are the variables which define the agricultural productivity of any region. These variables affect the crop differently during different stages of crop growth. During low and high rainfall years, the agro-meteorological parameters differ greatly. Moisture is a condition resulting from frequent rainfall and humid weather. It may affect other meteorological parameters, such as air temperature, soil temperature, atmospheric pressure and humidity, etc., and may reduce the risk of heat stress. The impact of agro-meteorological parameters on wheat yield needs to be investigated during low and high rainfall seasons using remote sensing (RS) and geographic information systems (GIS) in the Shaheed Bhagat Singh Nagar district of Punjab, which includes both assured irrigation (76,600 ha) and rainfed (49,200 ha) areas. The current study was planned with the objectives of studying variation in agro-meteorological parameters during low- and high-rainfall *Rabi* seasons and identifying their impacts on wheat production, and to compare the wheat production results predicted from agro-meteorological index models and spectral vegetation index models with those from historical data.

2.2 STUDY AREA

The study was conducted in the SBS Nagar (Nawanshehar) district of Doaba region of Punjab. With an extent reaching between 30 degrees 58 minutes to 31 degrees 14 minutes North latitudes and 75 degrees 47 minutes to 76 degrees 31 minutes East longitudes. It covers an area of 1,267 sq kilometers. It is the third smallest district of Punjab, after Fatehgarh Sahib and SAS Nagar districts. Earlier it was carved out from Hoshiarpur and Jalandhar districts, and river Sutlej separates it from Ludhiana district in the south. It is comprised of alluvial plains and a few small hills in its eastern side, which is part of the Kandi region. This undulating

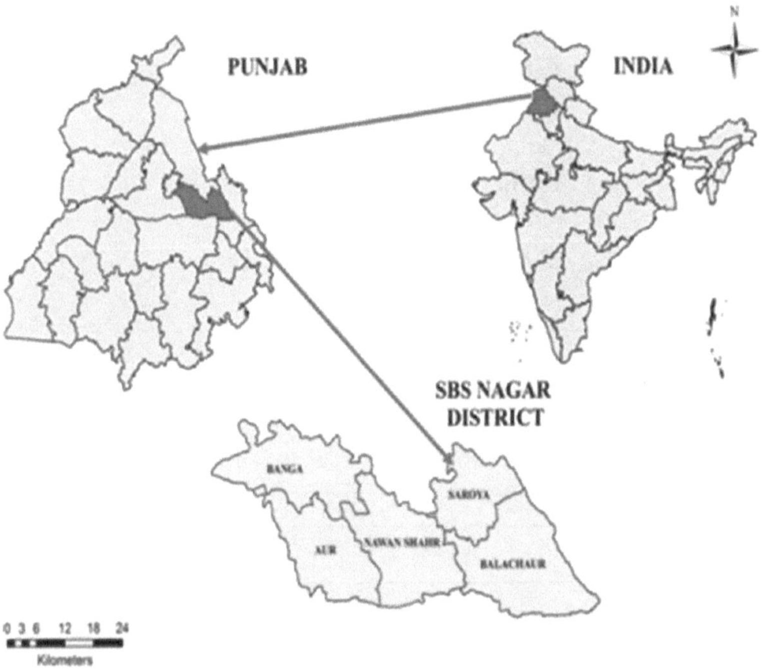

FIGURE 2.1 Location of study area.

area, which lies in the Saroya and Balachor blocks, has rainfed agriculture. All other parts of the district use canal and tubewell irrigation (Figure 2.1).

2.3 MATERIALS AND METHODS

2.3.1 Identification of Wet and Drought Years

Data on rainfall during the wheat-growing season (*Rabi* season – October to March) for the years 2005–06 to 2011–12 were tabulated (Table 2.1), and the years with minimum *Rabi* season rainfall (46 mm during 2007–08) and maximum *Rabi* season rainfall (239.5 mm during 2010–11) within this period were selected as drought and wet years, respectively, to study the impact of variation in agro-meteorological parameters on wheat production.

2.3.2 Data Used

2.3.2.1 Wheat Statistics

Historical district-level wheat yield statistics provided by the Statistical Abstract of Punjab from 1991 to 2012 (Anonymous 1991–2012), Punjab, were compiled and used in the study. District-level wheat yield, acreage and production are presented in Table 2.2.

TABLE 2.1
***Rabi* Season Rainfall Recorded at PAU Research Station for *Kandi* Area, Ballowal Saunkhri During the *Rabi* Season (October to March)**

Year	*Rabi* Season Rainfall (mm)
2005–06	96.4
2006–07	143.2
2007–08	46
2008–09	91.9
2009–10	54.4
2010–11	239.5
2011–12	95.4

TABLE 2.2
Wheat Area, Yield and Production in SBS Nagar District from 1991 to 2012

Year	Area ('00ha)	Production ('00 t)	Yield (kg/ha)
1991–1992	830	2770	3336
1992–1993	835	2835	3393
1993–1994	840	2900	3450
1994–1995	840	2870	3412
1995–1996	600	2260	3866
1996–1997	580	2370	4080
1997–1998	570	2000	3567
1998–1999	620	2650	4282
1999–2000	670	3080	4597
2000–2001	700	3130	4463
2001–2002	710	3240	4566
2002–2003	750	3110	4149
2003–2004	750	3110	4149
2004–2005	720	3030	4203
2005–2006	720	3030	4203
2006–2007	720	3000	4163
2007–2008	720	3280	4561
2008–2009	760	2970	3913
2009–2010	740	3160	4271
2010–2011	740	3480	4708
2011–2012	750	3870	5162

2.3.2.2 Meteorological Data

Keeping in view the cropping calendar of wheat (October to March), the historical meteorological data (2005 to 2012), namely maximum and minimum temperature, rainfall, maximum and minimum relative humidity and number of sunshine hours, of the district were collected from PAU Research Station for the *Kandi* area, Ballowal Saunkhri and the School of Climate Change and Agricultural Meteorology, PAU Ludhiana, and compiled to determine agro-meteorological indices such as Growing Degree Days (GDD), Accumulated Growing Degree Days (AGDD), Helio Thermal Units (HTU), Photo Thermal Units (PTU), Temperature Difference (TD) and Vapour Pressure Deficit (VPD).

2.3.2.3 Satellite Data

Single-date cloud-free IRS LISS-III data (Figure 2.2) at the maximum vegetation crop growth stage of wheat (February) and multi-date AWiFS digital data (November to March) were analyzed to determine the acreage under wheat and spectral indices, respectively for the years 2007–08 and 2010–11.

2.3.3 Computation of Indices

2.3.3.1 Spectral Indices

After classification of the images, gray-scale images of indices were generated, in which DN values varied from low to high. NDVI, RVI, IPVI and DVI profiles

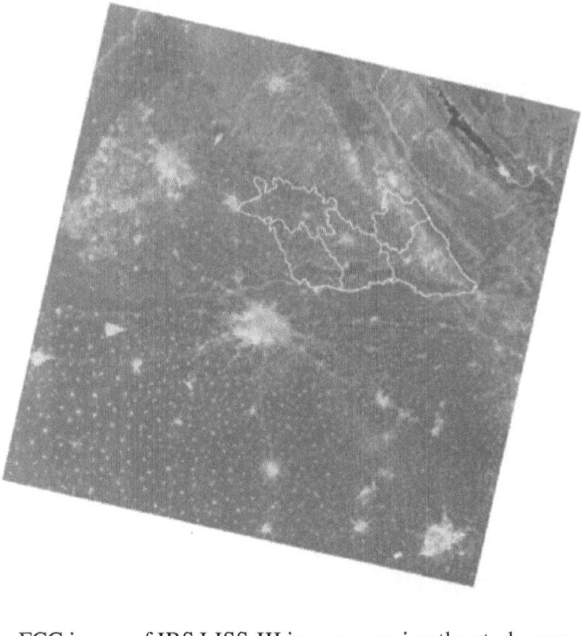

FIGURE 2.2 FCC image of IRS LISS-III image covering the study area.

TABLE 2.3
Equations for Calculation of Vegetation Indices

Sr. No.	Index Name	Equation	Reference
1.	Ratio Vegetation Index, RVI	NIR / Red	Jordon (1969)
2.	Normalized Difference Vegetation Index, NDVI	(IRR–Red) / (NIR+Red)	Rouse et al. (1974)
3.	Infrared Percentage Vegetation Index, IPVI	NIR / (NIR+Red)	Crippen (1990)
4.	Difference Vegetation Index, DVI	NIR–Red	Tucker (1979)

were generated from AWiFS data. Equations used for the computation of vegetation indices are given in Table 2.3.

2.3.3.2 Agro-Meteorological Indices

2.3.3.2.1 Accumulated Growing Degree-days (AGDD)

AGDD is used to estimate the growth and development of plants during the growing season. AGDD was computed as

$$\mathrm{AGDD} = \Sigma \frac{T_{\max} + T_{\min}}{2} - \mathrm{T_b} \tag{2.1}$$

where AGDD = accumulated growing degree-days (°C day); T_{max} = daily maximum temperature (°C); T_{min} = daily minimum temperature (°C); T_b = base temperature (°C)

2.3.3.2.2 Temperature Difference (TD)

$$\mathrm{TD} = \Sigma(\mathrm{T_{max}} - \mathrm{T_{min}}) \tag{2.2}$$

where TD = temperature difference (°C); T_{max} = daily maximum temperature (°C); T_{min} = daily minimum temperature (°C)

2.3.3.2.3 Heliothermal Units (HTU)

HTU is the product of growing degree-days and its corresponding actual number of sunshine hours, and it is expressed as:

$$\mathrm{HTU} = \mathrm{GDD} \times \mathrm{H_A} \tag{2.3}$$

where HTU = heliothermal units (°C day hours); AGDD = accumulated growing degree-days (°C day); H_A= actual sunshine hours.

2.3.3.2.4 Photothermal Units (PTU)

PTU is expressed as:

$$\mathrm{PTU} = \mathrm{GDD} \times \mathrm{D_L} \tag{2.4}$$

where PTU = photothermal units (°C day hours); AGDD = accumulated growing degree-days (°C day); D_L = maximum possible sunshine hours (daylength) (List 1951).

2.3.3.2.5 Vapour Pressure Deficit (VPD)

$$e_a = (RH_{mean} \times e_s) / 100 \quad (2.5)$$

$$VPD = e_s - e_a \quad (2.6)$$

where e_a = actual water vapour pressure (millibar); RH_{mean}= mean relative humidity (%); e_s= saturated water vapour pressure (millibar) as a function of mean air temperature (Michael 1978); VPD = vapour pressure deficit (millibar).

2.3.3.3 Wheat Yield Modeling

Agro-meteorological indices and agro-meteorological parameters were used, along with the historical yield data, to develop different yield prediction models. Multiple regression/regression analysis of each selected subset was carried out and regression coefficients were determined. The final yield prediction equation was selected based on the highest simple or multiple correlation coefficient (r).

2.3.3.4 Agro-meteorological indices model

Agro-meteorological parameters, namely rainfall, number of sunshine hours, maximum and minimum relative humidity, and maximum and minimum temperature, were used to compute various agro-meteorological indices for wheat. Simple and multiple correlation analysis were performed between the computed agro-meteorological indices and wheat yield provided by Bureau of Economics and Statistics (BES) in the appropriate years.

2.3.3.5 Agro-meteorological Multiple Regression Model

The regression equations between the wheat yield, given by BES for the years 2007–08 and 2010–11, and rainfall, maximum and minimum temperatures were determined.

2.3.4 Model Performance Evaluation

Wheat yield estimated by using the agro-meteorological indices model and the agro-meteorological multiple regression model were compared with the historical yield record of BES. The per cent relative deviation (% RD) was computed as a measure of accuracy as follows:

$$RD = \frac{(Y_m - Y_h)}{Y_m} \times 100 \quad (2.7)$$

where Y_m = model-estimated wheat yield (kg/ha); Y_h = historical wheat yield as per BES data (kg/ha).

2.4 RESULTS AND DISCUSSION

2.4.1 Variation in Agro-Meteorological Parameters

The average weekly agro-meteorological parameters (*Rabi* season – 1 Oct to 31 March) in the study area during the drought year 2007–08 and the wet year 2010–11 are presented in Table 2.4. The variation in different agro-meteorological parameters is also shown in Figures 2.2– 2.5. The maximum cumulative seasonal rainfall for the drought (2007–08) and wet (2010–11) years were 46 mm and 239.5 mm, with a coefficient of variation of 0.084 and 0.081, respectively. The average maximum seasonal temperature for the drought year was 25.7°C and for the wet year was 24.7°C. The average minimum seasonal temperature for years 2007–08 and 2010–11 was 9.2°C and 10.0°C, respectively. The average minimum and maximum seasonal relative humidity values for year 2007–08 were 43.3% and 81.5%, respectively, and for year 2010–11 were 53.4% and 89.9%, respectively. The average number of sunshine hours per day for the drought year was 8.0 hours, compared with 6.7 hours for the wet year. The standard deviation (SD) and coefficient of variation (CV) for all agro-meteorological parameters in the SBS Nagar district are given in Table 2.4.

There was no rainfall during the first eight weeks of the *Rabi* season of the drought year (Figure 2.3), which affected sowing of the wheat crop in the rainfed areas of the SBS Nagar district. But, in the case of the wet year, there was rainfall of 45.5 mm in the first four weeks of the *Rabi* season. The highest rainfall in the drought year was 14.2 mm in the 15th week of the *Rabi* season during the drought year, compared with 75.6 mm in the 20th week of the *Rabi* season during the wet year. The total rainfall values during the drought and the wet years were 46 mm and 239.5 mm, respectively.

The variability in maximum temperature during the drought year and the wet year did not indicate any trend (Figure 2.4). However, at the initial stage of crop growth, the temperature decreased gradually from 33.7°C to 19.7°C during the 1st to 11th week of the *Rabi* season in the drought year and from 33.5°C to 12.3°C during 1st to 14th week of the *Rabi* season in the wet year. The minimum temperature was 0.3°C in the 17th week of the *Rabi* season in the drought year and 2.9°C in the 15th week of the *Rabi* season during the wet year. The maximum temperature was 33.7°C and 33.5°C, both in the 1st week of *Rabi* season in both the drought and the wet years. The average maximum temperature was 1°C higher during the drought year as compared with the wet year. However, the average minimum temperature was 0.8°C higher during the wet year than the drought year.

The maximum relative humidity during the drought year was 96.7% in the 17th week of *Rabi* in the drought year and was 99.1% in the 15th week of *Rabi* during the wet year, but the trend was similar during the two years (Figure 2.5). Minimum relative humidity in the 14th week of 2010–11 was quite high (85.0%). The minimum relative humidity values were 25.0% and 38.0% in the 25th week of *Rabi* season during the drought year and wet year, respectively.

TABLE 2.4
Variation of Agro-Meteorological Parameters in the Study Area During the Drought (2007–08) and Wet Years (2010–11)

	2007–08						2010–11					
Week	T-max (°C)	T-min (°C)	RH max (%)	RH min (%)	Sunshine Hours	Rainfall (mm)	T-max (°C)	T-min (°C)	RH max (%)	RH min (%)	Sunshine Hours	Rainfall (mm)
1	33.7	15.6	76.3	44.3	10.6	0	33.5	19.6	91.3	58.3	8.2	8
2	33.2	14.3	69.6	28.6	10.3	0	33.0	18.9	88.7	57.1	8.1	0.5
3	31.5	14.7	74.7	41.3	9.4	0	32.8	19.0	90.6	62.0	6.4	0
4	32.2	14.4	74.1	40.4	9.5	0	30.1	13.6	88.4	47.7	7.7	37
5	30.2	12.4	76.3	46.6	9.3	0	28.9	12.2	91.0	56.0	8.2	0
6	29.3	12.8	72.3	41.3	8.5	0	29.2	12.4	92.4	46.6	9.4	0
7	27.6	10.0	83.9	42.7	8.6	0	28.4	12.2	79.7	47.1	8.1	0
8	25.8	8.1	74.1	31.7	8.6	0	25.8	10.3	94.4	50.7	7.6	0
9	24.9	9.5	81.3	47.4	6.7	4.2	24.4	8.6	94.9	44.9	7.1	0
10	22.4	6.0	94.0	50.3	5.9	0.4	23.7	7.4	92.0	46.6	7.9	0
11	19.7	7.5	89.7	60.0	4.6	12.8	21.2	5.5	95.3	57.9	6.1	0
12	22.5	4.5	92.1	42.6	7.9	0	21.0	4.2	98.7	47.7	7.4	0
13	21.7	3.6	93.9	37.0	8.8	0	19.8	4.9	95.1	56.0	4.0	61.4
14	20.6	3.0	83.6	38.9	5.9	0	12.3	5.3	97.4	85.0	1.4	0
15	20.8	8.0	92.3	56.4	5.1	14.2	14.1	2.9	99.1	77.6	2.1	3.8
16	18.9	6.6	95.4	56.9	4.5	7	17.8	4.4	90.7	58.6	4.7	9.6
17	16.5	0.3	96.7	37.4	8.0	0	20.3	4.2	93.1	45.0	8.7	0
18	17.1	2.3	88.3	44.6	6.3	3.8	22.0	5.5	90.6	53.1	6.8	0
19	17.2	5.7	95.3	60.3	5.0	3.6	23.7	9.6	84.6	52.7	6.1	17.9
20	22.4	4.0	76.4	34.4	9.4	0	20.6	9.5	86.3	63.7	3.0	75.6

(Continued)

TABLE 2.4 (CONTINUED)

Variation of Agro-Meteorological Parameters in the Study Area During the Drought (2007–08) and Wet Years (2010–11)

	2007–08						2010–11					
Week	T-max (°C)	T-min (°C)	RH max (%)	RH min (%)	Sunshine Hours	Rainfall (mm)	T-max (°C)	T-min (°C)	RH max (%)	RH min (%)	Sunshine Hours	Rainfall (mm)
21	24.6	10.1	83.0	43.7	7.4	0	21.7	8.0	92.6	50.1	5.5	0
22	27.4	9.1	84.0	40.1	9.8	0	21.5	10.1	81.3	56.3	4.0	9.8
23	29.8	13.1	76.7	42.7	8.8	0	25.7	10.0	91.1	48.6	8.9	0
24	30.3	13.4	82.9	59.7	8.8	0	29.3	13.0	91.0	42.7	8.8	0
25	33.7	14.4	58.1	25.0	10.3	0	30.8	14.5	74.3	38.0	8.0	0
26	32.9	14.7	55.1	30.8	9.7	0	31.7	15.3	72.3	39.4	9.2	15.9
Avg	25.7	9.2	81.5	43.3	8.0	-	24.7	10.0	89.9	53.4	6.7	-
Min	16.5	0.3	55.1	25.0	4.5	-	12.3	2.9	72.3	38.0	1.4	-
Max	33.7	15.6	96.7	60.3	10.6	14.2	33.5	19.6	99.1	85.0	9.4	75.6
SD	5.7	4.6	11.0	9.6	1.9	3.9	5.8	4.8	6.7	10.6	2.2	19.5
CV (%)	22.1	49.8	13.5	22.2	23.7	8.5	23.3	48.3	7.5	19.8	33.5	8.1

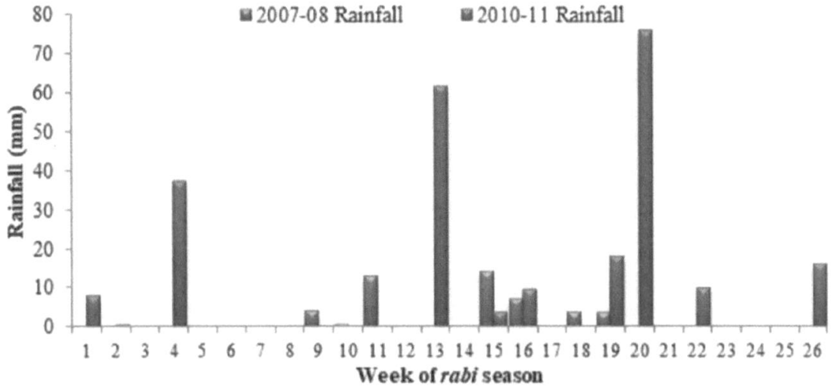

FIGURE 2.3 Variation of rainfall during drought and wet years.

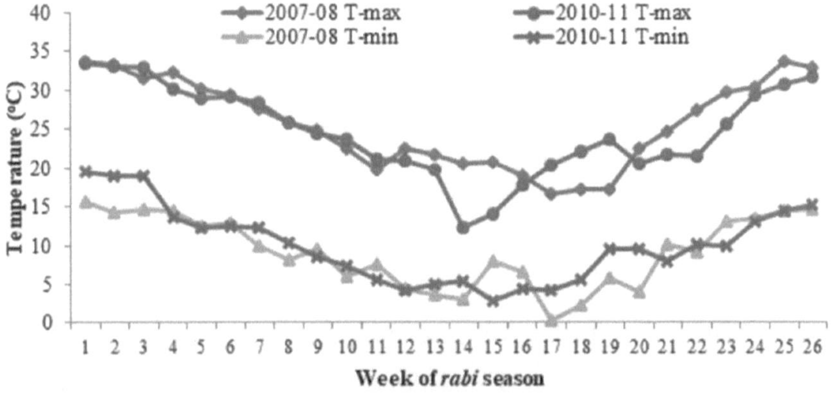

FIGURE 2.4 Variation of maximum and minimum temperature during drought and wet years.

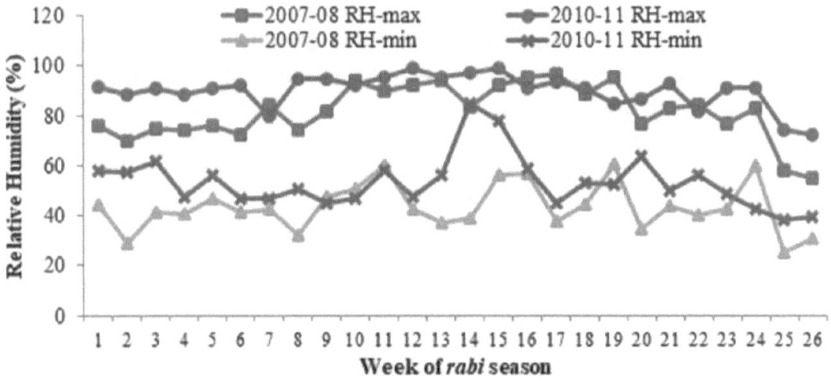

FIGURE 2.5 Variation of maximum and minimum relative humidity during drought and wet years.

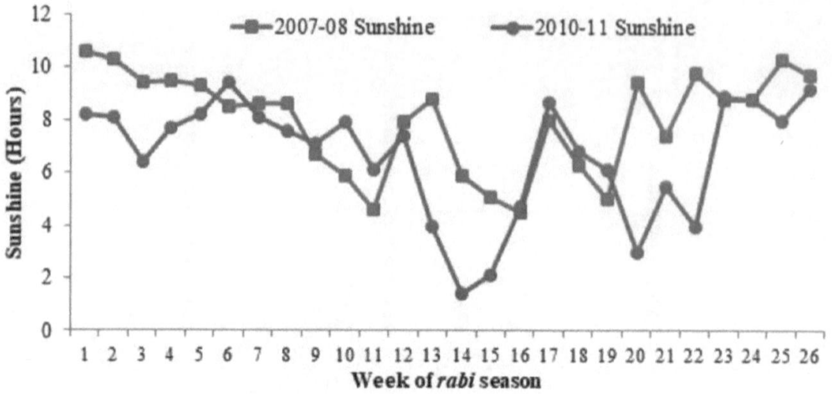

FIGURE 2.6 Variation of the number of sunshine hours during drought and wet years.

During the initial stage of the crop, up to the 5th week, the number of sunshine hours per day during the drought year was higher than in the wet year. During the 14th week, the number of sunshine hours was 1.4 hours during the wet year (Figure 2.6). The average number of sunshine hours during the drought year was 8.0 hours, compared with 6.7 hours during the wet year. The minimum and maximum numbers of sunshine hours were 4.5 hours and 10.6 hours during the drought year, respectively, and 1.4 hours and 9.4 hours, respectively, during the wet year.

2.4.2 Wheat Yield Prediction through Models

2.4.2.1 Agro-Meteorological Index Models

Six meteorological indices, namely GDD, AGDD, HTU, PTU, TD and VPD, were computed from meteorological data. Results indicated a negative linear relationship between yield and accumulated growing degree-days, AGDD (r=0.696) and GDD (r=0.666). Fischer and Kohn (1966) also found that, under higher temperature conditions, a decrease in grain yield is associated with a decrease in either vegetative growth or post-flowering plant water status. HTU showed a significantly negative relationship with yield (r = 0.779). PTU also showed a significant negative relationship (r = 0.723). Similarly, Angus et al. (1981) and Bazgeer (2005) reported significant negative relationships between PTU and yield. The correlation between wheat yield and VPD was not significant, although it suggested a negative linear relationship. Scatter plots of meteorological indices against yield are shown in Figure 2.7.

Wheat yield prediction by the use of different agro-meteorological indices and the estimated yield was compared from BES, as shown in Table 2.5. The results revealed that the yield predicted in years 2007–08 (RD = 3.18%) and 2010–11 (RD = 0.78%), following modeling by GDD, was greater than the yield

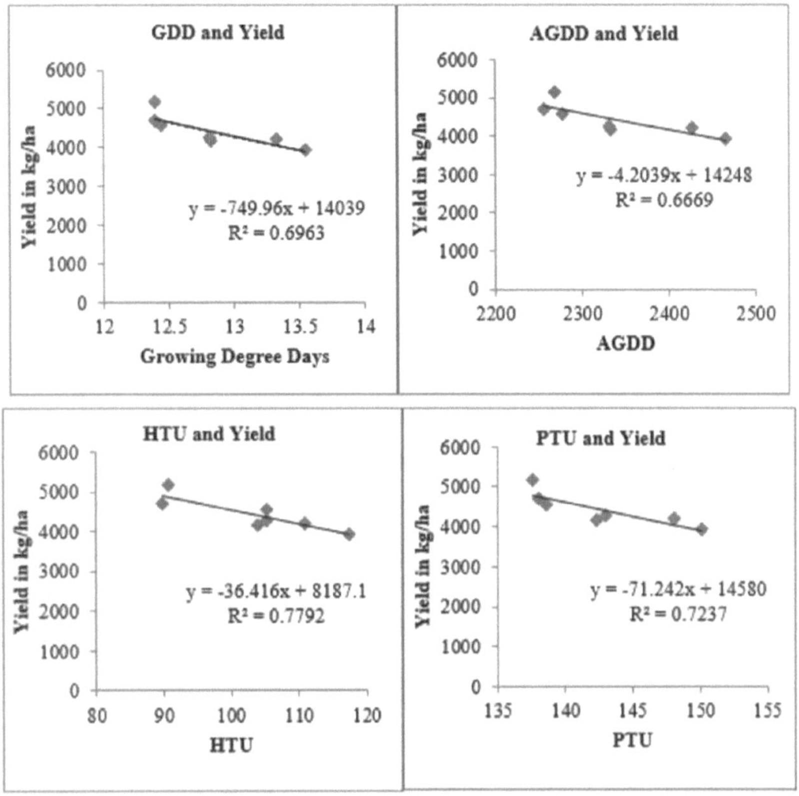

FIGURE 2.7 The relationship between wheat yield and agro-meteorological indices (GDD, AGDD, HTU, PTU) for SBS Nagar district.

TABLE 2.5
Prediction of Wheat Yield by Different Agro-Meteorological Indices and Relative Deviation* from BES Yield Estimate For the Years 2007–08 and 2010–11 in SBS Nagar District

		Prediction by Agro-meteorological Indices (kg/ha)			
Year	BES (kg/ha)	GDD	AGDD	HTU	PTU
2007–08	4561	4711 (3.18)	4680 (2.54)	4355 (–4.73)	4705 (3.06)
2010–11	4708	4745 (0.78)	4767 (1.23)	4922 (4.34)	4745 (0.77)

* Figures in parentheses show the relative deviation from the BES estimate

given by BES. Also, yield predicted by AGDD was higher than that of BES in both 2007–08 (RD = 2.54%) and 2010–11 (RD = 1.23%). Similarly, the yield predicted by HTU was lower than the BES estimated yield in year 2007–08 (RD = −4.73%) but higher in year 2010–11 (RD = 4.34%). The yield predicted by PTU overestimated BES in year 2007–08 (RD = 3.06%) and underestimated in the year 2010–11 (RD = −0.77%). However, the predicted yields by GDD, AGDD, PTU were in closer agreement with the BES-estimated yields during 2010–11, than with 2007–08. This may be due to the fact that 2007–08 was a drought year.

2.4.2.2 Agro-Meteorological Multiple Regression Model

Regression equations were developed to predict the yield using different agro-meteorological parameters, as follows

For year 2007–08:

$$Y = 5700.794 + 1.501 * RF + 421.812 * T_{max} - 124.267 * T_{min} \left(R^2 = 0.667\right) \quad (2.8)$$

For year 2010–11:

$$Y = 5013.704 + 0.400 * RF + 375.968 * T_{max} - 49.443 * T_{min} \left(R^2 = 0.635\right) \quad (2.9)$$

where RF = rainfall (mm); T_{max} = maximum temperature (°C); T_{min} = minimum temperature (°C).

Actual and predicted wheat yields using agro-meteorological model (Equation 2.8 and Equation 2.9) for the verification periods (2007–2008 and 2010–2011) are shown in Table 2.6.

The results revealed that the prediction of yield for year 2010–11 (RD = 0.08 %) was closer to the BES estimates than was the value for 2007–08 (RD = −12.50 %).

Estimates of wheat yield predicted by the use of spectral vegetation indices were closer to the actual yield, compared with results predicted by agro-meteorological indices (Table 2.7).

TABLE 2.6
Prediction of Wheat Yield through the Agro-meteorological Multiple Regression Model and Relative Deviation from the BES Estimate for the Years 2007–08 and 2010–11 in SBS Nagar District

Year	BES (kg/ha)	Prediction by Model (kg/ha)	Relative Deviation (%)
2007–08	4561	4054	−12.50
2010–11	4708	4712	0.08

TABLE 2.7
Comparison of Wheat Yield Predicted from Different Spectral Indices, Vegetation Indices and Relative Deviation* from BES Estimate for the Years 2007–08 and 2010–11 in SBS Nagar District

		Prediction by Vegetation Indices (kg/ha)				Prediction by Agro-meteorological Indices (kg/ha)			
Year	BES (kg/ha)	NDVI	RVI	IPVI	DVI	GDD	AGDD	HTU	PTU
2007–08	4561	4458	4520	4551	4487	4711	4680	4355	4705
		(−2.31)	(−0.90)	(−0.21)	(−1.64)	(3.18)	(2.54)	(−4.73)	(3.06)
2010–11	4708	4609	4662	4680	4615	4745	4767	4922 (4.34)	4745
		(−2.14)	(−0.99)	(−0.59)	(−2.01)	(0.78)	(1.23)		(0.77)

* Figures in parentheses show the relative deviation from the BES estimate

2.5 CONCLUSION

The area under wheat was 4.43% higher in the wet year than that achieved during the drought year. This increase in wheat area may be attributed to the wetter conditions in the area during 2010–11 (239.5 mm rainfall) than in 2007–08 (46 mm rainfall). The increase in the wheat area in the wet year was mostly in the rainfed Shivalik foothills and was negligible in the lower parts of the district, with assured irrigation. The wheat yield predicted by were closer to the actual yields for both the drought and wet years than were the values predicted using agro-meteorological index models. The per cent relative deviation for both the years was less than 5 per cent. The lowest per cent relative deviation occurred in the case of AGDD (2.54%) and PTU (0.77%) for the drought year and the wet year, respectively. The per cent relative deviation was maximum in the case of HTU (−4.73% and 4.34%) for both years. Agro-meteorological multiple regression model prediction of wheat yield for the wet year (RD = 0.08%) was closer to actual yield than for the dry year (RD = −12.50%).

REFERENCES

Angus J F, Mackenzie D H, Morton R and Schafer C A (1981) Phasic development in field crops, II. Thermal and photoperiodic responses of spring wheat. *Field Crop Res* **4**: 269–83.

Anonymous (1991–2012) *Statistical Abstracts of Punjab.* The Economic Advisor, Punjab.

Anonymous (2014a) http://www.indiastat.com/table/agriculture/2/foodgrains/17786/379015. (Accessed on 8.08.2016).

Anonymous (2014b) http://www.esopb.gov.in/static/PDF/glance%202014.pdf. (Accessed on 9.08.2016).

Anonymous (2015) http://www.indiastat.com/table/agriculture/2/wheat/17195/968258/data. (Accessed on 8.08.2016).

Bazgeer S (2005) *Land Use Change Analysis in the Submountainous Region of Punjab Using Remote Sensing, GIS, and Agrometeorological Parameters.* Ph.D. Dissertation, Punjab Agricultural University, Ludhiana.

Crippen R E (1990) Calculating the vegetation index faster. *Remote Sens Environ* **34**: 71–73.

Fischer R A and Kohn G D (1966) The relationship of grain yield to vegetative growth and post-flowering leaf area in the wheat crop under conditions of limited soil moisture. *Aust J Agric Res***17**: 281–95.

Jordan C F (1969) Derivation of leaf area index from quality of light on the forest floor. *Ecology* **50**: 663–66.

Michael A M (1978) *Irrigation: Theory and Practice*, pp. 754–61, Vikas Publishing House Pvt. Ltd, New Delhi.

Rao G S L H V P (2003) *Agricultural Meteorology*, pp. 95–112, Director of Extension, Kerala Agricultural University, Thrissur.

Rouse J W, Haas R H, Deering D W, Schell J A and Harlan J C (1974) *Monitoring the Vernal Advancement and Retrogradation (Green Wave Effect) of Natural Vegetation*, p. 371, NASA/GSFC Type III Final Report, Greenbelt, MD.

Tucker C J (1979) Red and photographic infrared linear combinations for monitoring vegetation. *Remote Sens Environ* **8**: 127–50.

3 Geostationary Meteorological Satellite for Agro-Meteorological Applications

Rahul Nigam and Bimal K. Bhattacharya

CONTENTS

3.1 INTRODUCTION

Accurate high-resolution short- and medium-range weather forecasts are of paramount importance to the Indian economy and agriculture. The day-to-day weather fluctuations play an important role in the decision making of farmers for numerous farming practices, to curtail crop losses and reduce the cost of farming. An economic impact study over different parts of India showed that implementation of the agro-meteorological advisories issued from the India Meteorological Department, through the current framework, data quality and reliability, could reduce the cost of farming by 5–10% (Rathore and Maini, 2008) and increase crop yields by 10–25%. At present, point-scale agro-meteorological information at an irregular temporal domain is used to monitor the situation and issue agro-meteorological advice for a cluster of districts. Hence, there is need to provide the desired agro-meteorological information at an appropriate spatial and temporal scale to capture the variability within a district. There is an urgent need for high-resolution weather forecasting, especially at block levels, improvement in accuracy through better understanding and representation of land surface processes and high-resolution continuous agro-meteorological monitoring of different agro-meteorological parameters. The recent advances of remote sensing over a wide range of electromagnetic spectra in optical, thermal and microwave bands, in terms of spatial, temporal and radiometric resolution, is the key to having better agro-meteorological parameters for near-real-time assessment of crop condition and growth from low Earth orbiting (LEO) and geostationary (GEO) satellites. The synoptic observations of agro-meteorological variables from polar and geostationary platforms is key to the monitoring of agriculture at a regional to national scale. This will help to plan appropriate agricultural operations in order to offset adverse crop conditions.

The daily observations available from polar satellites, such MODIS, are unable to provide a synoptic view of the whole of India. MODIS data are available in tiled grids and with limited passes (one day and night pass) over a region. Hence, this may lead to having discontinuity of data, requiring extra computation to get data at a country scale. Moreover, the view and solar zenith of a polar satellite also changes with each overpass and hence it requires directional correction. The MODIS data are available to users, with a latency time of 16 days. However, geostationary satellite sensors provide data at multiple times (half-hourly) within a day, with a constant view direction, which gives better opportunities to obtain cloud-free observations (Fensholt et al., 2006) for continuous monitoring of vegetation. The country-scale synoptic coverage data at high temporal frequency (30-minute to one-hour intervals) from meteorological payloads (e.g., VHRR, Imager) and multi-spectral camera (e.g., a charge-coupled device, CCD) on board geostationary satellites (e.g., Kalpana 1, INSAT 3A, INSAT 3D and 3DR) are the potential sources to achieve these indicators on a country scale. Moreover, many satellite-based land surface variables, in addition to atmospheric variables, need to be assimilated on an operational basis to improve the accuracy of weather forecasting. The combination of advanced level products and weather forecasting can lead to the extraction of practical value-added information.

Quantified land surface variables are now available, using observations from Indian geostationary platforms at between 1- and 10-km spatial resolutions through an operational and automated data processing chain of IMDPS (INSAT Meteorological Data Processing System), with pre-defined algorithms. Through IMDPS, geo-registered products are generated within 10 minutes of image acquisition after basic corrections for sensor parameters, such as servo, ephemeris extraction and stagger correction. The geostationary satellite products are available for scientific users through the MOSDAC web portal (www.mosdac.gov.in). At present, many meteorological and agro-meteorological variables are available at half-hourly and daily scales, such as insolation, vegetation index, land surface temperature (LST), rainfall, hydro-estimator, fog, etc. These variables will further be used to derive value-added agro-meteorological parameters/indicators, such as progressive in-season crop area, sowing date, leaf area index and potential and actual evapotranspiration for regular monitoring of crops. These will be further used for crop monitoring, yield estimation and improvement of the high-resolution weather forecast.

These agro-meteorological indicators on a spatial scale at regular intervals and with accurate high-resolution weather forecasting (Joshi et al., 2011) at short, medium and extended ranges are essential components to generate value-added information on early-warning and forewarning for farmers' advisories. The suite of current and future geostationary sensors is summarized in Table 3.1. The retrieval of various meteorological and value-added agro-meteorological parameters, using geostationary satellites, are discussed in the following sections.

3.2 CORE AGROMET PRODUCTS FROM AN INDIAN GEOSTATIONARY SATELLITE

3.2.1 Normalized Difference Vegetation Index (NDVI)

Vegetation index is a mathematical representation of the spectral response of vegetation at different wavelengths, to assess the vigour and health of vegetation. The Normalized Difference Vegetation Index (NDVI) is widely used for monitoring vegetation growth. Vegetation reflects more in the green band (550 nm) and less in the red band (650 nm), due to strong absorption by the leaf pigments (chlorophylls) used for photosynthesis. Reflectance is greater in the near-infrared NIR (800 nm) band, due to the cellular structure of the leaves. For land surfaces dominated by vegetation, the NDVI values normally range from 0.25 to 0.9 during the vegetation growing cycle, the higher values being associated with greater vegetation vigour and greenness of the plant canopy. To obtain regular NDVI from INSAT 3A, CCD cloud screening and atmospheric correction were applied

TABLE 3.1
Characteristics of Indian Geostationary Satellite Sensors

Sensors	Channels	Wavelength (μm)	Spatial Resolution
		Satellite: Kalpana 1 (K 1)	
VHRR	Visible	0.55–0.75	2 km × 2 km
	Water vapour	5.70–7.10	8 km × 8 km
	Thermal IR	10.5–12.5	8 km × 8 km
		Satellite: INSAT 3A	
VHRR	Visible	0.55–0.75	2 km × 2 km
	Water vapour	5.70–7.10	8 km × 8 km
	Thermal IR	10.5–12.5	8 km × 8 km
CCD	Red	0.62–0.68	1 km × 1 km
	NIR	0.77–0.86	1 km × 1 km
	SWIR	1.55–1.69	1 km × 1 km
		Satellite: INSAT 3D & 3DR	
Imager	Visible	0.55–0.75	1 km × 1 km
	Shortwave infrared (SWIR)	1.55–1.70	1 km × 1 km
	Middle infrared (MIR)	3.80–4.00	4 km × 4 km
	Water vapour (WV)	6.50–7.00	8 km × 8 km
	Thermal infrared 1	10.2–11.3	4 km × 4 km
	Thermal infrared 2	11.5–12.5	4 km × 4 km
Sounder	19 channels		10 km × 10 km

*Future ISRO's Geostationary Satellite: GISAT, INSAT 3D-S

to the cross-calibrated Top of the Atmosphere (TOA) reflectances (Nigam et al., 2011). These were computed as:

$$\rho_{TOA(\lambda)} = \frac{\pi d^2 L_{TOA(\lambda)}}{E_{0(\lambda)} \cos\theta_s} \tag{3.1}$$

where $L_{TOA(\lambda)}$ = at-sensor cross-calibrated band radiances in $Wm^{-2}\mu m^{-1}sr^{-1}$ in a given band of CCD;

d = earth–sun distance correction factor, calculated as follows:

$$d = (1 - 0.01672 \times \cos((0.9856 \times (Julianday - 4))))$$

$E_{0(\lambda)}$ = exo-atmospheric bandpass irradiances weighted through CCD RSR (relative spectral response) at a fixed wavelength interval;

θ_s = solar zenith angle (degree).

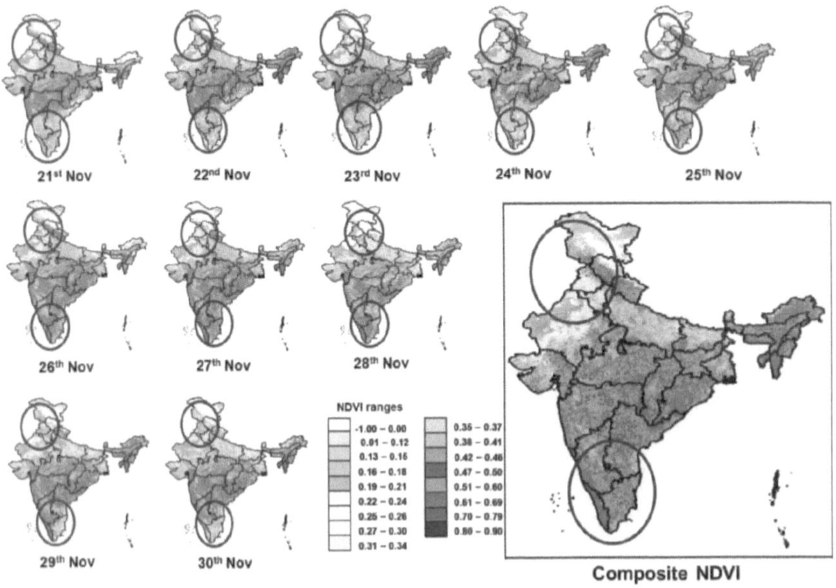

FIGURE 3.1 NDVI temporal dynamics from INSAT 3A CCD.

The atmospheric correction can be carried out using SMAC (Simple Model for Atmospheric Correction) (Rahman and Dedieu, 1994). The time-series of a CCD NDVI composite for a growing year (21 November to 30 November) is shown in Figure 3.1 (Nigam et al., 2012). The operational products were made available to users from the MOSDAC (Meteorological and Oceanographic Satellite Data Archival Centre) (*http://www.mosdac.gov.in*) site.

3.2.2 Surface Insolation

The amount of solar radiation flux or irradiance reaching the ground surface is known as surface insolation or global (direct + diffuse) insolation. There are two main approaches, the Kondratyev–Kubelka–Monk (KKM) (Rosema, 1993) or the Tanahashi physical model (TPM) (Tanahashi et al., 2001), to obtain insolation. The KKM is based on the "bottom-up" approach, first determining the albedo, the atmospheric transmissivity and then the insolation. The TPM is basically a "top-down" approach that starts with exo-atmospheric irradiance in the shortwave and stepwise subtracts the attenuation by atmospheric constituents and cloud till it reaches the ground. The basic steps in both approaches are: (i) generation of solar zenith angle, (ii) cloud mask generation, differing under (iii) clear or cloudy sky conditions, to achieve the computation of insolation.

3.2.2.1 Computation of Clear Sky Insolation (Rs_{clear})

$$Rs_{clear} = S.\,fj.Cos\theta.t_{clear}$$

For KKM,

$$t_{clear} = (b-a)/[(b(1-A)-(a-bA)e((a-b)\tau)]$$

where

t_{clear} = clear sky transmissivity;

τ= integrated atmospheric optical depth

For TPM,

$$t_{clear} = t_{direct} + t_{Rayleigh} + t_{aerosol}$$

$$t_{direct} = \tau_{o*}\tau_{R*}(1-a(u))*\tau_A$$

$$t_{Rayleigh} = \tau_{o*}(0.5(1-\tau_R))*\tau_A$$

$$t_{aersol} = \tau_o * \tau_{R*}(1-a(u))*F_c*Wo*(1-\tau_a)$$

τ_o = transmittance due to absorption by ozone, estimated by Lacis and Hansen (1974);

τ_R = transmittance due to Rayleigh scattering, which is wavelength dependent;

τ_a = transmittance due to attenuation by aerosols, estimated according to Macher (1983). β = The seasonal representative angstrom turbidity parameter are given by Iqbal (1983).

F_c = ratio of forward scattering to total scattering as a function of solar zenith angle

W_o= single-scattering albedo

3.2.2.2 Computation of Cloudy Sky Insolation (Rs_{cloudy})

For KKM

$$Rs_{cloudy} = S.* f_{j.} * Cos\theta.\ t_{cloudy}$$

where

t_{cloudy} = transmission through cloud, estimated using the Kubelka-Monk theory, relating cloud transmission to planetary albedo for different land-cover types

For TPM,

$$Rs_{cloudy} = Rs_{clear} * (1-aA_c)$$

where

A_c = cloud-top albedo;

a = cloud attenuation coefficient from a lookup table (LUT) generated for different cloud classes, based on cloud-top albedo and cloud-top temperature. By

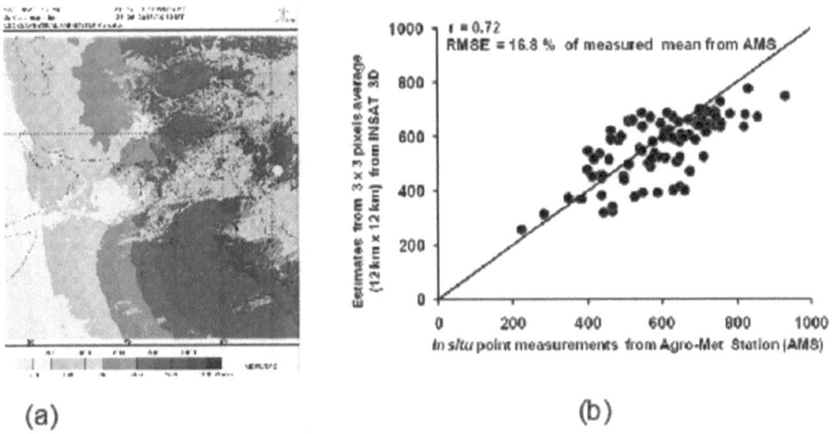

FIGURE 3.2 (a) Instantaneous insolation from INSAT 3D and (b) validation from AMS stations.

applying the above-mentioned algorithm, instantaneous insolation from INSAT 3D (Bhattacharya et al., 2013) was generated as shown in Figure 3.2. The solar hot spot over India is also computed using the long-term INSAT-derived insolation data (Vyas et al., 2016).

3.2.3 Land Surface Temperature (LST)

Crop surface (or skin) temperature may be more directly related to crop growth and canopy water loss in the form of actual evapotranspiration than to air temperature. The difference between the two is a measure of crop stress. However, it should be noted that the satellite-derived "skin temperature" and "crop canopy temperature" are equivalent only when the satellite field of view (FOV) is filled with vegetation. Otherwise, it is the temperature of bare soil, crop, etc. A Single-Channel (SC) method was developed for retrieval of land surface temperature (LST) from K 1 VHRR TIR data (Pandya et al., 2010) through a radiative transfer (RT) model. The basis of the algorithm depends on transmissivity, upwelling and downwelling radiances of the atmosphere. The transmissivity, upwelling and downwelling radiances were computed for the VHRR sensor through the RT simulations by the MODTRAN model, based upon varying atmospheric and surface inputs. The atmospheric transmissivity, upwelling and downwelling radiances were fitted with the atmospheric columnar water vapour content and then a set of coefficients was derived for LST retrieval. The diurnal LST over the Indian landmass is shown in Figure 3.3.

3.2.4 Rainfall

The use of only land-based techniques of rainfall estimation are inadequate for global rainfall assessment as about 70% of the Earth's surface is covered with

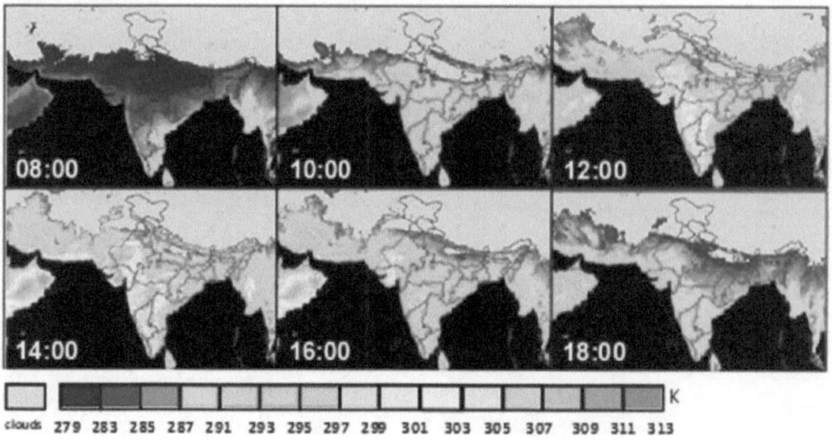

FIGURE 3.3 LST images generated from the Kalpana 1 VHRR data.

water. Moreover, the cost of installation of instruments, their operation and longevity are also not feasible on temporal and spatial scales. The meteorological satellites have shown their potential for improved identification and quantification of precipitation (rainfall) with advances in spectral bands and retrieved algorithms. The primary objective of satellite rainfall monitoring is to provide spatial rainfall data in high temporal domains. Around the globe, the limited and unequal distribution of rain gauges and weather radar over land and the scarcity of rainfall data over the oceans have significantly limited the use of these datasets in numerical prediction, hydrological and crop models. Precipitation is one of the most variable quantities in space and time, due to its uneven distribution over land and sea. Current meteorological geostationary satellites provide visible (VIS) and infrared (IR) spectral data at high temporal resolution, able to capture the growth and decay of precipitating clouds. Microwave sensing has an ability to provide interaction of radiation with hydrometeors but at a coarser resolution and with limited swath from polar satellites, such as Tropical Rainfall Measuring Mission (TRMM) (Kummerow et al., 1998). At present, three algorithms are used to estimate rainfall from Indian geostationary satellites, namely (i) GOES Precipitation Index (GPI), (ii) INSAT Multispectral Rainfall Algorithm (IMSRA) and (iii) the Hydro-estimator. The spatial estimation of rainfall from INSAT is shown in Figure 3.4.

3.2.4.1 GPI Algorithm

The GOES Precipitation Index (GPI) algorithm uses IR images of observed radiant energy from the Earth's atmosphere or from the land surface and water bodies. The strength of this radiated energy is integrated over all spectral bands, using the Stefan–Boltzmann law. If the atmosphere, land or water body emits spectral energy, according to some temperature less than its thermal temperature, then medium emissivity is accounted for in the computation. The emissivity of a

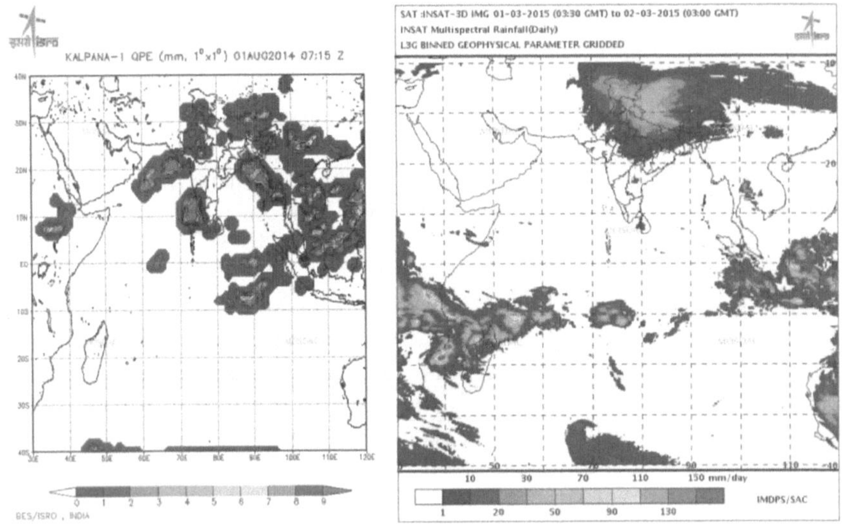

FIGURE 3.4 Rainfall estimation from K 1 VHRR and INSAT 3D (*www.mosdac.gov.in*).

medium determines its emission efficiency. Hence, the brightness temperature of a medium or body can be defined as T_b, and it is related to the physical temperature T of medium or body by the following equation:

$$T_b = \varepsilon T$$

where ε is emissivity of the medium or body. If $\varepsilon = 1$, $T_b = T$, it is a black body; if $\varepsilon = 0$, $T_b = 0$, it is a white body; if $0 < \varepsilon < 1$, $T_b < T$, and it is a grey body.

Here, IR refers to the thermal IR band, that is 10.0–12.5 μm for rainfall computation. This is also an atmospheric window where all earthly bodies radiate maximum spectral radiation. This specific spectral band has strong absorption for different cloud types and land surface/water bodies and minor absorption for atmospheric gases. The opaque land surfaces do not transmit radiation, so measured intensity is closely approximated by the fourth power of the temperature. For surfaces, such as clouds, which are not opaque, the measured intensity is close to effective emissivity times the fourth power of the temperature. Hence, sometimes effective emissivity will be referred more often as cloud emissivity. According to the principle, the cloud emissivity cannot be measured from observations from a single spectral channel, so it is simply assumed to be unity. In that case, the temperature calculated from the observed spectral intensity of radiation is termed the `brightness temperature'. As already mentioned, when $\varepsilon = 1$, black body temperature equals brightness temperature.

The estimates of rainfall using the GOES precipitation Index (GPI) in the 1°×1° grid is computed. The high correlation between the fractional coverage of "cold" clouds and observed rainfall is computed for all grids (Mishra et al., 2011). A linear regression method is developed using all grid data. On the basis of regression of the multi-dataset, the following equation evolves:

$$R = \left(\mathrm{mm\,h^{-1}}\right) * (\mathrm{frac}) * (\mathrm{hours})$$

where R is the rainfall estimate in mm; frac is the fractional coverage of cloud-top temperature < 235 K for each grid; and hours represents the number of hours during the period of measurement (Gairola and Varma, 2015).

3.2.4.2 IMSRA Algorithm

The clouds with the highest brightness in VIS, and with low cloud-top temperature in the IR channel, are expanding in nature and produce more rainfall. The clouds with high cloud-top temperatures generate little or no rainfall, as we know that maximum rainfall occurs during the upwind portion of a convective structure. Moreover, cloud microphysics indicates that the effective radius of cloud-top temperatures need to be known for better rainfall estimation, using the IR channel. The data are classified between pixels having rainy clouds as per the aforementioned statement, then these pixel data are used to pre-calibrate mean rain rate for cloud-top brightness temperature from IR with a Satellite Passive Microwave Radiometer (e.g., TRMM-TMI) and/or Active Microwave Radar (TRMM-PR) for the same geolocation and time. Then, the estimated rainfall is further calibrated and validated with ground-based Doppler Weather Radar data at limited locations over India. At present, the rainfall products are available at 30-minute and daily intervals (Prakash et al., 2010).

3.2.4.3 Hydro-Estimator (H-E) Algorithm

The Hydro-Estimator (H-E) algorithm provides improved rainfall estimates in near-real-time, as compared with the rainfall estimates described in Sections 3.2.4.2 and 3.2.4.3. In this, first of all, brightness temperature at spectral channel 10.7 μm ($T_{b=}10.7$) and rain rate (mm h^{-1}) (R) is related as: $R = a*exp(-bT_b{}^{12}{}_{10.7})$

This method has an ability to provide rainfall estimates in both convective and non-convective systems by using different scaling function R–T_b relationships. Hence, we are able to observe more rainfall during the convective system. During the convective system, a similar equation, with a different set of coefficients for precipitable water, is generated, using the numerical weather prediction (NWP)-based model output. Thus, this will give a higher water vapour content below the cloud during the convective system. The relationship is generated in such a way that 0.5 mm h^{-1} of rain rate at 240 K, and a precipitable water (NWP model) – dependent value at 210 K of Tb 10.7. In this algorithm, the water vapour content below the cloud limits the rainfall at each grid. In this manner, the rainfall at a pixel is dependent on the columnar water content below the cloud. Hence, in this algorithm, both cloud-top temperature as well as columnar water content is considered. For example, if the rain rate is higher at a particular pixel but a higher cloud-top temperature is achieved at an adjacent pixel, then the colder pixel/grid will get more rainfall (Kumar and Varma, 2016). To run this algorithm at an operational relationship between the NWP model-derived columnar water content, temperature and humidity profiles up to cloud height and rainfall (*R*) (in mm h^{-1}) were generated and applied in this algorithm. A further few add-on corrections for wet and dry conditions, warm top modification, orography and parallax correction were also adopted in this algorithm (Varma, 2018).

3.2.5 Fog

Fog occurs due to the difference between air temperature and dew point exceeding some threshold, leading to condensation of atmospheric water droplets. Fog is disastrous in the case of air and road transport and is hazardous in highly polluted areas, in terms of human health. The current Indian geostationary satellite spectral bands provide an opportunity for real-time detection of fog at a country-wide scale. Many workers have tried to detect fog, using different satellite platforms at local and regional scale (Bendix et al., 2003). According to the satellite sensor specification and their saturation limits for different spectral channels, dedicated threshold radiances or brightness temperatures have to be computed to detect fog, using a multi-temporal dataset. In general, fog can be categorized based on its method of formation, such as radiative or advective fog, but, at present, only one fog product can be detected from a satellite. Further categorization can be achieved on the basis of a numerical weather prediction model, with the data being reanalyzed. Due to fog, the radiative energy received by different channels of the satellite will change, compared to energy received from non-fog areas, and this has been taken as the basis by which to discriminate fog from a clear sky.

To detect fog, the difference in brightness temperature between 3.9 μm (Mid Infrared) and 10.8 μm (thermal or IR) is used primarily. The emissivity properties of fog droplets in these spectral channels (MIR and IR) are the basis for the identification of fog. Through different studies, it was found that the small droplets in fog have lower emissivity at 3.9 μm than at 10.8 μm. For larger droplets, emissivity is approximately the same for both spectral channels (Hunt, 1973). Ellord (1995) computed that, when the difference in brightness temperature values is less than 0.5 K, it is a cloud-free sky, whereas a difference of more than 2.5 K is an indication of opaque clouds. By using this concept and a multi-dataset of a particular satellite, the threshold values will be evaluated for difference between brightness temperatures of these channels to distinguish between clear, foggy and cloudy skies. In the case of INSAT 3D, the different threshold values and the difference between brightness temperatures were revisited according to the sensor specification and its ability to detect fog. The new threshold was generated for Indian regions specific to INSAT 3D satellite spectral channels MIR (5.5 K) and IR (7.0 K) by Chaurasia et al. (2011). This threshold will change according to the central spectral band and width of MIR and IR channels, along with the spatial footprint of the satellite. An example of country-scale fog detection over the Indian landmass from INSAT 3D is shown in Figure 3.5.

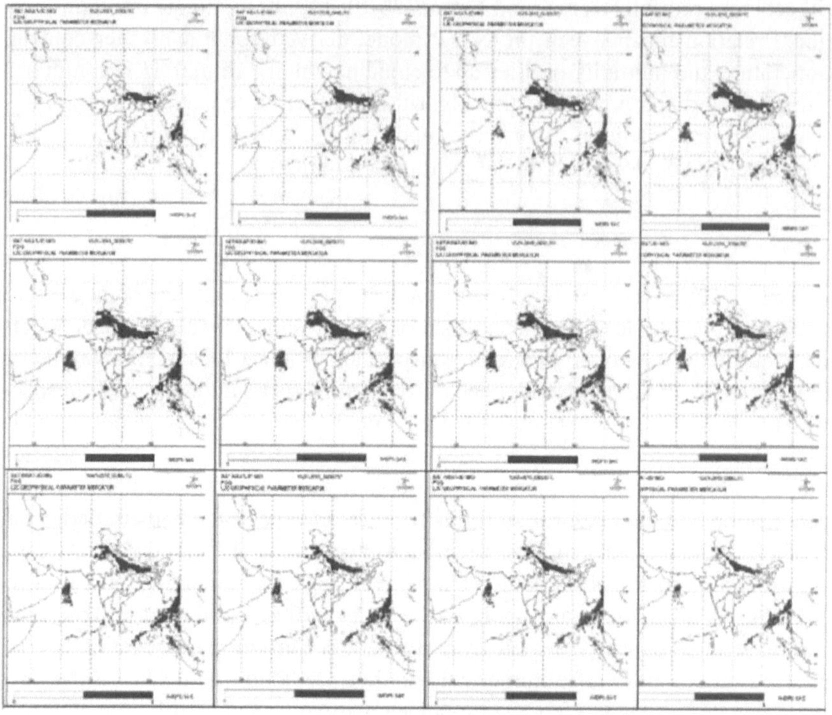

FIGURE 3.5 Diurnal variation of fog as detected by INSAT 3D for 10 January 2015 from 0330 UTC to 0900 UTC (*www.mosdac.gov.in*).

3.2.6 Land Surface Albedo

The ratio of hemispherically reflected shortwave radiation to incident shortwave radiation (0.4–4 μm) or surface insolation, is termed surface albedo. There are three main approaches to obtaining surface albedo, using K 1 VHRR, namely the (i) Kondratyev (Rosema, 1993), (ii) Mossou et al. (1989) and (iii) Gautier et al. (1980) approaches. Among all three approaches, the Kondratyev approach was found to provide least bias (−0.002 to 0.002) for K 1 data. In the Kondratyev approach, the minimum 10-year planetary (earth–atmosphere)

albedo (A'_{min}) was generated in a time-series of K 1 VHRR noon time (1430 IST) VIS band data:

$$A'_{min} = \pi . L_{VIS} \,/\, I_{OVIS} Cos\theta$$

L_{VIS} = radiance (Wm^{-2}) = [(grey count)/(maximum grey count)] L_{max};

L_{max} = saturation radiance for VHRR VIS band (=524 Wm^{-2});

I_{OVIS} = exo-atmospheric bandpass irradiance (Wm^{-2}) for VHRR VIS;

The spectral range (=1384 Wm^{-2}) is determined by weighting, through the relative spectral response (RSR) curve of VHRR;

θ = solar zenith angle computed from the time of acquisition, calendar day of the year, latitude and longitude, using universal astronomical functions;

A'= Planetary (earth–atmosphere) albedo.

Then, A'_{min} values were collected from dark dense forest targets. Typical albedo values of dark dense forests are approximately 0.07. Then, the atmosphere optical depth was obtained by inverting the atmosphere radiation transmission function that relates surface (A) and planetary albedo (A'_{min}) through optical depth. The function form is given below:

$$A = \left(A'b - a\right) - a\left(A' - 1\right)e^{\left((a-b)\tau'_o\right)} / \left(A'b - a\right) - b\left(A' - 1\right)e^{\left((a-b)\tau'_o\right)}$$

where $b = 2\beta$= backscatter coefficient, typically 0.1;

$a = \beta / Cos\theta$, τ'_o = atmospheric optical depth.

Applying this optical depth uniformly, the whole-image albedo could be obtained from the planetary albedo. The temporal and composite albedo generated for K 1 VHRR (Bhattacharya et al., 2009) is shown in Figure 3.6. For generation of composite albedo, the minimum value from 11 November to 19 November is computed to minimize the perturbation by atmospheric components, such as cloud.

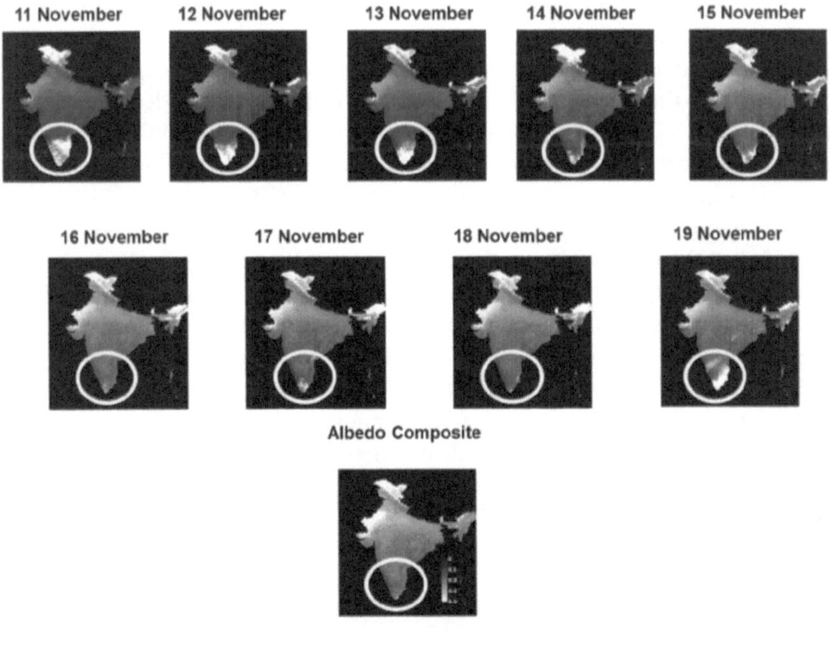

FIGURE 3.6 Albedo compositing from a geostationary satellite.

3.3 VALUE-ADDED AGROMET PRODUCTS FROM A INDIAN GEOSTATIONARY SATELLITE

3.3.1 Leaf Area Index (LAI)

LAI is defined as the single-sided area of green, functioning leaves per unit ground area. It is essential in agricultural applications to estimate LAI, as it exhibits a major control on energy (transpiration) and mass (CO_2) exchange between canopy and the atmosphere and is hence an indicator of vegetation growth. LAI can play a vital role for determining vegetation physiological state and health and was found to be useful for the detection of crop biotic and abiotic stresses to model crop productivity (Boegh et al., 2002). Agricultural crop LAI from INSAT 3A CCD has been estimated by using the one- dimensional (1D) radiative transfer model *ProSail*. This model is the amalgamation of the *PROSPECT* and *SAIL* models. The *PROSPECT* model (Jacquemoud and Baret, 1990) simulates reflectances at leaf level, whereas *SAIL* addresses their directionality. To simulate reflectances for INSAT 3A CCD spectral bands with its specific band width sensor, spectral response function was integrated into the *ProSail* model. *ProSail* simulates a unique set of reflectances for all three INSAT 3A CCD bands, based on the values of input parameters. To simulate reflectances for maximum reasonable possible conditions, according to input parameters, forward simulation was designed. Input parameters were divided into a reasonable interval within a specific tolerance range of all input parameters, to cover the entire dynamic range of *rabi* agricultural crops (sown in winter, harvested in spring) (Houborg et al., 2007). For different agro-climatic zones, the soil spectral library (Saxena et al., 1997) was used as an input for running the model. The model was run in forward mode to generate synthetic reflectance for the three INSAT 3A CCD spectral bands over all scenarios for respective soil layers of the region. Reflectances obtained for all CCD bands by running the model in forward mode correspond to unique input parameters. To invert the satellite-observed surface reflectance of CCD, an inversion technique of least square distance (LSD) was used to obtain the unique LAI for a given three-band reflectance of a pixel for a *rabi* agriculture crop. An agriculture crop map was superimposed over each date for the CCD data to obtain the agriculture crop pixel for LAI retrieval. Surface reflectance in all three bands was compared with the entire synthetic database of reflectances. By applying the inversion technique, a unique LAI was estimated for a unique set of three band CCD reflectances. Spatial 10-day LAI composites from INSAT 3A CCD are shown in Figure 3.7 (Nigam et al., 2016). Ground-measured LAI data at three field sites over different crop phenological stages was compared with INSAT 3A CCD-retrieved LAI data. The CCD LAI was able to capture seasonal fluctuations in LAI similar to the *in-situ* data during the crop growth cycle. The LAI value retrieved from INSAT 3A CCD showed a root-mean-square-error (RMSE) of 0.62 (n = 20).

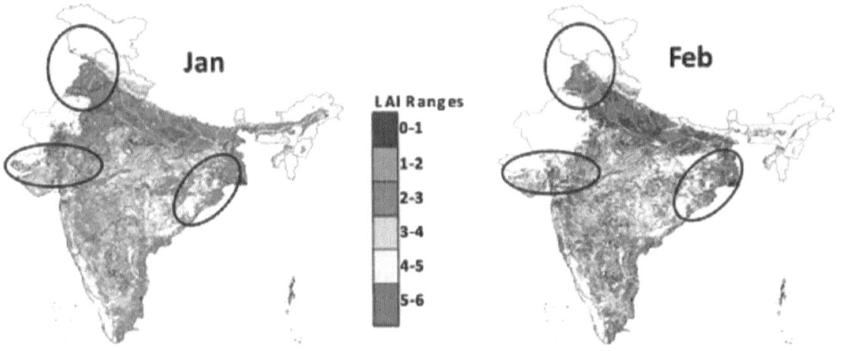

FIGURE 3.7 Crop LAI from INSAT 3A CCD, using the *ProSail* radiative transfer model.

3.3.2 Evapotranspiration

3.3.2.1 Potential Evapotranspiration

Surface evapotranspiration (ET) is one of the potential drivers of the carbon and hydrological (or water) cycles. ET plays key roles in energy and water exchange between land surface and the atmosphere. The potential or reference evapotranspiration (ET_0) represents the atmospheric water demand over vegetative surface. The rate of ET_0 is influenced by various meteorological parameters, such as air temperature, wind speed, solar radiation and vapour pressure deficit. Among all meteorological parameters, solar radiation or insolation is found to be the most responsive parameter for ET_0 in various agro-climatic regions. Due to its theoretical background, potential evapotranspiration (ET_0) is expressed in terms of amount of water loss per unit time to the atmosphere from a non-limiting moist surface covered with a uniformly and actively growing short grass such as alfalfa. ET_0 represents the evaporative demand of the atmosphere for a given atmospheric state. The deficiency to fulfill the requirement of the atmosphere leads to moisture stress. The daily ET_0 is computed using daily insolation from INSAT (K 1 and 3D) and three-hourly weather forecast from the WRF (Weather Research Forecast) model (Nigam and Bhattacharya, 2015). The FAO56 model is customized for INSAT and weather data to generate daily ET_0. The ET_0 generated over the Indian region is shown in Figure 3.8 (Vyas et al., 2016b). Presently, the daily operational product of ET_0 is available from the MOSDAC portal for the user community.

3.3.2.2 Actual Evapotranspiration

Actual evapotranspiration (AET) can be estimated from latent heat fluxes (Eλ or LE) and latent heat (L) of evaporation (Brutsaert and Chen, 1996). The satellite-based surface latent heat flux (Eλ) estimation is generally accounted for from the residual of the surface energy balance (Kustas et al., 1994; Moran et al., 1994;

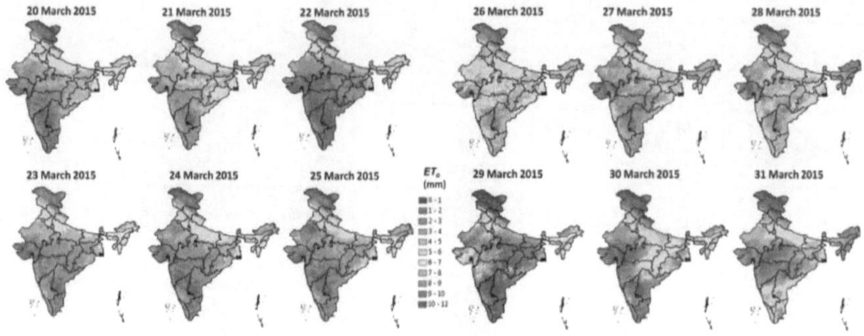

FIGURE 3.8 *ET_0* over the Indian landmass from a geostationary satellite.

Mallick et al., 2007). In the single source (soil–vegetation complex as a single unit) surface energy balance approach, the energy balance formulation can be written as:

$$R_n = H + G + \lambda E + M$$

The energy component for metabolic activities (M) is very small (Samson and Lemeur, 2001) and hence can be ignored. Equation (3.1) can be rewritten as:

$$\lambda E = R_n - G - H$$

Where R_n = net radiation (Wm^{-2}), H = sensible heat flux (Wm^{-2}), G = ground heat flux (Wm^{-2}), and $R_n - G$ = net available energy (Q) in Wm^{-2}. Computation of H requires explicit characterization of many physical parameters, which are difficult to determine at a spatial scale. The combination of the evaporative fraction ($\lambda E/Q$) and Q results in λE estimates. By using the evaporative fraction (Λ), the computational complexities associated with the determination of H (Bastiaanssen et al., 1998; Cleugh et al., 2007) can also be avoided. Therefore, Equation (3.2) can also be written as,

$$\lambda E = Q.\wedge = (R_n - G).\wedge \quad (3.1)$$

The AET can also be estimated using the Priestley–Taylor (P–T) equation developed from the Penman–Monteith (P-M) equation, by assuming that the aerodynamic term can be replaced by a parameter, φ. The simplified form of P–T approach is as follows:

$$\lambda \boldsymbol{E} = \phi[\boldsymbol{Rn}-)*(\Delta / \Delta = \gamma)]$$

where Δ = slope of saturated vapour pressure ($KpaK^{-1}$), γ = psychrometric constant ($kpaK^{-1}$). The expression of the evaporative fraction is as follows:

$$\wedge = \phi(\Delta / (\Delta + \gamma))$$

Therefore, the net available energy (Q) and evaporative fraction (as a function of the Priestley–Taylor parameter, ϕ, air temperature) serve as key determinants of latent heat flux (λE). The fraction of night-time net available energy is much less than the daytime component. Moreover, the transpiration element of evapotranspiration occurs during daylight hours, as a result of stomatal opening and closure during photosynthesis. Night-time soil evaporation contributes a negligible fraction to evapotranspiration throughout the day. The daytime average latent heat flux ($\lambda \boldsymbol{E}_d$) was computed as a product of $\boldsymbol{Q}_d$ and Λ_d, and the $\lambda \boldsymbol{E}_d$ was converted to evapotranspiration (ET) or actual ET (AET) rate in terms of mm d^{-1}, as given below

$$ET = (\lambda E(dL / 24) * C$$

Here, dL is daylight hours, which is computed from latitude, longitude, solar declination and day of the year. C is the conversion factor from energy to water units.

The parameter, dL or daylight hours, in Equation (3.1) is highly dynamic in response to varying soil moisture and surface evaporative conditions at space and time scales. On a regional domain (country-scale), pixel-by-pixel dynamic (ϕ_i) was computed from upper (ϕ_{max}) and lower (ϕ_{min}) limits of ϕ, using warm and wet edges of the thermal inertia-NDVI triangle (Stisen et al., 2008). Here, thermal inertia is closely represented by morning sunrise in terms of surface temperature ($\Delta T = dT_s$), between 1.5 hours and 5.5 hours after sunrise. The morning rise in surface temperature generally remains linear within this four-hour slot and coincides with a linear increase in sensible heat flux during the daytime hours on a clear day (Stisen et al., 2008). Scaling of sensible heat flux with respect to total heat flux requires the determination of a scalar, such as leading to the determination of the evaporative fraction. The actual evapotranspiration has been generated using Kaplan 1 (INSAT) data over the Indian landmass, using the above- mentioned algorithm (Bhattacharya and Nigam, 2015). The monthly generated ET is shown in Figure 3.9. With the use of ET_0 and ET, the relative evapotranspiration (RET) is computed and shown in Figure 3.9. The RET provides information on the dryness of the surface.

3.4 APPLICATION OF AGROMET PRODUCTS TO AGRICULTURAL MONITORING AND WEATHER FORECASTING

3.4.1 Crop Sowing Date

The crop sowing date acts as a key input to initialize the crop conditions within a dynamic crop growth model. To generate crop yield at a spatial scale, the crop sowing date of a particular crop in space is needed to set the phenological

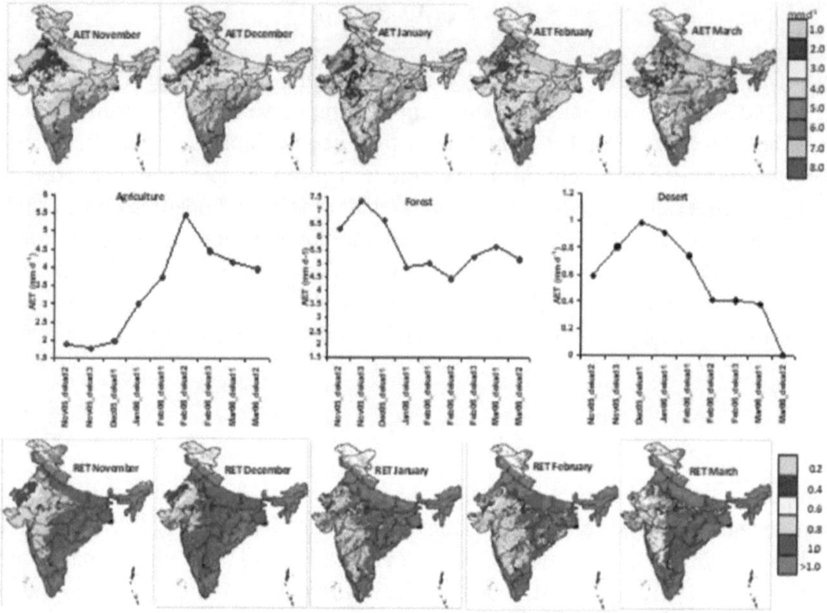

FIGURE 3.9 Monthly Actual and Relative Evapotranspiration during the *rabi* season.

calendar in any crop model. In the general crop growth model used, sowing date is the starting date for generating biomass until its physiological maturity stage is reached. This will decide the growing window for the crop and set the boundary conditions for crop yield modeling and other agronomic inputs, such as irrigation. To estimate wheat crop sowing date over a larger part of India, the daily INSAT 3A CCD NDVI data were used. From the daily CCD NDVI data, the 7-day maximum composite was generated from September to December to plot the temporal profile of NDVI over the development of the wheat crop. The maximum value composite of seven days was prepared to minimize the cloud and atmospheric perturbations in the NDVI data set. In general, a wheat crop takes 5–7 days after sowing to achieve emergence in India, so a 7-day composite has been selected over the conventional 10-day NDVI composite. The 7-day composite is able to capture the fluctuation in NDVI temporal profile within the dynamicity of the 7-day period. In season, the wheat crop mask was prepared from AWiFS (56 m × 56 m) and re-sampled at parent spatial resolution of INSAT 3A CCD (1 km × 1 km) through linear aggregation, and it was applied over each 7-day NDVI composite to obtain the wheat-only pixel. The model was developed to model the NDVI temporal profile in such a way that it was able to pick the date of the inflection point, where NDVI decreased due to the harvest of the *kharif* (monsoon; from late May to early October) crop and, after that, the NDVI profile started to increase with a persistent positive gradient due to emergence of the *rabi* wheat crop (Vyas et al., 2011a). The change in the temporal pattern of the NDVI profile

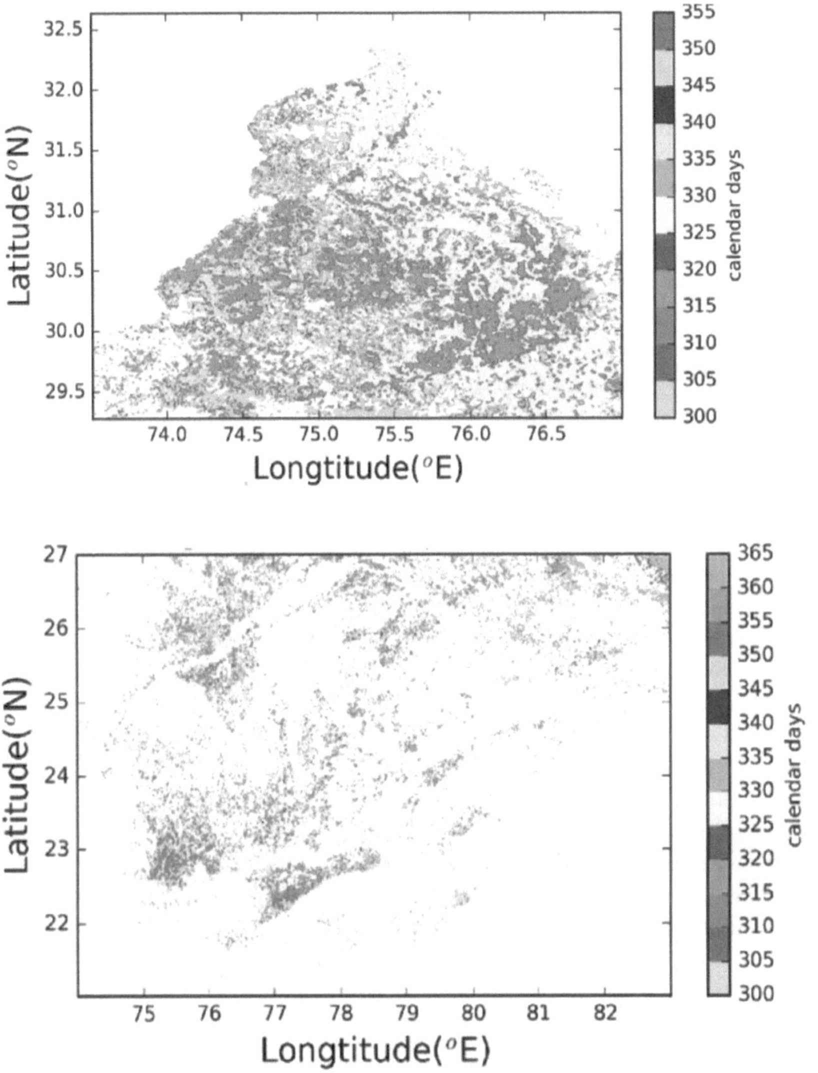

FIGURE 3.10 Spatial sowing map for wheat crops, using geostationary satellite data over Punjab and Madhya Pradesh State.

was detected on the basis of change in the NDVI minimum threshold value, with the persistent positive tangent and sowing date being computed for the wheat crop by deducting seven days from the inflection point date as being due to the coarse footprint of INSAT 3A CCD; the spectral emergence was observed with a lag of 4–5 days. The spatial map of wheat sowing for two wheat- dominant states in 2010–11 is shown in Figure 3.10.

3.4.2 In-Season Crop Area Progress

To monitor the vegetation and to address full resolution of crop growth to capture phenological dynamics, temporal data are required. But, due to frequent cloud over tropical and sub-tropical regions, daily data are required with a wide swath. The high temporal repeativity of INSAT 3A CCD data, along with data over a sub-continent in a single snapshot, is able to minimize the cloud problem for optical remote sensing. In India, different methodologies have evolved over the past two decades to monitor crop area and production, using Indian polar satellite data (Oza et al., 2008). But, as the polar satellite has limitations of temporal resolution, and periodic swath monitoring of crops at regular intervals is difficult at a country level. It can be overcome by using geostationary satellite data. Also, the data obtained from a geostationary satellite can add to the advantage of viewing the same area with changing solar zenith and azimuth angles. This will provide a need to explore geostationary satellite data to monitor the progress of crops at regular intervals under Indian conditions. A methodology has been developed for monitoring progress of crops during the *rabi* (October to April) season, using Indian geostationary satellite data (Nigam et al., 2015). To monitor the progress of *rabi* crops, detailed analysis of NDVI derived from the INSAT-3A CCD sensor during the crop season has been carried out. To monitor the progress of *rabi* crop area, decision tree-based classification was used in this study. This classification technique is widely applied around the world for a range of studies (Ghose et al., 2010). The technique is particularly advantageous for remote-sensing data classification problems relative to the other classification methods, such as artificial neural network (ANN), supervised, unsupervised, etc., due to its flexibility, intuitive simplicity and computational efficiency. Decision tree-based classification has gained increased acceptance for many studies, particularly on a continental or global scale. The progress of *rabi* area on a decade-scale is shown in Figure 3.11 for two consecutive years.

3.4.3 Cold and Heat Waves

In recent years, the impact of climatic variability has assumed increased significance, as warm winters have turned out to be a major impediment, whereas cold wave conditions have a significant positive impact on crop productivity in these areas. Thus, the possible impact assessment of cold wave on wheat productivity can be carried out either by analyzing productivity data, in relation to historical meteorological data, or by using thermal infra-red data from satellites. An attempt was made to capture cold wave conditions using night-time ((1800 GMT and 2350 IST)) thermal infrared brightness temperature (BT_{night}) from Kalpana 1 VHRR during 2008 in the northwestern part of India.

A minimum air temperature model from nighttime Kalpana 1 VHRR (Aich Bhowmick et al., 2008) was applied to generate air temperature from LST. The minimum air temperature, estimated from the Kalpana 1 thermal signatures,

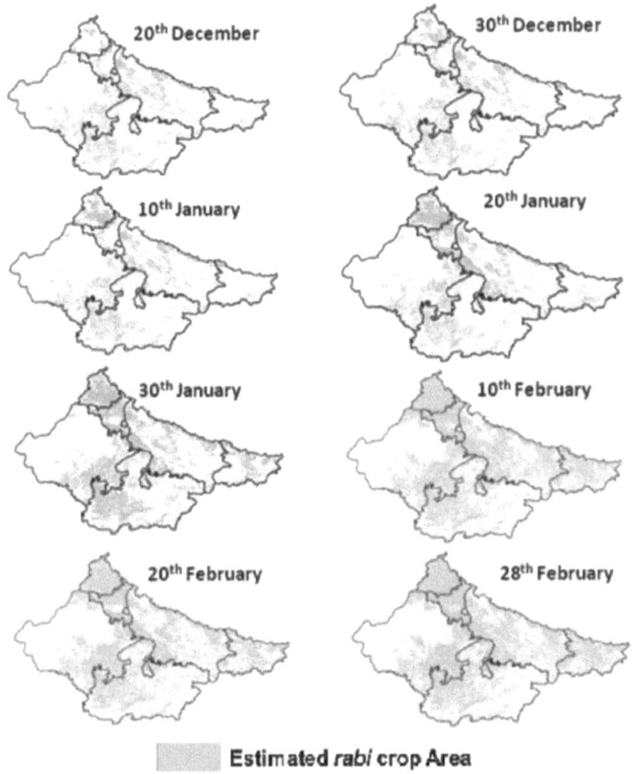

FIGURE 3.11 Near-real-time-cropped progress from INSAT 3A CCD in the *rabi* agricultural season.

clearly indicated a sharp drop in minimum air temperature during January and February 2008 as shown in Figure 3.12.

Similarly, heat waves during summer may also have an impact on milk production by bovine animals. The noontime LST at 1300 IST showed significantly higher surface temperature of the order of 5 to 10°C over north-west India during the third week of May in 2010, as compared with that of 2009, as shown in Figure 3.12. It should also be mentioned that heat waves were followed by heavy monsoon rains in north-west India in 2010, as compared with 2009.

3.4.4 Crop Yield Estimation

The radiation use efficiency (RUE)-based crop yield estimation can be carried out by following the Monteith (1972) approach, using geostationary data. In this approach, maximum radiation use efficiency (RUE_{max}) (in terms of maximally efficient CO_2 assimilation) was constrained by moisture and thermal scalars. The daily surface insolation data from the Indian geostationary satellite (INSAT) is taken and converted to photosynthetically active radiation (PAR)

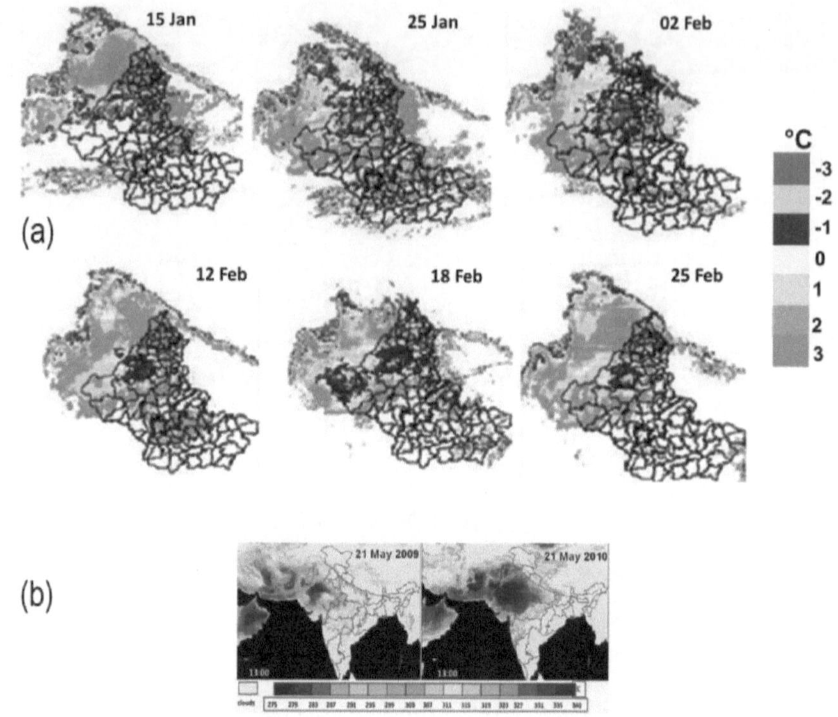

FIGURE 3.12 Zone of influence of (a) cold wave and (b) heat wave from LST.

by applying the conversion factor 'ε', generally taken to be 0.45 by the scientific community. The fractional absorbed PAR (fAPAR) was computed from the LAI retrieved from INSAT with PAR and the extinction coefficient (K_c). The K_c is a model for different wheat growth stages, using *in-situ* measurements of LAI and PAR components, and from reported literature over the study area. To generate an in-season wheat crop map, high-resolution AWiFS data (Oza et al., 2006) are used, as mentioned in Section 3.4.2. The moisture scalar (W_i) was used to determine water-limiting effects on biomass accumulation (Liu et al., 2010). Here, Land Surface Water Index (LSWI) for moisture is derived from NIR and SWIR spectral bands of INSAT 3A CCD, and the bias-corrected LST for temperature was used as a surrogate for moisture- and thermal-limiting factors, respectively (Nigam et al., 2016). All generated input is used to determine crop yield at a spatial scale, using an efficiency-based approach. The advantage of using this approach is that it uses the maximum amount of satellite data and provides more spatial heterogeneity in crop yield. The estimated spatial crop yield, using the RUE approach and geostationary data, is shown in Figure 3.13.

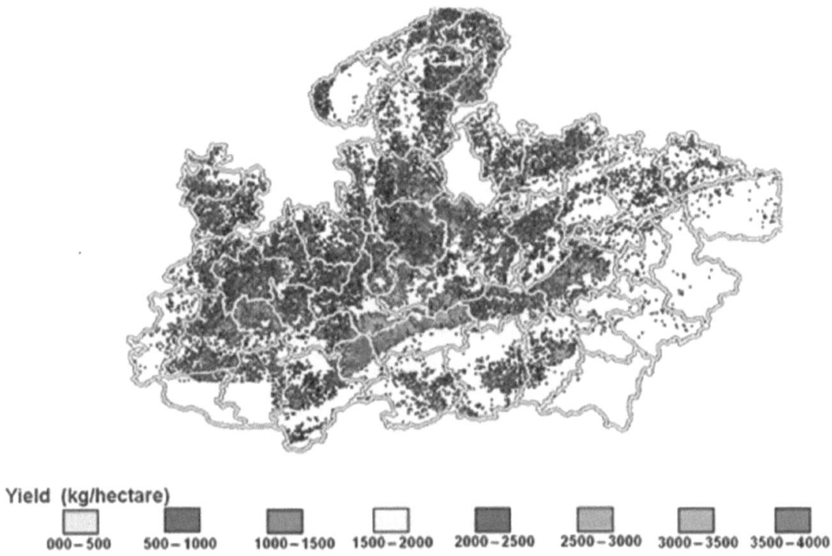

FIGURE 3.13 Spatially distributed wheat yield map, using geostationary data.

3.4.5 Improvement in Quality of Weather Forecast through Land Product Assimilation

To improve high-resolution weather forecasts, near-real-time land products, such as vegetation index, albedo and leaf area index, have to be assimilated in numerical weather prediction (NWP) models. Earlier, Kumar et al. (2013, 2014) assimilated the vegetation fraction (VF) from an operational vegetation index product (e.g., NDVI) and land surface albedo from the Indian geostationary satellites INSAT 3A CCD and K1 VHRR, respectively, in NWP models.

The weather forecast showed that the assimilation of the updated vegetation fraction from INSAT 3A CCD improved the low-level 24-hour temperature (18%) and moisture (10%) forecast in comparison with the control run. Similarly, real-time K 1 VHRR albedo improved the surface temperature, specific humidity and wind speed forecasts as compared with control run experiments.

3.5 CONCLUSION

The meteorological and value-added agro-meteorological products and indicators retrieved from Indian geostationary satellites (INSAT, K 1) appear to be valuable tools for crop growth and condition assessment in near-real-time. This will help planners to make decisions on a regional to national scale. This will also help to maintain timely demand and supply of different agriculture inputs at regional to national scales. In future, high-resolution and multi-spectral data from ISRO geostationary missions, such as INSAT 3D-S and GISAT, will promise more and accurate agro-meteorological parameters to guide agriculture decision-support systems in a better way.

ACKNOWLEDGEMENTS

The authors are grateful to Dr. Markand P. Oza, Dr. R. M Gairola, Dr. Atul Verma, Dr. (Mrs) S. Chaurasia and Dr. Mehul R. Pandya for providing valuable inputs for the manuscript. The authors would also like to thank Deputy Director of Earth, Ocean, Atmosphere, Planetary Sciences and Applications Area and Shri D. K. Das, Director of the Space Applications Centre, Ahmedabad, Gujarat, India, for their constant encouragement and motivation.

REFERENCES

Aich Bhowmick, S., Bhattacharya, B.K., Mallick, K. and Nigam, R. (2008). Retrieval of near surface air temperature in clear skies using Indian geostationary satellite data. *Journal of Agrometeorology (Special Issue)*, 2, 545–556.

Bastiaanssen, W.G.M., Menenti, M., Feddes, R.A. and Holstlag, A.A.M. (1998). A remote sensing surface energy balance algorithm for land (SEBAL). 1: Formulation. *Journal of Hydrology*, 213, 198–212.

Bendix, J., Reudenbach, C.H. and Rollenbeck, R. (2003). The Marburg satellite station. *Proceedings 2002 Met. Sat. Users' Conf. Dublin, 2–6 September 2002*, EUMETSAT, 139–146.

Bhattacharya, B.K. and Nigam (2015). Vegetation – atmosphere interaction: characterization and modelling of energy-mass exchange. *Scientific Report* SAC/EPSA/BPSG/IGBP- EMEVS/SR/01/2015.

Bhattacharya, B.K., Mallick, K., Padmanabhan, N., Patel, N.K. and Parihar, J.S. (2009). Retrieval of land surface albedo and temperature using data from the Indian geostationary satellite: a case study for the winter months. *International Journal of Remote Sensing*, 30, 3239–3257.

Bhattacharya, B.K., Padmanabhan, N., Mahammed, S., Ramakrishnan, R. and Parihar, J.S. (2013). "Assessing solar energy potential using diurnal remote-sensing observations from Kalpana-1 VHRR and validation over the Indian landmass." *International Journal of Remote Sensing*, 34, 7069–7090. doi:10.1080/01431161.2013.811311.

Boegh, E., Soegaard, H., Broge, N., Hasager, C.B., Jensen, N.O., Schelde, K., et al. (2002). Airborne multispectral data for quantifying leaf area index, nitrogen concentration and photosynthetic efficiency in agriculture. *Remote Sensing of Environment,* 81, 179–193.

Brutsaert, W. and Chen, D. (1996). Diurnal variation of surface fluxes during thorough drying (or severe drought) of natural praire. *Water Resources Research*, 32, 2013–2019.

Chaurasia, S., Sathiyamoorthy, V., Shukla, B.P., Simon, B., Joshi, P.C. and Pal, P.K. (2011). Nighttime fog detection using MODIS data over Northern India. *Meteorological Applications*, doi:10.1002/met.248.

Cleugh, H.A., Leuning, R., Mu, Q. and Running, S.W. (2007). Regional evaporation estimates from flux tower and MODIS satellite data. *Remote Sensing of Environment*, 106(3), 285–304.

Ellord, G.P. (1995). Advances in the detection and analysis of fog at night using GOES multispectral infrared imagery. *Weather and Forecasting*, 10, 606–619.

Fensholt, R., Sandholt, I., Stisen, S. and Tucker, C.(2006). Analysing NDVI for the African continent using the geostationary meteosat second generation SEVRI sensor. *Remote Sensing of Environment*, 101, 212–229.

Gairola, R.M. and Varma, A.K. (2015). Quantitative Precipitation Estimation (QPE) from GPI and IMSRA techniques. INSAT 3D algorithm and theoretical basis development document. www.mosdac.gov.in.

Gautier, C., Diak, G. and Masse, S. (1980). A simple physical model to estimate incident solar radiation at the surface from GOES satellite data. *Journal of Applied Meteorology*, 19, 1005–1012.

Ghose, M.K., Pradhan, R. and Ghose, S.S. (2010). Decision tree classification of remotely sensed satellite data using spectral separability matrix. *International Journal of Advanced Computer Science and Applications*, 1, 93–101.

Houborg, R., Soegaard, H. and Boegh, E. (2007). Combining vegetation index and model inversion methods for the extraction of key vegetation biophysical parameters using Terra and Aqua MODIS reflectance data. *Remote Sensing Environment*, 106, 39–58.

Hunt, G.E. (1973). Radiative properties of terrestrial clouds at visible and infra-red thermal window wavelengths. *Quarterly Journal of Royal Meteorological Society* 99: 346–369.

Iqbal, M. (1983). *An Introduction to Solar Radiation*. Canda: Academic Press, p. 390.

Jacquemoud, S. and Baret, F. (1990). PROSPECT: A model of leaf optical properties spectra. *Remote Sensing of Environment*, 34, 75 –91.

Joshi, P.C., Simon, B. and Bhattacharya, B.K. (2011). Advanced INSAT data utilization for meteorological forecasting and agrometeorological applications. In *Challenges and Opportunities in Agrometeorology*, Eds. Attri, S.D., Rathore, L.S., Sivakumar, M.V.K. and Dash, S.K., Springer, (ISBN 978-3-642-19359-0), pp. 273–286.

Kumar, P., Bhattacharya, B.K., Nigam, R., Kishtawal, C.M. and Pal, P.K. (2014). Impact of Kalpana-1 derived land surface albedo on short-range weather forecasting over the Indian sub-continent. *Journal of Geophysical Research-Atmospheres*. (American Geophysical Union). 119, doi:10.1002/2013JD0202534.

Kumar, P., Bhattacharya, B.K. and Pal, P.K. (2013). Impact of vegetation fraction from Indian geostationary satellite on short-range weather forecast. *Agricultural and Forest Meteorology*, 168, 82–92.

Kumar, P. and Varma, A.K. (2016). Assimilation of INSAT-3D hydro-estimator method retrieved rainfall for short-range weather prediction. *Quarterly Journal of the Royal Meteorology Society*, https://doi.org/10.1002/qj.2929.

Kummerow, C.D., Barnes, W., Kozu, T., Shiue, J. and Simpson, J. (1998). The Tropical Rainfall Measuring Mission (TRMM) sensor package. *Journal of Atmospheric and Oceanic Technology*, 15, 809–817.

Kustas, W.P., Moran, M.S., Humes, K.S., Stannard, D.I., Pinter, P.J., Hipps, L.E., Swiatek, E. and Goodrich, D.C. (1994). Surface energy balance estimates at local and regional scales using optical remote sensing from an aircraft platform and atmospheric data collected over semiarid rangelands. *Water Resources Research*, 30(5), 1241–1260.

Liu, J., Pattey, E., Miller, J.R., McNairn, H., Smith, A., and Hu, B. (2010). Estimating crop stresses, above ground dry biomass and yield of corn using multi-temporal optical data combined with a radiation use efficiency model. *Remote Sensing of Environment,* 114, 1167–1177.

Lacis, A.A. and Hansen, J.E. (1974). A parameterization for absorption of solar radiation in cloudy atmosphere. *Journal of the Atmospheric Sciences,* 33, 789–805.

Macher, M. (1983). *Parameterization of Solar Irradiation Under Clear Skies.* M.A.Sc. Thesis. University of British Columbia, Vancouver.

Mallick, K., Bhattacharya, B.K., Chourasia, S., Dutta, S., Nigam, R., Mukherjee, J., Banerjee, S., Kar, G., Rao, V.U.M., Gadgil, A.S. and Parihar, J.S. (2007). Evapotranspiration using MODIS data and limited ground observations over selected agroecosystems in India. *International Journal of Remote Sensing*, 28 (10), 2091–2110.

Mishra, A.K., Gairola, R.M., Varma, A.K. and Agarwal, V.K. (2011). Improved rainfall estimation over the Indian region using satellite infrared technique. *Advances in Space Research*, 48, 49–55.

Monteith, L. (1972). Solar radiation and productivity in tropical ecosystems. *Journal of Applied Ecology*, 9, 747–766.

Moran, M.S., Kustas, W.P., Vidal, A., Stannard, D.I., Blanford, J.H. and Nichols, W.D. (1994). Use of ground-based remotely sensed data for surface energy balance evaluation of a semiarid rangeland. *Water Resources Research*, 30, 1339–1350.

Mossou, G., Diabate, L., Obrecht, D. and Wald, L. (1989). A method for the mapping of the apparent ground brightness using visible image from geostationary satellite. *International Journal of Remote Sensing,* 10, 1207–1225.

Nigam, R., Bhattacharya, B.K., Gunjal, K.R., Padmanabhan, N. and Patel, N.K. (2011). A continental scale vegetation index from Indian geostationary satellite. *Current Science*, 100, 1184–1192.

Nigam, R., Bhattacharya, B.K., Gunjal, K.R., Padmanabhan, N. and Patel, N.K. (2012). Formulation of time series vegetation index from Indian geostationary satellite and comparison with global product. *Journal of Indian Society of Remote Sensing*, 40(1), 1–9.

Nigam, R. and Bhattacharya, B.K. (2014). Potential evapotranspiration (PET) from INSAT 3D insolation Product and short range forecasts. *Algorithm Theoretical Basis Document.* https://mosdac.gov.in/data/doc/INSAT_3D_ATBD_MAY_2015.pdf.

Nigam, R., Vyas, S.S, Bhattacharya, B.K., Oza, M.P. and Manjunath, K R. (2016). Retrieval of regional LAI over agricultural land from an Indian geostationary satellite and its application for crop yield estimation. *Journal of Spatial Science*, http://dx.doi.org/10.1080/14498596.2016.1220872.

Nigam, R., Vyas, S.S, Bhattacharya, B.K., Oza, M.P., Srivastava, S.S., Bhagia, D., Dhar, D. and Manjunath, K.R. (2015). Modeling temporal growth profile of vegetation index from Indian geostationary satellite for assessing in-season progress of crop area. *GIScience & Remote Sensing.* https://doi.org/10.1080/15481603.2015.1073036.

Oza, M.P., Rajak, D.R., Bhagia, N., Dutta, S., Vyas, S.P., Patel, N.K. and Parihar, J.S. (2006). Multiple production forecasts of wheat in india using remote sensing and weather data. *Proceedings of SPIE*, 6411, 102–108.

Oza, M.P., Pandya, M.R. and Rajak, D.R. (2008). Evaluation and Use of ResourceSat-I Data for Agricultural Applications. *International Journal of Applied Earth Observation and Geoinformation*, 10, 194–205. doi:10.1016/j.jag.2008.02.006

Pandya, M.R., Darji, N.P., Shah, D.B., Trivedi, H.J., Ramakrishnan, R., Panigrahy, S., Parihar, J.S. and Kirankumar, A.S. (2010). *Development of An Algorithm to Retrieve Land Surface Temperature from the INSAT-VHRR Data.* Paper presented at ISRS National Symposium, Lonavala.

Prakash, S., Mahesh , C., Gairola, R.M. and Pal, P.K. (2010). Estimation of Indian summer monsoon rainfall using Kalpana-1 VHRR data and its validation using rain gauge and GPCP data. *Meteorology and Atmospheric Physics*, 110, 45–57. doi:10.1007/s00703-010-0106-8.

Rahman, H. and Dedieu, G. (1994). SMAC: A simplified method for the atmospheric correction of satellite measurements in the solar spectrum. *International Journal of Remote Sensing*, 15, 123–143.

Rathore, L.S. and Maini, P. (2008). *Economic Impact Assessment of Agro-Meteorological Advisory Service of NCMRWF*. Report no. NMRF/PR/01/2008, NCMRWF, Ministry of Earth Sciences, Government of India.

Rosema, A. (1993). Using METEOSAT for operational evapotranspiration and biomass monitoring in the Sahel region. *Remote Sensing of Environment,* 46, 27–44.

Samson, R. and Lemeur, R. (2001). Energy balance storage terms and big-leaf evapotranspiration in a mixed deciduous forest. *Annals of Forest Sciences*, 58, 529–541.

Saxena, R.K., Srivastava, R. and Verma, K.S. (1997). Spectral library of Indian soils. NATP mission mode programme code no. 27(2)/97/NATP/MM-III-2.

Stisen, S., Sandholt, I., Nørgaard, A., Fensholt, R. and Jensen, K.H. (2008). Combining the triangle method with thermal inertia to estimate regional evapotranspiration applied to MSG-SEVIRI data in the Senegal River basin. *Remote Sensing of Environment*, 112, 1242–1255.

Tanahashi, S., Kawamura, H., Matsuura, T. and Yusa, H. (2001). A system to distribute satellite incident solar radiation in real-time. *Remote Sensing of Environment,* 75, 412–422.

Varma, A.K. (2018). Measurement of precipitation from satellite radiometers (visible, infrared, and microwave): Physical basis, methods, and limitations. remote sensing of aerosols, clouds, and precipitation, Eds. Ahmed, Tanvir, Hu, Y., Kokhanovsky, A. and Wang, J., (ISBN:978-0-12-810437-8), pp. 223–248.

Vyas, S.S., Bhattacharya, B.K. and Nigam, R. (2016a). Assured solar energy hot-spots over Indian landmass detected through remote sensing observations from geostationary meteorological satellite. *Current Science*, 111, 836–842.

Vyas, S.S, Nigam, R., Bhattacharya, B.K. and Kumar, P. (2016b). Development of real-time reference evapotranspiration at the regional scale using satellite-based observations. *International Journal of Remote Sensing*, 37, 6108–6126.

Vyas, S.S., Nigam, R., Patel, N.K. and Panigrahy, S. (2011a). *Estimation of Wheat Crop Sowing Date Using CCD Data from Indian Geostationary Satellite INSAT 3A*. Paper presented at ISRS National Symposium, Bhopal.

Vyas, S.S., Nigam, R., Patel, N.K., Panigrahy, S. and Parihar, J.S. (2011b). Monitoring of *rabi* crop progress using geostationary satellite INSAT 3A CCD data. *Scientific Note*, ABHG/SAC/FASAL/SN/03/2011.

4 Prediction of Urban Expansion and Identifying Its Impacts on the Degradation of Agricultural Land

A Machine Learning-Based Remote-Sensing Approach in Rajshahi, Bangladesh

Abdulla-Al Kafy, Md. Nazmul Huda Naim, Md. Hasib Hasan Khan, Md. Arshadul Islam, Abdullah Al Rakib, Abdullah-Al-Faisal, Md. Hasnan Sakin Sarker

CONTENTS

4.1 INTRODUCTION

Changes in the land use/land cover (LULC) is a notable environmental problem nowadays (Bajocco et al. 2012, Comarazamy et al. 2013). Such transitions happen due to changes in different climatic and physical conditions, along with several socio-economic influences. In addition, human actions, combined with natural hazards, may have a significant effect on indigenous people's livelihood and the economic development of the country (Huang et al. 2008, Kaliraj et al. 2017, Muttitanon and Tripathi 2005). Moreover, the sustainability of existing and potential ecosystems is strongly dependent on human actions on a global level (Bucx et al. 2010).

Situated on the coast of the Bay of Bengal, in South Asia, Bangladesh houses the most endangered coastal regions of the world, being under the constant threat of sea-level rise and increasing global temperatures. Bangladesh is impacted by numerous natural calamities each year on a regular basis, which hampers the prosperity of the nation (Zimmermann, Glombitza, and Rothenberger 2009). Several factors operating in combination, which include geographical location, soil features, access to fresh water, shoreline features, tropical monsoon climate, etc., create a naturally hazardous situation in Bangladesh. Over the past four decades, a notable transition in the LULC trend has been observed in the southern coastal areas of the country. In this region, cultivatable land fell by 20%, and the area of water bodies increased by around 10% in the period 1980–2009 (Rahman and Begum 2011). This situation is worse in the north-western region of the country, where people are much more exposed to natural disasters, i.e. drought, flood, storm surge, etc. (Murad and Islam 2011, Zimmermann, Glombitza, and Rothenberger 2009), which greatly hampers the socio-economic development of the region. On average approximately 1% of cultivable land of the country transforms into non-agricultural land each year, which is raising concerns of national food security (Stevens 1972, Chang-Gil 2009, Ahmad 2011).

Identification of LULC transitions is an effective approach to addressing the problems regarding unregulated urbanization and environmental degradation (Beevi, Sivakumar, and Vasanthi 2015, Hadeel, Jabbar, and Xiaoling 2011, Mubea, Ngigi, and Mundia 2010). In addition, to ensure proper management and sustainable distribution of environmental resources, a detailed analysis of the evolving trend of LULC change is essential, which can be achieved through the implementation of remote-sensing (RS) techniques (Fanelli and Piazza 2020,

Kafy et al. 2018, Lu et al. 2003, Turner, Lambin, and Reenberg 2007). Remote-sensing techniques facilitate modeling of upcoming LULC change patterns and help to develop an understanding of the continuous LULC transitions (Verburg et al. 2004). It further helps to analyze and develop appropriate land management regulations. Different topographical features, such as demographic information, past patterns of LULC transition, positions of various facilities, etc., are considered to model the possible location of the LULC transition. These models harness the potential to detect changes in LULC in different dimensions (Freier, Schneider, and Finckh 2011, Guan et al. 2011, Halmy et al. 2015, Veldkamp and Lambin 2001, Wickramasuriya et al. 2009).

A Markov chain approach has proven to be the most frequently-used model when it comes to modeling and forecasting land use patterns (Xiong et al. 2012). It is a probabilistic model that represents a series of potential occurrences, where the likelihood of each occurrence relies upon the condition reached in the preceding case. This approach is best suited to short-term forecasts (Baker 1989, Bell and Hinojosa 1977, Muller and Middleton 1994, Sinha and Kimar 2013, Weng 2002). Incorporating Markov chain techniques with the probabilistic approach of geographic cellular automata can help to forecast any changes in multi-dimensional land usage. It works more effectively than models which use regression techniques. Earlier models only interacted with unidirectional changes (Eastman et al. 2013, Guan et al. 2011, Pontius and Malanson 2005, Roy et al. 2015, Theobald and Hobbs 1998, Ye and Bai 2008). Software, such as DINAMICA EGO, CA-MARKOV, CLUE-S, and Land Change Modeler, are appropriate for predicting upcoming LULC dynamics, using analytical approaches based on previous LULC trends (Mas et al. 2014). Recent quantum geographic information system (QGIS) plug-ins include MOLUSCE, which can study the LULC transition and forecast upcoming LULC transitions. This plug-in can use the Markov chain method to create transition potential or the probability matrix, and develop a prediction model focused on four separate frameworks i.e. multi-criteria evaluation (MCE), weights of evidence (WoE), artificial neural networks (ANN), and logistic regression (LR). Lastly, the plug-in predicts the future LULC map, using a Cellular Automata (CA) modeling framework (Nadoushan, Soffianian, and Alebrahim 2015, Balogun and Ishola 2017, Kafy, Rahman, et al. 2020, Ullah et al. 2019).

For spatial and temporal LULC modeling, CA-Markov modeling has proved to be quite useful, according to many studies, as planning of land use, development of urban policy, and climate change research of various areas can be done promptly (Santé et al. 2010, Arsanjani et al. 2013, Al-sharif and Pradhan 2014, Ozturk 2015, Nadoushan, Soffianian, and Alebrahim 2015, Vázquez-Quintero et al. 2016, Balogun and Ishola 2017). Land-use transition has been predicted by many researchers, using CA-artificial neural networks (ANN) and Monte Carlo approaches (Sejati, Buchori, and Rudiarto 2019, Liu et al. 2020). Prediction of Dhaka city's future LULC pattern (Ahmed and Ahmed 2012, Islam and Ahmed 2011), including the trend of LULC variation in the hilly terrains of the south-eastern region of Bangladesh, have also been conducted through the CA-Markov approach (Roy et al. 2015). The entire future LULC pattern of Bangladesh has been forecasted for the period 2010–2030. Prioritizing environmental protection and economic development were taken as the baseline for this model. In order to model the land use variations with respect to the driving factors of the variation in land cover in the south-western regions of Bangladesh, the earlier Dynamics of Land Systems (DLS) model was used (Hasan et al. 2017). Only a handful of studies have been conducted to model this region's future LULC patterns up to this time.

In this study, the north-western region of Bangladesh, i.e. the city of Rajshahi, is kept in focus, as it is prone to LULC changes due to environmental mishaps and unregulated human activities. Agricultural land in this region is rapidly being transformed into urban landscapes and other forms of land use. Considering these issues, the study focuses mainly on: (i) assessing the spatial and temporal variation of LULC in Rajshahi 1999–2019; (ii) predicting the future LULC pattern for the year 2029, with the help of an CA-ANN approach; and (iii) displaying how rapidly urbanization is impacting agricultural lands in the study area.

4.2 STUDY AREA

Rajshahi District is the social, economic, and administrative hub of northern Bangladesh, situated on the North bank of the Padma River (RDA 2008, BBS 2013). The study is situated between 24° 12′ to 24° 42′ N latitude and 88° 15′ to 88° 50′ E longitude in the north-western region of Bangladesh. The location is nearly flat, with a surface elevation between 10 and 47 m (Figure 4.1). Rajshahi District territory is around 2,428 km^2, consisting of nine upazilas (subdistricts), four thanas (Thana was a subdistrict in the administrative geography of Bangladesh which is controlled by a police station), 13 municipalities, and 147 wards (BBS 2013, Kafy, Faisal, et al. 2020). Rajshahi District is primarily a dry-wet tropical area. The maximum temperature is 30–35 °C, with 1,448 mm rainfall (Ferdous and Baten 2011, Kafy, Rahman, et al. 2020, Kafy, Faisal, et al. 2020). The study land area consists of agriculture, infrastructure, and other areas, totaling 394,986.32 ha, 117,615.42 ha, and 63829.56 ha, respectively (Clemett et al. 2006, BBS 2013, Kafy, Faisal, et al. 2020).

Historically, Rajshahi has been recognized as a rural agrarian region. However, after construction of the Jamuna bridge and the spread of industrialization, this region is experiencing significant urbanization. Rapid urbanization has drastically increased the duration of the winter and summer seasons in the past few years, which is harmful to this district's climate, livelihood, and agricultural development (Ferdous and Baten 2011, Kafy et al. 2019, Kafy, Faisal, et al. 2020). Urban growth is the fastest-growing and most overwhelming problem in Bangladesh. The expansion of Rajshahi is also haphazard and largely unplanned. The district has a total population of around 2.6 million, with a density of 1,070 individuals per km^2, whereas the district had a population of about 2.3 million people in 2001, with a density of about 950 per km^2 (BBS 2013). Large-scale rural-to-urban migration and rapid urbanization are among the main reasons for the study region's population growth. Land-use history in this area shows that over 10% of

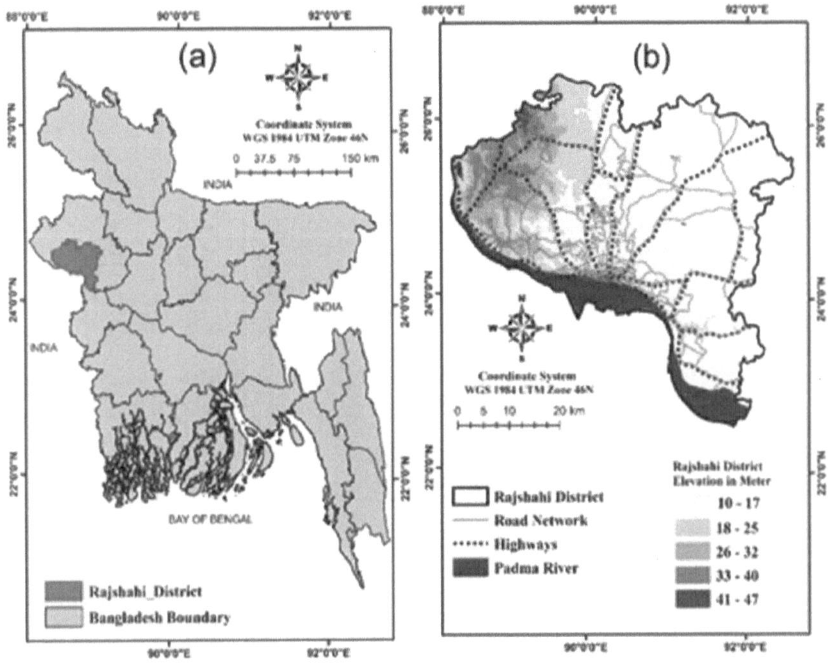

FIGURE 4.1 Location map of the study area: a) Rajshahi District in Bangladesh b) Rajshahi District.

the agricultural area has been lost in the past 20 years, due to rapid urbanization (RDA 2003, 2008, Kafy et al. 2019, Kafy, Faisal, et al. 2020).

4.3 MATERIALS AND METHODS

4.3.1 Data Description

For this research, the study period was selected to be the years 1999, 2009, and 2019. Over these two decades, three sets of multi-spectral Landsat Satellite data (for the years 1999, 2009, and 2019) were collected from United States Geological Survey (USGS) domain for estimating the changes in LULC and land surface temperature (LST) of the study area.

All of these images had been captured by sensors within April and May, to prevent undue influences of seasonal variations. In the image downloading process, maximum cloud coverage was set to less than 10% for ensuring accurate measurements from the images. However, across the study region, cloud coverage was close to zero percent (0 %). No additional geo-correction or image processing was required for the preprocessing of the images since the Landsat Satellite data are free of radiometric and geometric distortions. Images details were gathered from the USGS repository (Table 4.1).

TABLE 4.1
Description of the Images Acquired From Landsat Satellite Sensors

Satellite Data	Date of Acquisition	Sensor	Path/Row	Band No.	Spectral Range (Wavelength, μm)	Spatial Resolution (m)
Landsat 4-5	9 May 1999; 5 May 52009	TM	138/43	1	0.45–0.52	30
				2	0.52–0.60	30
				3	0.63–0.69	30
				4	0.76–0.90	30
				5	1.551.75	30
				6	10.4012.50	120 resampled to 30
				7	2.082.35	30
Landsat 8	16 May 2019	OLI	138/43	1	0.43–0.45	30
				2	0.45–0.51	30
				3	0.64–0.67	30
				4	0.53–0.59	30
				5	0.85–0.88	30
				6	1.57–1.65	30
				7	2.11–2.29	30
				8	1.36–1.38	15
				9	0.50–0.68	30
				10 (TRIS1)	10.60–11.19	100 resampled to 30
				11 (TRIS 1)	11.50–12.51	100 resampled to 30

Source: (https://earthexplorer.usgs.gov)

TABLE 4.2
Descriptions of LULC Groups

LULC Classes	Description
Built-up area	Residential, commercial and industrial services, transportation network.
Agricultural land	Trees, grassland, cropland, and fallow land.
Water bodies	River, wetlands, lakes, ponds, and reservoirs.
Bare land	Vacant land, open space, sand, bare soils, and landfill sites.

4.3.2 Land Use/ Land Cover (LULC) Classification

The satellite images obtained from Landsat sensors were enhanced in ERDAS IMAGINE V.14 software, by the 3×3 majority filtering technique, to achieve greater visibility (Kafy, Faisal, et al. 2020, Kafy, Rahman, et al. 2020,). True Color Composites (TCC) were generated, using the correct band combinations for all images, to choose training samples of various LULC classes (Kafy, Rahman, et al. 2020, Trolle et al. 2019). The collected Landsat images were classified into four LULC classes: built-up area, agricultural land, water bodies, and bare land for the years of 1999, 2009, and 2019 (Table 4.2). The Maximum Likelihood Supervised Image Classification technique was used for the LULC classifications. In the process of creating LULC maps, about 35 training samples were taken for each LULC class.

4.3.2.1 Accuracy Assessment

Accuracy of land cover maps is measured from available field data and Google Earth images through 180 ground truths. Kappa coefficients and confusion matrices, considered among the best indicators of image classification accuracy were used in the study of accuracy assessment of the classified LULC maps (Foody 2002, Story and Congalton 1986, Pontius Jr and Millones 2011, Congalton and Green 2008).

4.3.3 Land Use/ Land Cover (LULC) Transformations

Analysis of changes in different LULC classes is carried out in with respect to study area shifts. Three LULC change maps, from 1999 to 2009, 2009 to 2019, and from 1999 to 2019 were prepared in QGIS software, using the MOLUSCE plug-in. The transformation between different LULC classes and the transformation within the same LULC class were calculated.

4.3.4 Prediction of LULC Changes in 2029

The cellular automata (CA) model was used, with the help of the MOLUSCE plug-in in QGIS software, to predict the future changes in the study area. Land-use models, such as the CA model, include static and dynamic aspects of LULC class change. The CA model helps to predict future land-use patterns because of its high level of accuracy (Wickramasuriya et al. 2009, Yang, Zheng, and Lv 2012, Ozturk 2015, Vázquez-Quintero et al. 2016). The predictions of LULC are based on dependent variables (LULC maps produced for 1999, 2009, and 2019) and independent variables (distances of roads, elevation, water bodies, and slopes). Roads and water body vector data were used in the Euclidian distance function of ArcMap software to calculate proximity distances of roads and water bodies. For the estimation of elevation and slope data, SRTM DEM data were used in ArcMap software. These variables mentioned were used to produce a transition potential matrix. The random sampling mode was used for this research. The transition matrix analysis creates an empirical probability image, estimating the likelihood of changes in different LULC classes (Rahman et al. 2018, Kafy et al. 2019, Kafy, Rahman, et al. 2020). The maximum number of iterations was 1500, and the neighborhood pixel was set at 3×3 cells for the model.

4.3.5 Validation of the CA Model

The CA model is used to predict the future LULC map for the year 2029 after the transition potential matrix is modeled in QGIS software. The CA model is used, with the help of the MOLUSCE plug-in in QGIS software. To ensure that the model is reliable at forecasting LULC changes for a predicted year, it must be validated using current datasets. The CA model has been checked for its accuracy in predicting the 2019 LULC map, by being compared with the estimated LULC map of 2019 from the Landsat image of the study area. IDRISI Taiga Software is used to validate the model. For this validation, several Kappa (K) parameters were generated: K-no, K-location, and K-standard Kappa parameters were used to measure the accuracy of the model (Pontius Jr and Millones 2011, Pontius and Malanson 2005, Congalton and Green 2008). Overall, Kappa coefficients and %-correctness were calculated, using the QGIS MOLUSCE validation module between classified and predicted LULC maps for the year 2019. The two approaches mentioned were used to assess the accuracy of the CA model before using it to simulate the LULC map.

4.4 RESULTS AND DISCUSSION

The LULC patterns over the past two decades, the spatial distribution of the LULC transitions, and the prediction of the future distributions of LULC for 2029 were performed for the study area, and the results are presented in the section below.

4.4.1 Analysis of LULC Change

To determine the LULC change patterns over the past two decades, the maximum likelihood supervised classification (MLSC) algorithm was used. Table 4.3 describes the overall accuracy with the Kappa coefficient, where the overall accuracy for all three years tested was greater than 80%.

TABLE 4.3
Accuracy Assessment of LULC Maps

	User's Accuracy (%)				Producer's Accuracy (%)					
Year	Water Body	Built-up Area	Vegetative	Bare Soil	Water Body	Built-up Area	Vegetative	Bare Soil	Classification Accuracy	Kappa Statistics
1999	87	85	85	80	91	82	81	82	89%	0.824
2009	86	89	86	85	93	84	93	86	91%	0.856
2019	91	83	81	80	87	89	84	92	87%	0.851

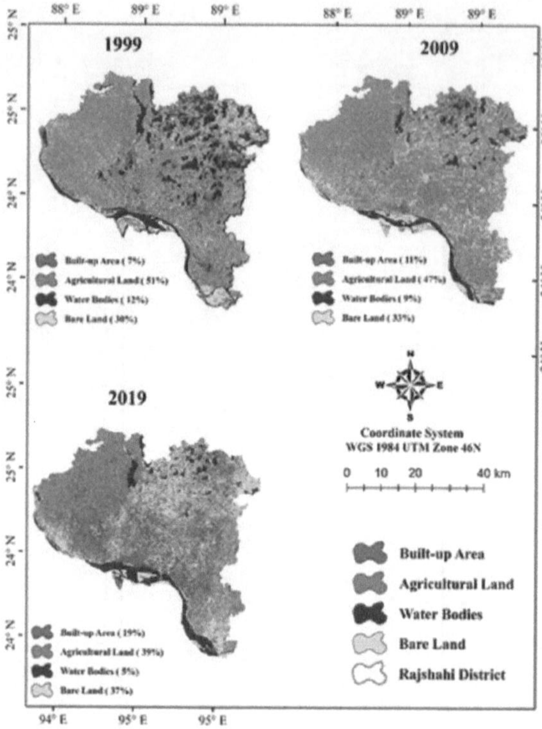

FIGURE 4.2 Classified LULC maps of the study area.

Two significant patterns of changes clearly stood out in the classification results throughout the entire two decades. These are the gradual increase in built-up area and bare land and a gradual decrease in area of agricultural lands and water bodies. The results clearly show that agricultural and vegetated lands and water bodies are being continuously replaced with built-up areas and bare lands. As shown graphically (Figure 4.2) and statistically (Table 4.4) over the past two decades, there were increases of 7.81% of the built-up area and 5.20% of the bare land from 1999 to 2019, associated with 8.27% and 4.75% decreases in agricultural land and water bodies areas, respectively, in that period in the study area. Several influential factors, such as random growth, migration pull factors, better study opportunities, and being the divisional sector pushed such LULC changes all over the study area (Kafy, Faisal, et al. 2020, Kafy, Rahman, et al. 2020, Regmi, Saha, and Balla 2014, Rahman 2016, Silva, da Silva, and Santos 2018, Fernando 2018, Anand and Oinam 2020).

Although agricultural/vegetated land still represents the classification with the highest land cover in the area, the proportion gradually decreased from 51.27% (1,221.50 km^2) in 1999 to 46.54% (1,108.91 km^2) in 2009 and 38.88% (926.18 km^2) in 2019 (Table 4.4). This gradual decrease resulted a decrease of 8.27% in the net

TABLE 4.4
Area Distribution of Different LULC Classes With the Net Change

	Area in km²			Net Change (%)
LULC	1999	2009	2019	1999–2019
Built-up area	161.78	270.23	440.70	7.81
Agricultural Land	1221.50	1108.91	926.18	–8.27
Water bodies	295.24	206.80	125.67	–4.75
Bare land	703.70	796.29	889.54	5.20

TABLE 4.5
Transformation Rate of Different LULC Classes

	1999–2019			
Area (km²)	Water Body	Built-Up Area	Agricultural Land	Bare Land
Area change to	20.36	331.45	29.56	308.52
Area converted from	190.28	26.34	708.25	225.45
Increasing rate (%)	0.85	13.91	1.24	12.95
Decreasing rate (%)	–7.98	1.10	–29.73	9.46

amount of agricultural land in 1999. On the other hand, there was only 6.79% of built-up areas in 1999.

The Urban area increased to 11.34% in 2009 and eventually to 18.5% in 2019, which was almost three times the area in 1999 (Figure 4.2). This marked change led to a 7.81% increase in the net area of the built-up areas. As it stands beside the bank of the River Padma, this can be considered to be a significant decrease in the area of water bodies, which fell to 8.68% in 2009 and 5.28% in 2019, the latter being less than the half of the area represented in Rajshahi in 1999 (12.39%). As the water bodies decreased, the area of bare lands increased but not at the same rate as that of the built-up areas. In 2009, the area of bare land was 33.43%, and it increased to 37.34% in 2019, compared with 29.53% in 1999.

4.4.2 Urban Expansion Analysis

Analyzing the urban expansion occurring in the study area will help explain changes in urbanization patterns in 16 different spatial directions. The spatial pattern of urban expansion between 1999, 2009, and 2019 shows that the area underwent rapid urbanization, resulting in significant changes to the LULC over the 20-year period. The study shows that the area in the north-northeasterly direction experienced the most rapid urban expansion (Figure 4.3). In contrast, development

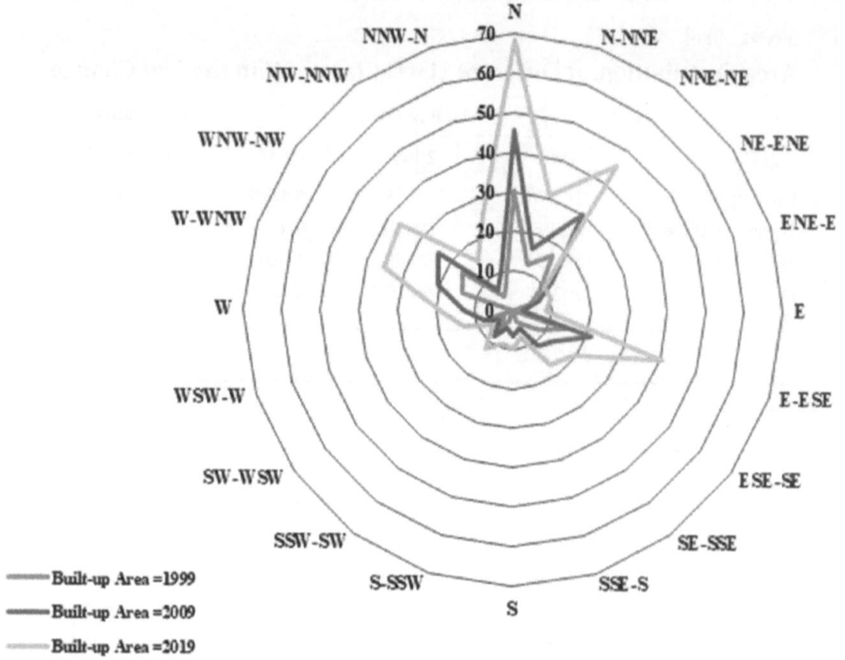

FIGURE 4.3 Spider diagram showing urban expansion in 16 directions from 1999 to 2019 in the study area.

in the southwesterly and southeasterly regions remained relatively small, where agricultural and bare land dominated, limiting the growth of urban expansion. It has been estimated that urban growth to the north and along the Dhaka–Rajshahi highways had been of the order of between 35 and 40% between 1999 and 2019. The northeasterly and southeasterly regions and along the Rajshahi–Naogaong and Rajshahi–Chapai highways experienced relatively little urban expansion over the past 20 years, ranging from 25–30%, 22–25%, respectively. The urban growth toward the northwest was significant because residential and commercial operations were readily accessible. A number of projects, such as new highways, new commercial buildings, new job opportunities, and 10 km^2 of rail, bus stations, and airports also render the northwest a region more oriented toward urban development.

Rapid urban development significantly influences the LULC change, converting agricultural land to built-up areas. It can be seen from Table 4.5 that the majority of LULC units have been modified in the past 20 years, and the rate of change of built-up (13.91%) and bare land (12.95%) areas is both strong and positive. The largest transition rate for vegetated land (708.25 km^2) is reported for the whole study period, followed by bare land (225.45 km^2). Between 1999 and 2019, the majority of unchanged pixels (331.45 km^2) were detected in built-up areas. The most intense change dynamics were concentrated in the central contiguous urban area, which was continuously moving in an unplanned way and converting other LULC classes into impervious layers. Agricultural land and wetland within or adjacent to urban areas are being continuously changed into other classes by the construction of residential or commercial areas.

4.4.3 Transformation of Agricultural Land

A transition map of the conversion of agricultural land to other LULC classes is presented for the periods 1999–2009, 2009–2019, and 1999–2019 (Figure 4.4). This section briefly discusses the impact of expansion of built-up areas and of other LULC class influences on agricultural land conversion over past 20 years (Figure 4.5).

4.4.3.1 Transformation of Agricultural Land to Water Bodies

It was seen that 0.23% of agricultural lands had been transformed into water bodies in the time period between 1999 and 2009. The percentage increased to 0.34% from the year 2009 to 2019. If we observe the overall transformation of agricultural lands into water bodies from 1999 to 2019 we can see that approximately 0.44% of agricultural lands were transformed into water bodies. There was a gradual increase in the percentage of transformation from agricultural lands to water bodies throughout the study period.

4.4.3.2 Transformation of Agricultural Land Into Built-up Areas

Approximately 1.19% of agricultural land was transformed into built-up area from 1999 to 2009. This percentage doubled from the year 2009 to 2019 (Figure 4.4 and Figure 4.5). The overall percentage of transformation of agricultural land into built-up area from 1999 to 2019 represented a notable transformation of 6.78% of agricultural land into built-up area.

4.4.3.3 Transformation of Agricultural Land Into Bare Land

Approximately 10.5% of agricultural lands were transformed into bare lands from 1999 to 2009. This percentage increased remarkably to 17.89% from 2009 to 2019, with an overall percentage transformation of agricultural lands into bare lands of 16.54%.

4.4.3.4 Non-Transformation of Agricultural Land

Approximately 36.82% of agricultural land area remained untransformed from 1999 to 2009. This percentage decreased markedly to 27.81% from 2009 to 2019. From 1999 to 2019 the overall percentage of agricultural lands which remained untransformed was 26.21%.

4.4.4 Prediction of LULC Change for the Year 2029

During the study period 1999–2019, the LULC classes in the study area changed significantly. It is therefore important to predict the future dynamics of LULC

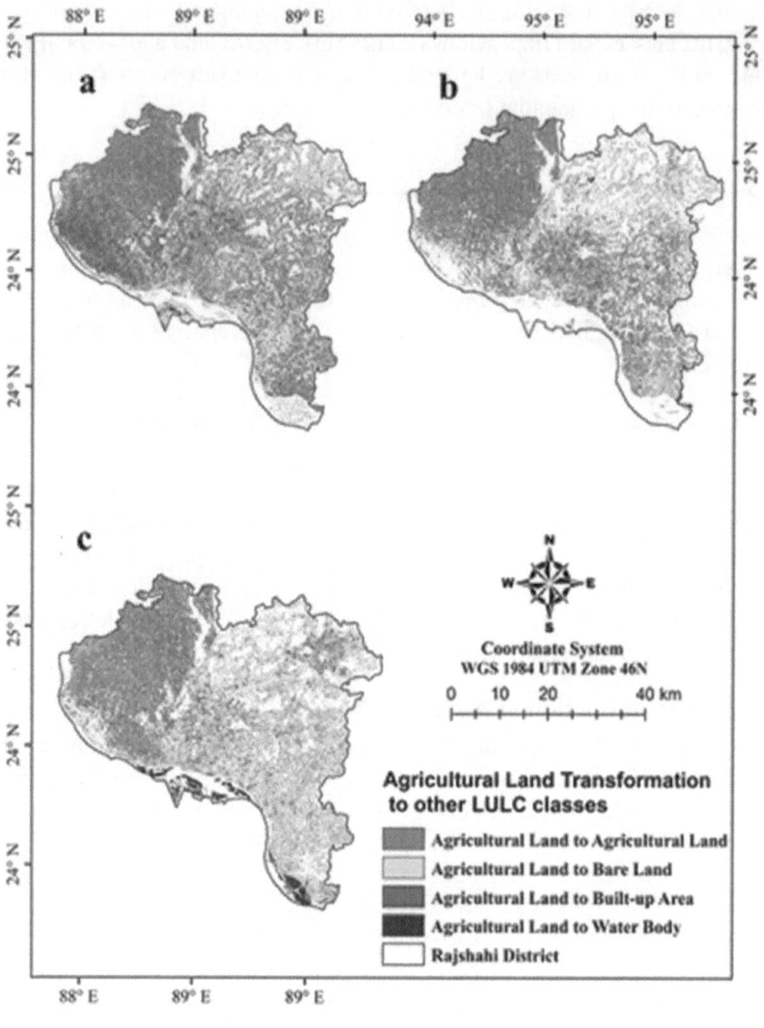

FIGURE 4.4 Transformation of agricultural land to different LULC classes: a) 1999 to 2009 b) 2009–2019, and c) 1999– 2019.

because, if past trends persist, the future changes in LULC classes will affect the biodiversity and micro-climate of the study area. Future LULC prediction will provide the framework for sustainable urban planning. The cellular automata (CA) model is used to predict the LULC changes for the year 2029. If a constant rate in LULC change is assumed, then the model is reliable in forecasting future change in LULC. The model is validated in the study area for the year 2019, using estimated and predicted LULC values. Overall Kappa values and the correctness of the predicted LULC model are 0.8 and 85.45%, respectively. The accuracy level is acceptable for LULC modeling (Table 4.6).

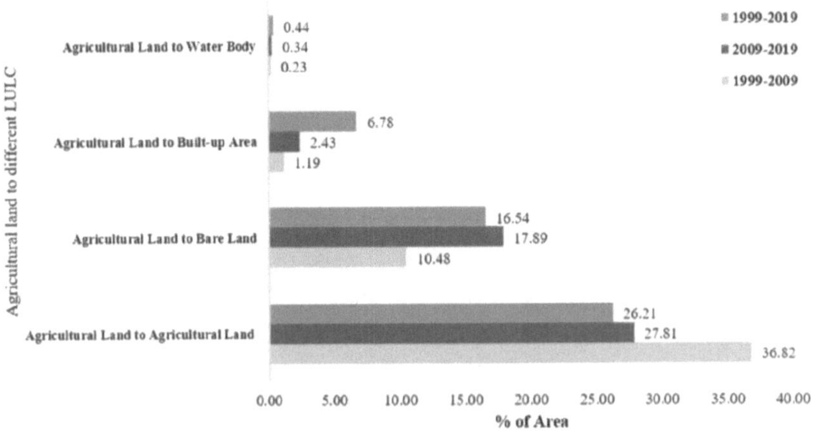

FIGURE 4.5 Agricultural land transformation to different LULC classes.

TABLE 4.6
Predicted LULC Model Validation in IDRISI Selva and QGIS

Prediction Year	CA Model Validation For LULC Prediction Using Two Modules					
2019	Kappa Parameters of IDRISI Selva Land Change Module				QGIS-MULUSCE Plug-in Module	
	K-location	K-no.	K-location Strata	K-standard	% correctness	Overall Kappa Value
	0.85	0.82	0.81	0.80	85.45	0.8

Estimated predicted maps for 2029 indicated that other land uses would be significantly lowered in the study region, except for built-up and bare land areas. Water body will decrease by 2% by 2029, compared with 2019. By 2029, the urbanized and bare land areas will increase by 3% and 5%, respectively. The agricultural land area will decrease dramatically by 6%, compared with 2019 LULC results. Considering the predicted year for prediction, 2029, agricultural land would decrease significantly (18%) in 30 years (1999–2029) (Figure 4.6), whereas urban areas (15%) will increase dramatically in the study area (Figure 4.7).

The trend toward increasing built-up area will significantly impact the area of agricultural land in the study area. Figure 4.8 describes the transformation pattern of agricultural land and its association with built-up area. The highest rate of built-up area was found in urban areas from 1999 to 2029 (15.06%), with this rate decreasing to 10.51% from 2009 to 2029.

From 2019 to 2029, the rate decreased markedly to 3.35%. The rates of decrease in agricultural land were 18.33%, 13.60%, and 5.93% from 1999–2029, 2009–2029, and 2019–2029, respectively. Finally, the impact of urban expansion

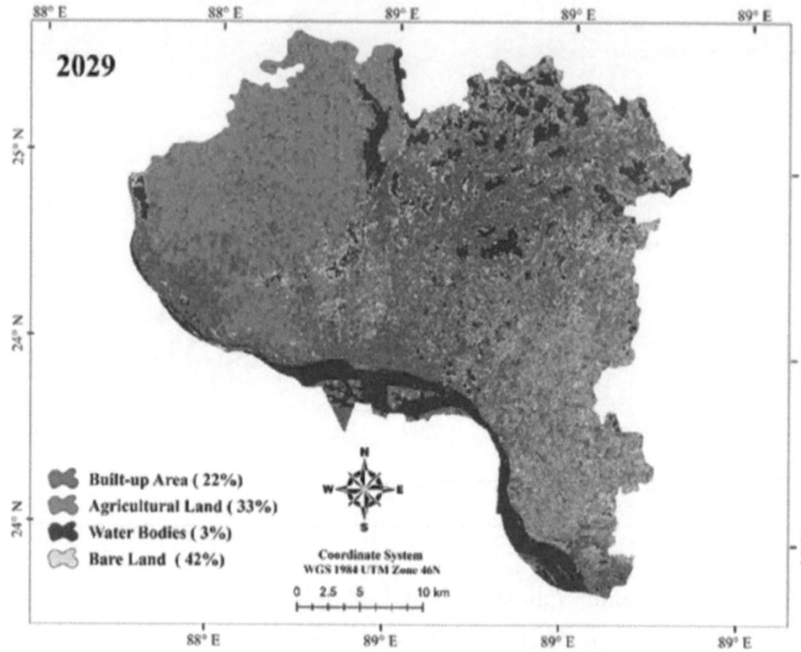

FIGURE 4.6 LULC-predicted map for the study area in 2029.

on agricultural land was determined by the rate of transformation of agricultural land into built-up areas, with the percentages of change being −9.77%, −8.45%, and −7.23% from 1999–2029, 2009–2029, and 2019–2029, respectively (Figure 4.8).

4.5 CONCLUSION

This study reveals that Rajshahi city has been going through rapid change in its land-uses, especially the agricultural/vegetated lands, over the past two decades. The agricultural lands were mostly converted into built-up area (more than 6%), compared to conversion into other land-uses. This change is alarming for the city as incremental increases in built-up area affects the environment negatively. In addition, decreases in agricultural land area will result in less crop production, causing food scarcity. The applied CA model also predicts that the decreasing trend of agricultural land will continue in the future and that these areas will be replaced alarmingly by the built-up area by 2029.

Despite the low spatial resolution (30 m) of the satellite images, the outcomes are very reliable, as confirmed by the high accuracy coefficients found in the study. Considering the outcomes, it is an essential task for the city planners, environmental engineers, and the local decision makers to take the necessary steps to mitigate these changes, to avoid the unexpected consequences of an excessive built-up area as well as a shortage of agricultural/vegetated land. An effective and

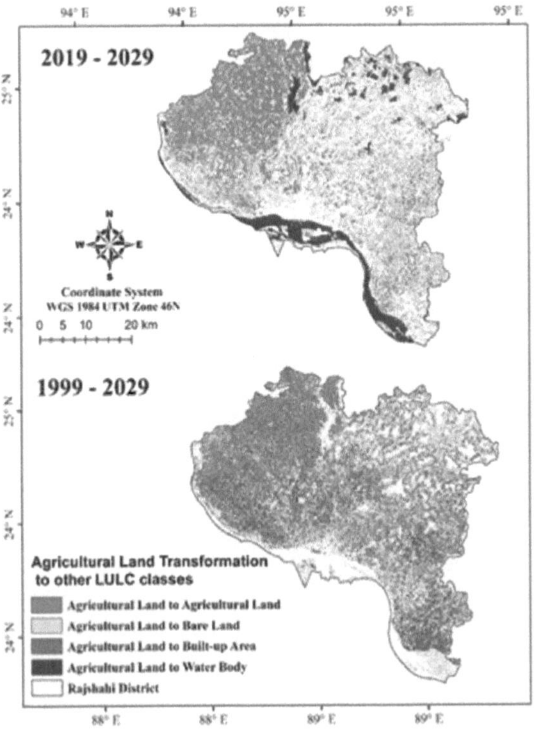

FIGURE 4.7 Transformation of agricultural land in different LULC classes from 2019 to 2029 and 1999 to 2029.

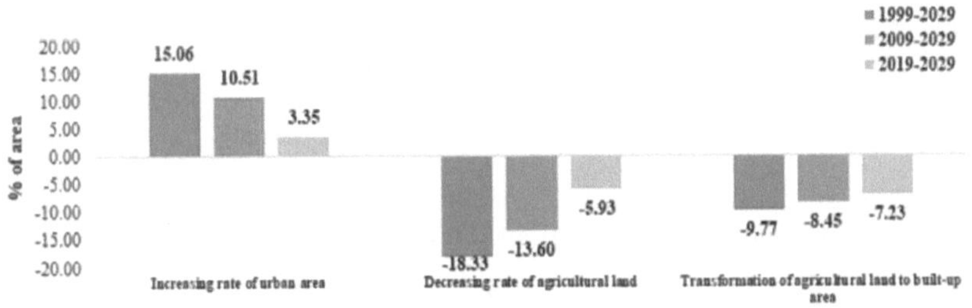

FIGURE 4.8 Increasing and decreasing trends of the urban area and agricultural land area from 1999–2029, 2009–2029, and 2019–2029.

comprehensive land-use plan for the city must be prepared, and all the subsequent development should follow the plan for the city. Moreover, as the urban expansion is currently happening in a disorderly unplanned manner, proper implementation of the comprehensive land-use plan can control this situation, to ensure sustainable urbanization into the future.

CONFLICT OF INTEREST

The authors declare no conflict of interest.

ACKNOWLEDGMENTS

Authors are thankful to the United States Geological Survey (USGS) for access to the Landsat archives.

REFERENCES

Ahmad, A. 2011. "Agricultural adjustment in flood porn areas in Comilla of Bangladesh: Geographical study." *Journal of Development and Agricultural Economics* 3 (12):602–609.

Ahmed, B., and R. Ahmed. 2012. "Modeling urban land cover growth dynamics using multi-temporal satellite images: A case study of Dhaka, Bangladesh." *ISPRS International Journal of Geo-Information* 1. doi:10.3390/ijgi1010003.

Al sharif, Abubakr A.A., and Biswajeet Pradhan. 2014. "Monitoring and predicting land use change in Tripoli Metropolitan City using an integrated Markov chain and cellular automata models in GIS." *Arabian Journal of Geosciences* 7 (10):4291–4301.

Anand, Vicky, and Bakimchandra Oinam. 2020. "Future land use land cover prediction with special emphasis on urbanization and wetlands." *Remote Sensing Letters* 11 (3):225–234.

Arsanjani, Jamal Jokar, Marco Helbich, Wolfgang Kainz, and Ali Darvishi Boloorani. 2013. "Integration of logistic regression, Markov chain and cellular automata models to simulate urban expansion." *International Journal of Applied Earth Observation and Geoinformation* 21:265–275.

Bajocco, S., A. De Angelis, L. Perini, A. Ferrara, and L.J. Salvati. 2012. "The impact of land use/land cover changes on land degradation dynamics: a Mediterranean case study." *Environmental Management* 49 (5):980–989.

Baker, W.L. 1989. "A review of models of landscape change." *Landscape Ecology* 2. doi:10.1007/BF00137155.

Balogun, I.A., and K.A. Ishola. 2017. "Projection of future changes in landuse/landcover using cellular automata/markov model over Akure city, Nigeria." *Journal of Remote Sensing Technology* 5 (1):22–31.

BBS, Bangladesh Bureau of Statistics. 2013. *District Statistics 2011, Rajshahi.* Edited by Statistics and Informatics Division. Ministry of Planning, Government of The People's Republic of Bangladesh.

Beevi, H.N., S. Sivakumar, and R. Vasanthi. 2015. "Land use / land cover classification of Kanniykumari Coast, Tamilnadu, India. Using remote sensing and GIS techniques." *International Journal of Engineering Research and Applications* 5.

Bell, E.J., and R.C. Hinojosa. 1977. "Markov analysis of land use change: Continuous time and stationary processes." *Socio-Economic Planning Sciences* 11. doi:10.1016/0038-0121(77)90041-6.

Bucx, Tom, Marcel Marchand, Bart Makaske, and Cees van de Guchte. 2010. *Comparative Assessment of the Vulnerability and Resilience of 10 Deltas: Work Document.* Deltares.

Chang-Gil, Kim. 2009. "Strategies for implementing green growth in agricultural sector." *Proceedings in Green Korea 2009-Green Growth and Cooperation.*

Clemett, Alexandra, Md Maksudul Amin, Sharfun Ara, Md Mashiur, and R. Akan. 2006. "Background information for Rajshahi City, Bangladesh." *WASPA Asia Project Report* 2:1–29.

Comarazamy, Daniel E., Jorge E. González, Jeffrey C. Luvall, Douglas L. Rickman, and D.J. Robert. 2013. "Climate impacts of land-cover and land-use changes in tropical islands under conditions of global climate change." *Journal of Climate Bornstein* 26 (5):1535–1550.

Congalton, Russell G., and Kass Green. 2008. *Assessing the Accuracy of Remotely Sensed Data: Principles and Practices*. CRC Press.

Eastman, J., Florencia Sangermano, Elia Machado, John Rogan, and Assaf Anyamba. 2013. "Global trends in seasonality of normalized difference vegetation index (NDVI), 1982–2011." *Remote Sensing* 5 (10):4799–4818.

Fanelli, Duccio, and Francesco Piazza. 2020. "Analysis and forecast of COVID-19 spreading in China, Italy and France." *Chaos, Solitons & Fractals* 134:109761.

Ferdous, M.G., and M.A. Baten. 2011. "Climatic variables of 50 years and their trends over Rajshahi and Rangpur division." *Journal of Environmental Science and Natural Resources* 4 (2):147–150.

Fernando, Gmts. 2018. "Identification of Urban Heat Islands & its relationship with vegetation cover: A case study of Colombo & Gampaha Districts in Sri Lanka." *Journal of Tropical Forestry and Environment* 8 (2).

Foody, Giles M. 2002. "Status of land cover classification accuracy assessment." *Remote Sensing of Environment* 80 (1):185–201.

Freier, K.P., U.A. Schneider, and M. Finckh. 2011. "Dynamic interactions between vegetation and land use in semi-arid Morocco: Using a Markov process for modeling rangelands under climate change." *Agriculture, Ecosystems and Environment* 140. doi:10.1016/j.agee.2011.01.011.

Guan, D., H. Li, T. Inohae, W. Su, T. Nagaie, and K. Hokao. 2011. "Modeling urban land use change by the integration of cellular automaton and Markov model." *Ecological Modelling* 222. doi:10.1016/j.ecolmodel.2011.09.009.

Hadeel, A.S., M.T. Jabbar, and C. Xiaoling. 2011. "Remote sensing and GIS application in the detection of environmental degradation indicators." *Geo-spatial Information Science* 14. doi:10.1007/s11806-011-0441-z.

Halmy, M.W.A., P.E. Gessler, J.A. Hicke, and B.B. Salem. 2015. "Land use/land cover change detection and prediction in the north-western coastal desert of Egypt using Markov-CA." *Applied Geography* 63. doi:10.1016/j.apgeog.2015.06.015.

Hasan, S.S., X. Deng, Z. Li, and D. Chen. 2017. "Projections of future land use in Bangladesh under the background of baseline, ecological protection and economic development." *Sustainability* 9.

Huang, Wenli, Huiping Liu, Qingzu Luan, Qingxiang Jiang, Junping Liu, and Hua Liu. 2008. "Detection and prediction of land use change in Beijing based on remote sensing and GIS." *The International Archives of the Photogrammetry, Remote Sensing and Spatial Information Science* 37:75–82.

Islam, M.S., and R. Ahmed. 2011. "Land use change prediction in Dhaka city using GIS aided Markov chain modeling." *Journal of Life and Earth Science* 6.

Kafy, Abdulla Al, Abdullah-Al Faisal, Soumik Sikdar, Mohammad Hasan, Mahbubur Rahman, Mohammad Hasib Khan, and Rahatul Islam. 2020. "Impact of LULC changes on LST in Rajshahi District of Bangladesh: A remote sensing approach." *Journal of Geographical Studies* 3:11–23. doi:10.21523/gcj5.19030102.

Kafy, Abdulla Al, Lamia Ferdous, Md Sazib Ali, and Pintu Sheel. 2018. "Using contingent valuation method to determine economic value of Padma River Wetland in Rajshahi District, Bangladesh." *1st National Conference on Water Resources Engineering (NCWRE 2018)*, Chittagong, Bangladesh.

Kafy, Abdulla Al, Muhaiminul Islam, Abdur Khan, Lamia Ferdous, and Md Hossain. 2019. "Identifying most influential land use parameters contributing reduction of surface water bodies in Rajshahi City, Bangladesh: A remote sensing approach." *Remote Sensing of Land*:87–95. doi:10.21523/gcj1.18020202.

Kafy, Abdulla Al, Md Shahinoor Rahman, Abdullah Al Faisal, Mohammad Mahmudul Hasan, and Muhaiminul Islam. 2020. "Modelling future land use land cover changes and their impacts on land surface temperatures in Rajshahi, Bangladesh." *Remote Sensing Applications: Society and Environment*. https://doi.org/10.1016/j.rsase.2020.100314.

Kaliraj, S., N. Chandrasekar, K.K. Ramachandran, and Y. Srinivas. 2017. "Coastal landuse and land cover change and transformations of Kanyakumari coast, India using remote sensing and GIS." *The Egyptian Journal of Remote Sensing Saravanan, and Space Science* 20 (2):169–185.

Liu, Zhihua, Pierre Magal, Ousmane Seydi, and Glenn Webb. 2020. "Predicting the cumulative number of cases for the COVID-19 epidemic in China from early data." *arXiv preprint arXiv:2002.12298.*

Lu, D., P. Mausel, E. Brondizio, and E. Moran. 2003. "Change detection techniques." *International Journal of Remote Sensing* 25. doi:10.1080/0143116031000139863.

Mas, J.F., M. Kolb, M. Paegelow, M.T.C. Olmedo, and T. Houet. 2014. "Inductive pattern-based land use/cover change models: A comparison of four software packages." *Environmental Modelling & Software* 51. doi:10.1016/j.envsoft.2013.09.010.

Mubea, K.W., T.G. Ngigi, and C.N. Mundia. 2010. "Assessing application of Markov chain analysis in predicting land cover change: A case study of Nakuru municipality." *Journal of Agriculture, Science and Technology* 12.

Muller, M.R., and J. Middleton. 1994. "A Markov model of land-use change dynamics in the Niagara Region, Ontario, Canada." *Landscape Ecology* 9.

Murad, Hasan, and A.K.M.S. Islam. 2011. "Drought assessment using remote sensing and GIS in north-west region of Bangladesh." *Proceedings of the 3rd International Conference on Water & Flood Management.*

Muttitanon, W., and N.K. Tripathi. 2005. "Land use/land cover changes in the coastal zone of Ban Don Bay, Thailand using Landsat 5 TM data." *International Journal of Remote Sensing* 26 (11):2311–2323.

Nadoushan, M.A., A. Soffianian, and A. Alebrahim. 2015. "Modeling land use/cover changes by the combination of Markov chain and cellular automata Markov (CA-Markov) models." *International Journal of Earth, Environment and Health* 1. doi:10.4103/2423-7752.159922.

Ozturk, Derya. 2015. "Urban growth simulation of Atakum (Samsun, Turkey) using cellular automata-Markov chain and multi-layer perceptron-Markov chain models." *Remote Sensing* 7 (5):5918–5950.

Pontius, G.R., and J. Malanson. 2005. "Comparison of the structure and accuracy of two land change models." *International Journal of Geographical Information Science* 19. doi:10.1080/13658810410001713434.

Pontius Jr, Robert Gilmore, and Marco Millones. 2011. "Death to Kappa: Birth of quantity disagreement and allocation disagreement for accuracy assessment." *International Journal of Remote Sensing* 32 (15):4407–4429.

Rahman, Md Shahinoor, Hossain Mohiuddin, Abdulla-Al Kafy, Pintu Kumar Sheel, and Liping Di. 2018. "Classification of cities in Bangladesh based on remote sensing derived spatial characteristics." *Journal of Urban Management.*

Rahman, M.R., and S. Begum. 2011. "Land cover change analysis around the Sundarbans Mangrove Forest of Bangladesh using remote sensing and GIS application." *Journal of Science Foundation* 9 (1–2):95–107.

Rahman, Muhammad. 2016. "Detection of land use/land cover changes and urban sprawl in Al-Khobar, Saudi Arabia: An analysis of multi-temporal remote sensing data." *ISPRS International Journal of Geo-Information* 5 (2):15.

RDA, Rajshahi Development Authority. 2003. Preparation of Structure Plan, Master Plan and Detailed Area Plan For Rajshahi Metropolitan City Government of the People's Republic of Bangladesh Ministry of Housing and Public works.

RDA, Rajshahi Development Authority. 2008. *Working Paper on Existning Landuse, Demographic and Transport (revised).* Edited by Mohakhali C/A Development Design Consultants Limited 47, Dhaka-1212. Government of The People's Republic of Bangladesh Ministry of Housing and Public Works.

Regmi, R., S. Saha, and M. Balla. 2014. "Geospatial analysis of land use land cover change predictive modeling at Phewa Lake Watershed of Nepal." *International Journal of Current Engineering and Technology* 4:2617–2627.

Roy, S., K. Farzana, M. Papia, and M. Hasan. 2015. "Monitoring and prediction of land use/land cover change using the integration of Markov chain model and cellular automation in the southeastern tertiary hilly area of Bangladesh." *International Journal of Sciences: Basic and Applied Research (IJSBAR)* 24.

Santé, Inés, Andrés M. García, David Miranda, and Rafael Crecente. 2010. "Cellular automata models for the simulation of real-world urban processes: A review and analysis." *Landscape and Urban Planning* 96 (2):108–122.

Sejati, Anang Wahyu, Imam Buchori, and Iwan Rudiarto. 2019. "The spatio-temporal trends of urban growth and surface urban heat islands over two decades in the Semarang Metropolitan Region." *Sustainable Cities and Society* 46:101432.

Silva, Janilci Serra, Richarde Marques da Silva, and Celso Augusto Guimarães Santos. 2018. "Spatiotemporal impact of land use/land cover changes on urban heat islands: A case study of Paço do Lumiar, Brazil." *Building and Environment* 136:279–292.

Sinha, P., and L. Kimar. 2013. "Markov land cover change modeling using pairs of time-series satellite images." *Photogrammetric Engineering & Remote Sensing* 79. doi:10.14358/PERS.79.11.1037.

Stevens, Robert D. 1972. *Rural Development Programs for Adaptation from Comilla, Bangladesh.* Department of Agricultural Economics, Michigan State University.

Story, Michael, and Russell G Congalton. 1986. "Accuracy assessment: A user's perspective." *Photogrammetric Engineering and Remote Sensing* 52 (3):397–399.

Theobald, D.M., and N.T. Hobbs. 1998. "Forecasting rural land-use change: A comparison of regression- and spatial transition-based models." *Geographical and Environmental Modelling* 2.

Trolle, Dennis, Anders Nielsen, Hans E. Andersen, Hans Thodsen, Jørgen E. Olesen, Christen D. Børgesen, Jens Chr Refsgaard, Torben O. Sonnenborg, Ida B. Karlsson, and Jesper P. Christensen. 2019. "Effects of changes in land use and climate on aquatic ecosystems: Coupling of models and decomposition of uncertainties." *Science of the Total Environment* 657:627–633.

Turner, B.L., E.F. Lambin, and A. Reenberg. 2007. "The emergence of land change science for global environmental change and sustainability." *Proceedings of the National Academy of Sciences* 104. doi:10.1073/pnas.0704119104.

Ullah, Siddique, Adnan Ahmad Tahir, Tahir Ali Akbar, Quazi K. Hassan, Ashraf Dewan, Asim Jahangir Khan, and Mudassir Khan. 2019. "Remote sensing-based quantification of the relationships between land use land cover changes and surface temperature over the lower Himalayan region." *Sustainability* 11 (19):5492.

Vázquez-Quintero, G., R. Solís-Moreno, M. Pompa-García, F. Villarreal-Guerrero, C. Pinedo-Alvarez, and A. Pinedo-Alvarez. 2016. "Detection and projection of forest changes by using the Markov chain model and cellular automata." *Sustainability* 8.

Veldkamp, A., and E.F. Lambin. 2001. "Predicting land-use change agriculture." *Ecosystems & Environment* 85. doi:10.1016/S0167-8809(01)00199-2.

Verburg, P.H., J.R.V. Eck, T.C.D. Hijs, M.J. Dijst, and P. Schot. 2004. "Determination of land use change patterns in the Netherlands." *Environment and Planning B: Urban Analytics and City Science* 31. doi:10.1068/b307.

Weng, Q. 2002. "Land use change analysis in the Zhujiang Delta of China using satellite remote sensing, GIS and stochastic modelling." *Journal of Environmental Management* 64. doi:10.1006/jema.2001.0509.

Wickramasuriya, R.C., A.K. Bregt, H.V. Delden, and A. Hagen-Zanker. 2009. "The dynamics of shifting cultivation captured in an extended constrained cellular automata land use model." *Ecological Modelling* 220. doi:10.1016/j.ecolmodel.2009.05.021.

Xiong, Yongzhu, Shaopeng Huang, Feng Chen, Hong Ye, Cuiping Wang, and Changbai Zhu. 2012. "The impacts of rapid urbanization on the thermal environment: A remote sensing study of Guangzhou, South China." *Remote Sensing* 4 (7):2033–2056.

Yang, Xin, Xin-Qi Zheng, and Li-Na Lv. 2012. "A spatiotemporal model of land use change based on ant colony optimization, Markov chain and cellular automata." *Ecological Modelling* 233:11–19.

Ye, B., and Z. Bai. 2008. "Simulating land use/cover changes of Nenjiang County based on CA-Markov model." *International Federation for Information Processing Publications (IFIP)* 258. doi:10.1007/978-0-387-77251-6_35.

Zimmermann, M., K.F. Glombitza, and B. Rothenberger. 2009. "Disaster adaptation programme for Bangladesh 2010–2012." In *Swiss Agency for Development and Cooperation (SDC)*.

Section II

Hydrology, Microclimates and Climate Change Impacts

5 Spatio-Temporal Variations in Water Surface Temperature and Its Effect on Microclimate of Sukhna Lake in Chandigarh (India)

Abhishek Pathania, Raj Setia, Sagar Taneja, Tapan Ghosh and Brijendra Pateriya

CONTENTS

5.1 INTRODUCTION

Lake water surface temperature plays an important role in controlling the aquatic ecosystem of lakes, which, in turn, affects both the chemical and biological processes occurring in the water body (Piccolroaz et al., 2013). The urban blue areas, either lakes or rivers, can potentially reduce the heat stress in cities and may play an important role in diminishing the urban heat island (UHI) effect and improving the air quality (Cheval et al., 2018). In general, an increase in river water temperatures is driven primarily by rising air temperatures (Seekell et al., 2011). Morrill et al. (2005) analyzed water temperature changes in the three rivers of Massachusetts and they found that the temperature of these rivers increased by 0.6–0.8°C for every 1°C increase in air temperature.

A number of studies in different parts of the world have shown an increase in the surface temperatures of lakes due to climate change, urbanization and other factors. For example, the surface temperature of Lake Baikal (the world's largest freshwater lake, in Siberia) increased by about 1.2°C since 1946 due to climate change (Hampton et al., 2008). It has been found that urbanization raises water and air temperatures, due to an increase in the area of impervious surfaces *via* the heat island effect (Mohan et al., 2013), but water bodies in urban areas have a cooling effect on the surrounding area, due to evaporation and heat absorption. Robitu et al. (2006) observed that a small pond (4 m × 4 m) had a cooling effect of about 1°C at a distance of up to 30 m. However, the magnitude of the cooling effect of water bodies is affected by water temperature, relative to the air temperature in adjacent urban areas.

Traditional methods for measuring lake water surface temperature include *in-situ* data loggers, which are labor intensive and time consuming. In recent years, thermal remote sensing has been used for the mapping and monitoring of lake water surface temperature. Song et al. (2016) studied the spatio-temporal variations in surface temperature of 56 lakes across the Tibetan Plateau, using the MODIS Land Surface Temperature (LST) product, and they found that 32% of the lakes showed a warming rate of 0.04°C $year^{-1}$, based on daytime MODIS measurements. Reinart and Reinhold (2008) and Bresciani et al. (2011) also used MODIS data to map the surface temperature of large lakes in Sweden and Italy, respectively. Most of the studies have been based on MODIS data with moderate spatial resolution sensors (pixel size of about 1 km). Nevertheless, the use of these sensors in smaller lakes is challenging because their spatial resolution is too coarse, but there are very few studies in which Landsat satellite data have been used to map the surface temperature of smaller lakes (Ji et al., 2015). The aim of the present study was to study the surface temperature of Sukhna Lake in Chandigarh (India), using Landsat satellite data. Specific objectives were: (i) to study the changes in land use/land cover around the lake, using spectral indices; (ii) to examine the lake water surface temperature, using Landsat 5, 7 and 8 satellite data from the years 2001, 2011 and 2018; and (iii) to study the effect of water temperature on changes in microclimate of the areas surrounding the lake.

5.2 MATERIALS AND METHODS

5.2.1 Study Area

Sukhna Lake in Chandigarh (India) is a reservoir in the foothills (Shivalik Hills) of the Himalayas. The area of the rainfed lake is 3 km². Orginally, there was heavy siltation in the lake due to seasonal flow coming down directly from the Shivalik Hills. In order to check the inflow of silt, 25. 4 km² of land were placed under vegetation in the catchment area of the lake.

In this study, temporal variations in water surface temperature of the surroundings of Sukhna Lake were monitored using Landsat data. The study area is 23.16 km² and it comprises Sukhna Lake in the middle (Figure 5.1), Chandigarh in the southwest Sukhna catchment area to the north/northeast and some villages and suburbs in the east and southeast.

5.2.2 Satellite Data

The Landsat satellite data from February of the years 2001, 2011 and 2018 were downloaded from the "earthexplorer.usgs.gov" website of the United States Geological Survey (USGS). The datum of the satellite data was WGS 84 and the projection was UTM. The archive satellite data were Landsat-7 ETM+ for the

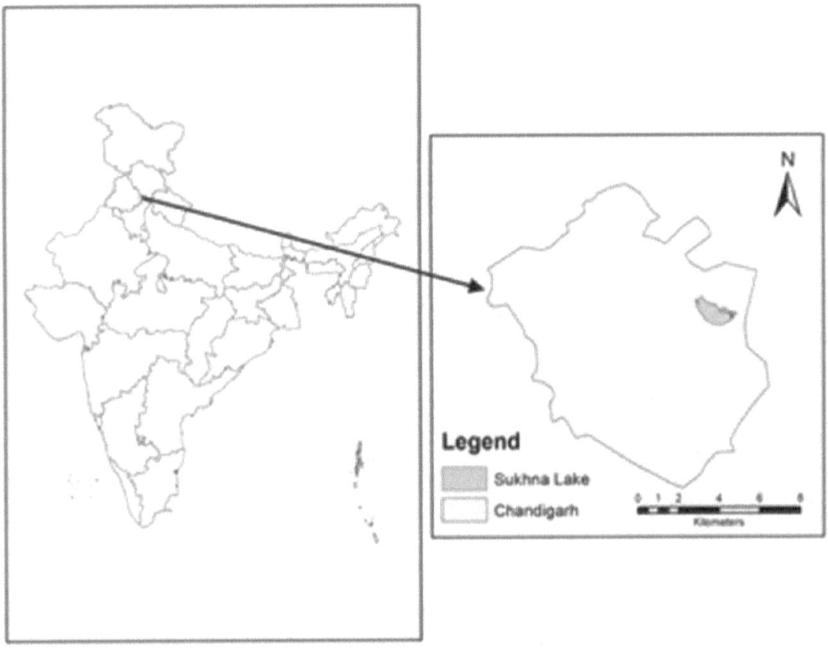

FIGURE 5.1 Location of the study area.

year 2001, Landsat-5 TM (Thematic Mapper) for the year 2011 and Landsat-8 for the year 2018.

5.2.2.1 Retrieval of Water Surface Temperature From Satellite Data

The following steps were used to obtain the water surface temperature from Landsat 5, 7 and 8.

(i) Pre-Processing of Satellite Data
In order to remove the effect of the source of the light and the atmosphere, the digital numbers of Landsat 5, 7 and 8 were converted to top of atmosphere reflectance

(ii) Use of Spectral Indices for Extracting Land Use/ Land Cover Features
The three main land use/ land cover features (vegetation, built-up areas, and water bodies) surrounding the lake were extracted using the following indices:
   (a) Normalized Difference Vegetation Index (NDVI) for extracting vegetation = (NIR-R) / (NIR+R)
   (b) Normalized Difference Built-up Index (NDBI) for extracting built-up features = (NIR – SWIR) / (NIR + SWIR)
   (c) Modified Normalized Difference Water Index (MNDWI) for extracting water features = (Green – MIR) / (Green + MIR),
   where G is green band, R is red band, NIR is near-infrared band and SWIR is short-wave-infrared band

(iii) Obtaining Water Surface Temperature From Landsat 5, 7 and 8
A number of algorithms have been developed to obtain LST, using sensor temperature and auxiliary data. In this study the Planck function has been used to convert brightness temperature to water surface temperature because atmospheric variables are not required during the satellite overpass time to obtain the surface temperature. The methodology to obtain the surface temperature from Landsat 5, 7 and 8 using the Planck function is described in Majumder et al. (2020).

5.2.2.2 Effect of Lake Water on Microclimate of the Surrounding Area

In order to study the cooling effect of water bodies on the microclimate of the surrounding area, buffers of 90-m to 900-m, at intervals of 90-m, were created around the lake, and the average temperature within each buffer zone was extracted for the years 2001, 2011 and 2018.

5.3 RESULTS AND DISCUSSION

5.3.1 Temporal Distribution of Land Use and Land Cover

There were substantial changes in land cover around the lake over the 17-year period (Figure 5.2). The built-up area increased by 2.73% from 2001 to 2011 and

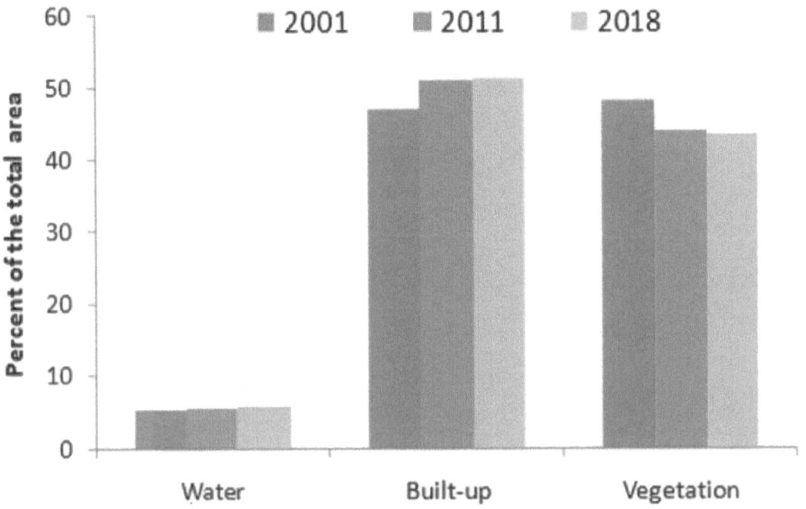

FIGURE 5.2 Temporal changes in area under the three main land use/land cover features in the study area.

by 1.18% from 2011 to 2018. The area of water bodies decreased by 0.03% from 2001 to 2011 but increased by 0.1% from 2011 to 2018. During the same period, the vegetated area decreased by 2.7% from 2001 to 2011 and by 1.28% from 2011 to 2018. The vegetated area lost 2.7% of its area to the built-up area between 2001 and 2011, and 1.18% between 2011 and 2018.

5.3.2 Spatial Distribution of Land Use and Land Cover

The spatial trends of land use and land cover (LULC) in the study area indicated that there was an expansion of the built-up area, mainly in the south-eastern and north-eastern parts of the area, between 2001 and 2011, and expansion continued in similar directions between 2011 and 2018 (Figure 5.3).

5.3.3 Temporal Distribution of Lake Water Surface Temperature

Of the different land cover features, the surface temperature was highest in built-up areas followed by vegetation and water bodies (Figure 5.4). The lake water surface temperature (LWST) was the lowest because water absorbs most of the radiation incident from the sun and distributes the energy throughout its depth. The average temperature of built-up areas increased by 0.8°C from 2001 to 2011 and by 2.91°C from 2011 to 2018. Similarly, the increase in temperature of vegetation was 0.66°C from 2001 to 2011 and 4.29°C from 2001 to 2018. The average surface temperature of the lake water also increased over the 17-year period, from 14.45°C in 2001 to 19.91°C in 2018.

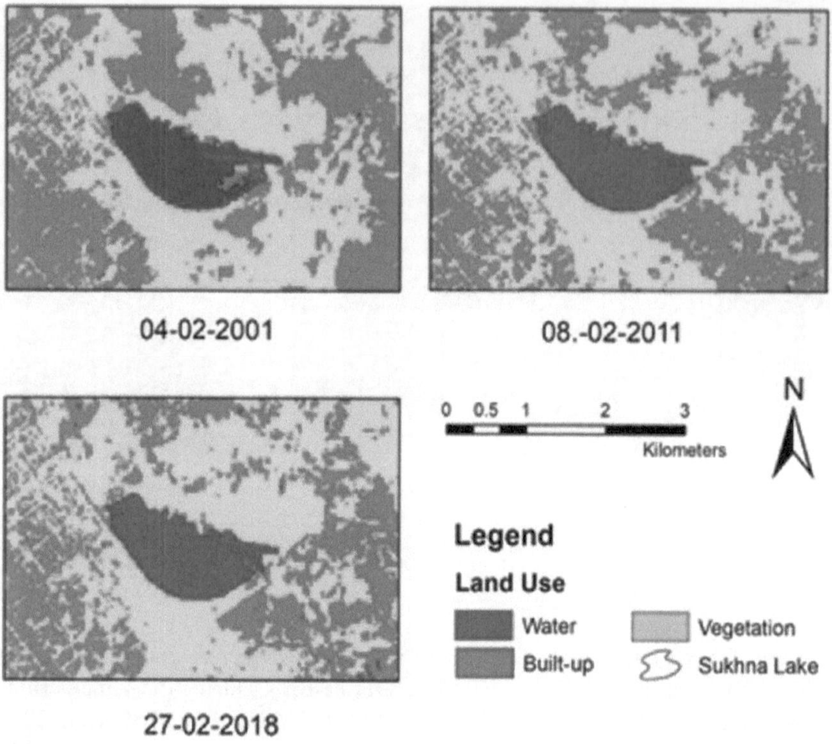

FIGURE 5.3 Spatio-temporal pattern of land use/land cover in the study area.

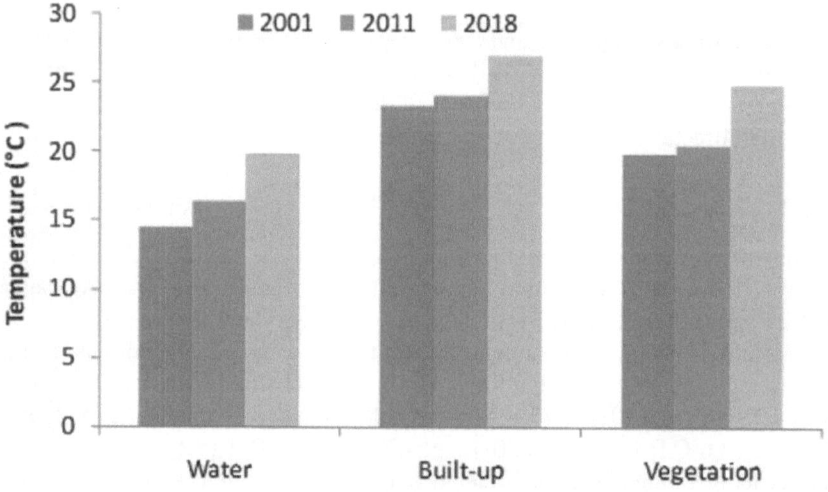

FIGURE 5.4 Temporal changes in average temperature of the three land use/land cover features in the study area.

The increase in built-up areas and the decrease in vegetated areas is one of the major reasons for increase in the water surface temperature of the lake. Another reason for the increase in the temperature of the lake was the rise in air temperature, which showed a maximum/minimum variation of 22.4/4.0°C in 2001 to 21.8/9.2°C in 2011 and 27.2/14.0°C in 2018. The depth of the lake had also decreased over the study years, which resulted in an increase in the water temperatures, because deeper water absorbs more energy, resulting in low temperatures.

It is pertinent to mention here that only diurnal changes in water surface temperature were measured from the Landsat satellites in this study. The water surface temperature is cooler during the night and warmer during the day due to the exceptionally large thermal capacity of water. The specific heat capacity values of water and air are 4.18 joule/g °C and 1.01 joule/g °C, respectively. The short-wave radiation absorbed during the day is released during the night and any increase in the diurnal temperature will also affect the night temperature, thereby significantly affecting various physical and biological phenomena. An increase in the average diurnal temperature of the lake water has affected the ecosystem of the lake, such as the number of migratory birds coming to the lake every year. A report from 2012 showed a staggering decrease of 42% in the total number of migratory birds compared with the previous year and that it had further affected the number of birds.

5.3.4 Spatial Pattern of Lake Water Surface Temperature

The degree of increase in the surface tempearture of the north-east area (Sukhna Reserved Forest) was higher compared with the other directions of the study area because vegetation cover was decreased in this direction but built-up increased in the direction which absorb more heat, thereby increasing the temperature in north-east direction (Figure 5.5).

The temperature of south-west and south-east areas were higher due to urban dominant area. The north-west and north directions of the areas include Sukhna reserved forest and villages, therefore, variations in temperatures in these areas are random with high temperatures in villages and lower temperatures in the forest-covered areas.

5.3.4.1 Effect of Water Temperature on the Microclimate Around the Lake

The average lake water surface temperature from 0 to 90 m decreased by 4.72, 3.32 and 4.06°C and in the range 90–180 m away decreased by 1.01, 0.66 and 0.72°C in the years 2001, 2011 and 2018, respectively (Figure 5.6). The effect of water bodies on affecting the land surface temperature was significant up to 360 m away, cooling the environment by 0.32, 0.17 and 0.14°C in the years 2001, 2011 and 2018, thereby respectively. The effect was variable beyond 360 m due to changes in land cover features. The contribution of water bodies to altering the temperature of the surrounding land is due to its cooling effect, either by evaporation or by transfer of heat between the air and the water. Furthermore, the

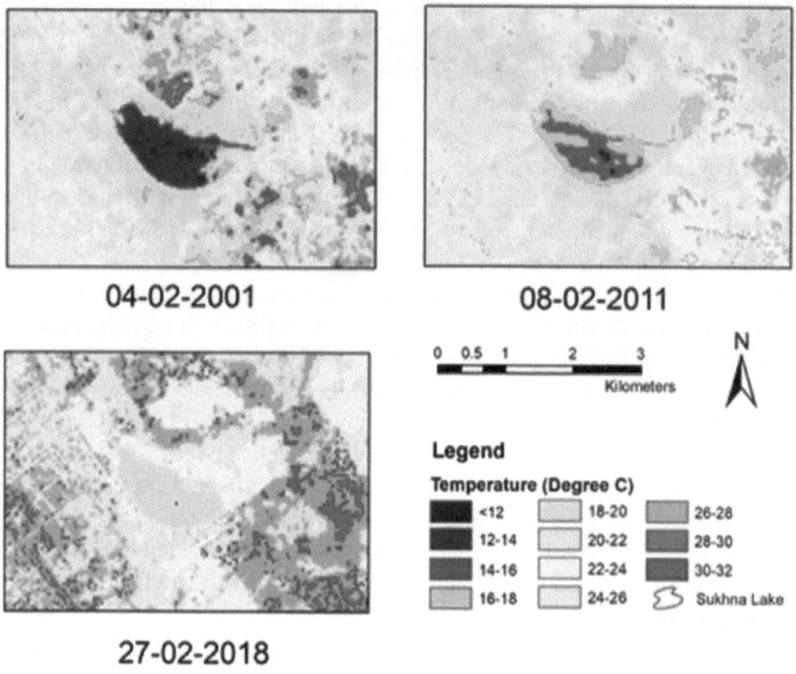

FIGURE 5.5 Spatio-temporal pattern of surface temperature in the study area.

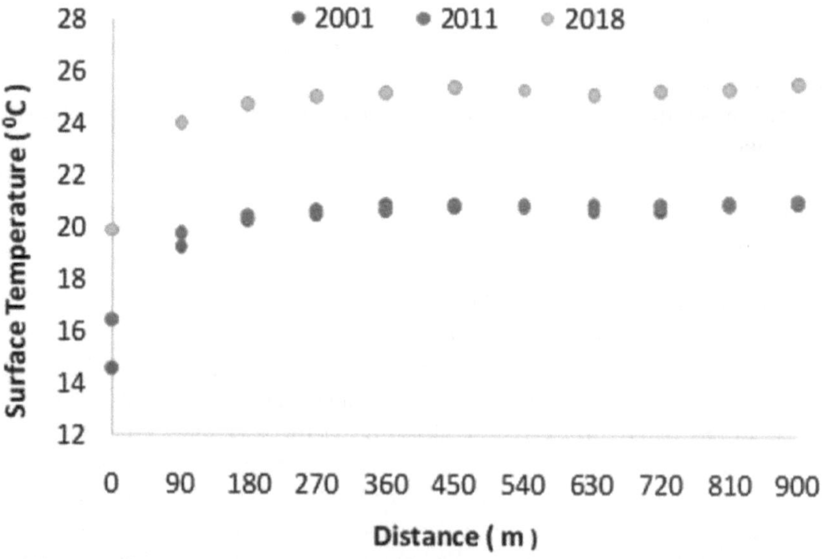

FIGURE 5.6 Changes in land surface temperature at various distances from Sukhna Lake due to the lake's microclimate effect.

water table near water bodies is close to the ground surface, and hence contributes toward lowering the surface temperature of surrounding land areas. Li et al. (2008) found that an independent 2 km^2 water body can cool the environment 1-km away by an average of 0.6°C.

5.4 CONCLUSIONS

Landsat satellite data is useful for assessing the diurnal changes in water surface temperature of the lake. The surface temperature increased over the years by 1.98°C from 2001 to 2011 and by 3.48°C from 2011 to 2018. This was mainly related to urbanization and the rise in average air temperature, which increased by 2.3°C from 2001 to 2011 and by 5.1°C from 2011 to 2018. Sukhna Lake cooled the surrounding environment by 0.32, 0.17 and 0.14°C up to 360 m away in the years 2001, 2011 and 2018, respectively, and the effect was variable beyond 360 m. More studies on diurnal and nocturnal changes in water surface temperature of Sukhna Lake are required to analyze the impact of temperature on the lake ecosystem, water budget, water storage and climate change.

REFERENCES

Bresciani, Mariano, Claudia Sotgia, Gian Luca Fila, Mauro Musanti, and Rossano Bolpagni. Assessing Common Reed Bed Health and Management Strategies in Lake Garda (Italy) by Means of Leaf Area Index Measurements. *Italian Journal of Remote Sensing* 43, no. 2 (2011): 9–22.

Cheval, Sorin, Sandric Ionut, Ioja Cristian, Dumitrescu Alexandru, Onose Diana, and Vanau Gabriel. The Influence of Urban Lakes on the Land Surface Temperature-a Remote Sensing Assessment. *EGUGA* (2018): 12582.

Hampton, Stephanie E, Lyubov R Izmest'eva, Marianne V Moore, Stephen L Katz, Brian Dennis, and Eugene A Silow. Sixty Years of Environmental Change in the World's Largest Freshwater Lake–Lake Baikal, Siberia. *Global Change Biology* 14, no. 8 (2008): 1947–58.

Ji, Luyan, Xiurui Geng, Kang Sun, Yongchao Zhao, and Peng Gong. Target Detection Method for Water Mapping Using Landsat 8 Oli/Tirs Imagery. *Water* 7, no. 2 (2015): 794–817.

Li, Shuyan, Chunyi Xuan, Wei Li, and H-bin Chen. Analysis of Microclimate Effects of Water Body in a City. *Chinese Journal of Atmospheric Sciences-Chinese Edition* 32, no. 3 (2008): 552.

Majumder, A, P Kingra, R Setia, SP Singh, and B Pateriya. Influence of Land Use/Land Cover Changes on Surface Temperature and Its Effect on Crop Yield in Different Agro-Climatic Regions of Indian Punjab. *Geocarto International* 35 (2020): 663–86.

Mohan, Manju, Yukihiro Kikegawa, BR Gurjar, Shweta Bhati, and Narendra Reddy Kolli. Assessment of Urban Heat Island Effect for Different Land Use–Land Cover from Micrometeorological Measurements and Remote Sensing Data for Megacity Delhi. *Theoretical and Applied Climatology* 112, no. 3–4 (2013): 647–58.

Morrill, Jean C, Roger C Bales, and Martha H Conklin. Estimating Stream Temperature from Air Temperature: Implications for Future Water Quality. *Journal of Environmental Engineering* 131, no. 1 (2005): 139–46.

Piccolroaz, Sebastiano, Marco Toffolon, and Bruno Majone. A Simple Lumped Model to Convert Air Temperature into Surface Water Temperature in Lakes. *Hydrology and Earth System Sciences* 17, no. 8 (2013): 3323.

Reinart, Anu, and Markus Reinhold. Mapping Surface Temperature in Large Lakes with Modis Data. *Remote Sensing of Environment* 112, no. 2 (2008): 603–11.

Robitu, Mirela, Marjorie Musy, Christian Inard, and Dominique Groleau. Modeling the Influence of Vegetation and Water Pond on Urban Microclimate. *Solar Energy* 80, no. 4 (2006): 435–47.

Seekell, David A, and Michael L Pace. Does the Pareto Distribution Adequately Describe the Size-Distribution of Lakes?. *Limnology and Oceanography* 56, no. 1 (2011): 350–56.

Song, Kaishan, Min Wang, Jia Du, Yue Yuan, Jianhang Ma, Ming Wang, and Guangyi Mu. Spatiotemporal Variations of Lake Surface Temperature across the Tibetan Plateau Using Modis Lst Product. *Remote Sensing* 8, no. 10 (2016): 854.

6 Assessing the Impact of Climate Variables on Soil Salinity Using Remote Sensing and GIS

Harkanwaljot Singh Sekhon, Som Pal Singh, Raj Setia, P. K. Kingra and Brijendra Pateriya

CONTENTS

6.1 INTRODUCTION

The advent of advanced crop production practices in the Indian Punjab has achieved unparalleled success in recent decades (Rangi and Sidhu 1998). Advances in agricultural technologies, including the use of chemical fertilizers, insecticides, pesticides and canal irrigation, enabled the state of Punjab to export surplus food grains (Bhatt et al. 2004), but it also caused the problems of waterlogging and poor drainage, resulting into soil salinity which affects normal plant growth and crop productivity (Singh 2013). The problem of salinity poses a serious threat to productive agricultural land around the world (Chatterjee et al. 2005). In addition to this, it causes deterioration of soil health and microbial diversity (Kulkarni and Shah 2013). The problem is further compounded by natural factors, such as the existence of topographic depressions, increases in temperature and persistent rains (Hundal and Abrol 1991).

Over recent decades, significant climatic variations have been observed under Punjab's conditions (Kingra et al. 2017; Kingra et al. 2018), with adverse effects on the water cycle and the recharging of groundwater, which are among the most important factors limiting crop production. In south-western Punjab, salinity is associated with waterlogging (Hundal and Kaur 1996). The salts dissolved in the water accumulate gradually at the soil surface, due to evaporation. Soil salinity patterns change with temperature variation and the frequency and intensity of rainfall. Salts mostly accumulate on the soil surface during the dry season and are leached down into the lower soil horizons by rainwater (World Development Report 2008). The alternate dry spells and intensive monsoon rainfalls in the few years of extreme climate events are very common, and are found to have significant impacts on variation in soil salinity in the region (Sontakke 1990).

Due to changes in climatic conditions, it is important to study the spatio-temporal variations in soil salinity. The spatial and spectral resolutions of satellite observations can provide consistent and repetitive global measurements of the Earth's surface (Xu 2007). The advantages of the spectral measurements are synoptic coverage, allowing the assessment of changes in dynamic phenomena, like soil salinity, before, during and after the monsoon seasons (Chatterjee et al 2003). Diagnosis, monitoring and mapping of saline areas are pre-requisites for the management of valuable land resources (Setia et al. 2013). Satellite remote sensing, integrated with geographic information systems (GIS), can better assess and monitor the waterlogged and saline areas (Kaul and Ingle 2011). In addition to slope, soil and other factors, the dynamic behavior of soil salinity is related to climate variables (Hingane et al. 1985). There are very few studies in which the impact of climate variables on salinity has been studied using the integration of remote sensing and GIS tools. Therefore, a study was carried out to assess the effect of climate variables on soil salinity, using Landsat satellite data in the GIS environment.

6.2 MATERIALS AND METHODS

6.2.1 Study Area

Four districts in south-western Punjab (Bathinda, Faridkot, Mansa and Muktsar) were selected for the study. The geographical area, comprising 3,385 km^2 of Bathinda district, 1,469 km^2 of Faridkot district, 2,171 km^2 of Mansa district and 2,615 km^2 of Muktsar district (Figure 6.1), was used for this study. Rice-wheat

FIGURE 6.1 Study area.

and cotton-wheat cropping patterns prevail in these districts. These parts of south-western districts are increasingly facing the problems of salinity over the past two decades. The area is nearly level and devoid of natural streams or gravity outlets causing salinity and associated problems. The area in Bathinda, Faridkot and Muktsar districts is irrigated by the Sirhind canal (with the Sirhind and Rajasthan feeders running North–South through this region). In Mansa district, the River Ghaggar enters at the southern end and flows through it for a distance of 15 km. The Bhakra mainline canal passes through the district, although the Sirhind drain runs across the entire district and has its out-fall in the River Ghaggar.

6.2.2 Satellite Data Used in the Study

The Landsat satellite images of pre-monsoon, monsoon and post-monsoon seasons for the years 2003, 2008 and 2013 were downloaded from "earthexplorer.usgs.gov" on the United States Geological Survey (USGS) website. The datum and projection of the satellite data were WGS 84 and UTM, respectively (Table 6.1). The archive satellite data of different seasons in the years 2003 and 2007 were Landsat-7 ETM+ (Launch date: 15 April 1999) but was Landsat-8 (Launch date: 11 February 2013) for all three seasons in the year 2013 (Table 6.2).

6.2.3 Spectral Indices Used to Extract the Salt-Affected Areas

Normalized Difference Salinity Index (NDSI) was used to extract the salt-affected areas from satellite images, and was calculated using the equation:

$$\text{NDSI} = (\text{Red} - \text{NIR})/(\text{Red} + \text{NIR})$$

where 'Red' is the red band (Band 3 of Landsat-7 and Band 4 of Landsat-8) and NIR is the near-infrared band (Band 5 of Landsat-7 and Band 5 of Landsat-8). The value of NDSI ranges from –1 to +1.

TABLE 6.1
Characteristics of the Satellite Data Used in the Study

Property	Landsat 7	Landsat 8
Spatial resolution	30 m for visible and IR, 15 m for panchromatic (PAN)* and 60 m for thermal	30 m for visible and IR, 15 m for PAN and 100 m for Thermal
Spectral resolution	8 bands (visible, IR, PAN and thermal bands)	11 bands (visible, IR, PAN and thermal bands)
Radiometric resolution	8 bit	16 bit
Temporal resolution	16 days	16 days
Spectral resolution (µm)	Band 1: 0.45–0.52 (Blue) Band 2: 0.53–0.61 (Green) Band 3: 0.63–0.69 (Red) Band 4: 0.76–0.90 (Near-IR) Band 5: 1.55–1.75 (SWIR-1) Band 6: 10.4–12.5 (Thermal) Band 7: 2.09–2.35 (SWIR-2) Band 8: 0.52–0.92 (PAN)	Band 1: 0.43–0.45 (Blue) Band 2: 0.45–0.51 (Blue-green) Band 3: 0.53–0.59 (Green) Band 4: 0.64–0.67 (Red) Band 5: 0.85–0.88 (Near IR) Band 6: 1.57–1.65 (SWIR-1) Band 7: 2.11–2.29 (SWIR-2) Band 8: 0.50–0.68 (PAN) Band 9: 1.36–1.38 (Cirrus) Band 10: 10.60–11.19 (Thermal IR) Band 11: 11.50–12.51 (Thermal IR)

*PAN: Panchromatic

TABLE 6.2
Season-wise Details of the Satellite Data Used in the Study

S. No	Satellite data	Year	Season
1	Landsat-7	2003	Pre-monsoon (May)
			Monsoon (September)
			Post-monsoon (December)
2	Landsat-7	2008	Pre-monsoon (April)
			Monsoon (August)
			Post-monsoon (December)
3	Landsat-8	2013	Pre-monsoon (April)
			Monsoon (August)
			Post-monsoon (December)

6.2.4 Data Observation and Analysis

Soil samples were collected during May 2003, using global positioning system (GPS), and analyzed for electrical conductivity. NDSI values corresponding to the soil sampling points were identified from the images of May 2003 classified according to land use, and correlated with the electrical conductivity of the soils. In addition, Pearson's correlation coefficient was calculated among values for rainfall, temperature and NDSI values at the soil sampling points. The significant differences in salt-affected areas among various seasons were determined, using t-tests.

6.3 RESULTS AND DISCUSSION

6.3.1 Spatio-Temporal Changes in Rainfall

During the pre-monsoon, monsoon and post-monsoon seasons (Table 6.2) of year 2003, the average rainfall was higher in Bathinda and Mansa than in Faridkot and Muktsar districts (Table 6.3). During the post-monsoon season, the rainfall was negligible in Muktsar and Faridkot districts. The rainfall was higher in Bathinda and Mansa districts than Faridkot and Muktsar districts during the pre-monsoon season of year 2008.

The amount of rainfall was relatively higher in Mansa district during the monsoon season but it was higher in Faridkot, Mansa and Muktsar districts during the post-monsoon season. The rainfall was relatively lower during the pre-monsoon and post-monsoon seasons of the year 2013 but it was higher during the monsoon seasons in all four districts (Figure 6.2).

6.3.2 Spatio-Temporal Changes in Maximum Temperature

In all the years tested, maximum temperature was significantly lower during the post-monsoon season than in the pre-monsoon and monsoon seasons (Table 6.4). Among the pre-monsoon seasons of the different years, temperature

TABLE 6.3

Temporal Changes in Rainfall (mm) in Bathinda, Faridkot, Mansa and Muktsar Districts in the Pre-Monsoon, Monsoon and Post-Monsoon Seasons of 2003, 2008 and 2013

Year/	2003			2008			2013		
Season	Pre-monsoon	Monsoon	Post-monsoon	Pre-monsoon	Monsoon	Post-monsoon	Pre-monsoon	Monsoon	Post-monsoon
Bathinda	5.5	84.7	2.7	39.6	133	0	0.5	305	0
Faridkot	0.6	32.4	0.2	2.1	163	2.1	6	368	1
Mansa	5.1	50.7	2.3	18.9	231	1.5	0	79	0
Muktsar	0.4	28.5	0	2.3	160	2.1	1	352	0

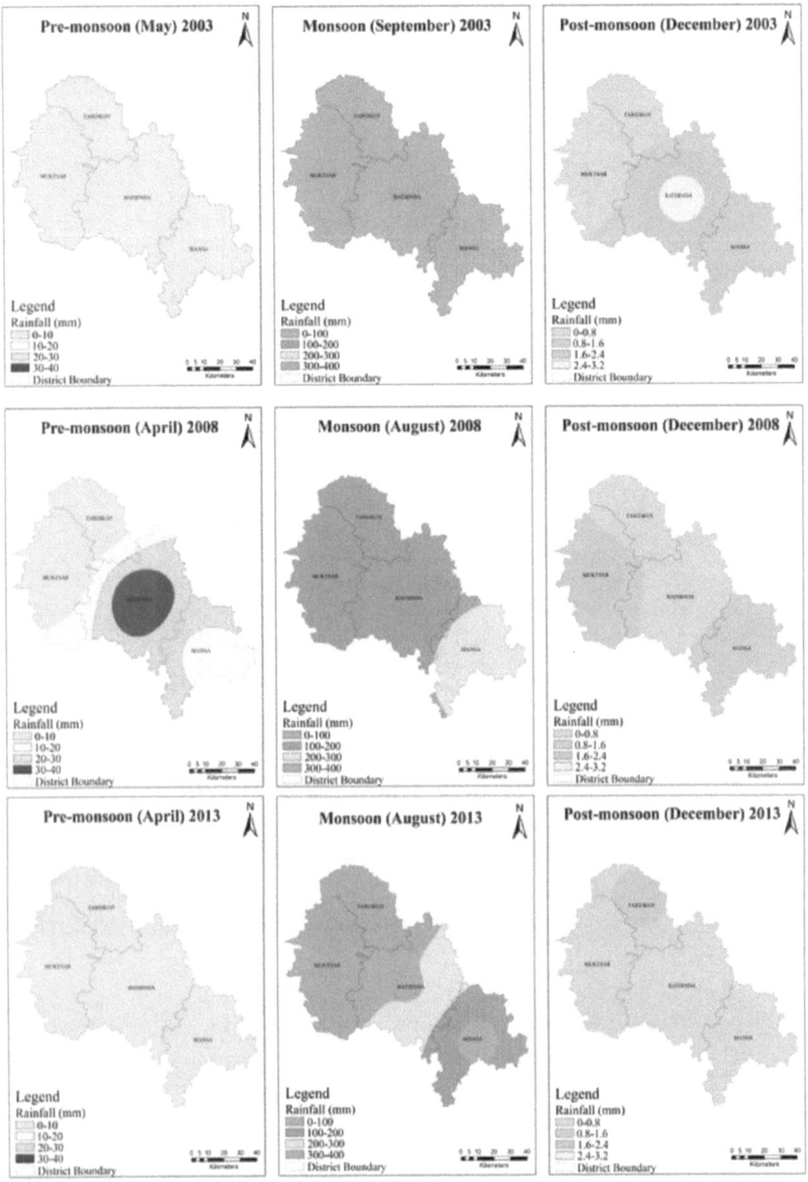

FIGURE 6.2 Spatial distribution of rainfall during 2003, 2008 and 2013.

was significantly lower in the year 2013. Maximum temperature during the monsoon season (July to September) of different districts could not be compared due to temperature data taken from August and September. For the post-monsoon seasons, maximum temperature was lower in Bathinda district than in the other districts during 2003, but it was lower in Faridkot district during the year 2008 and in Mansa and Muktsar districts during the year 2013 (Figure 6.3).

TABLE 6.4
Maximum Temperature (°C) in the Study Area During 2003, 2008 and 2013

	Year								
	2003			2008			2013		
District	Pre-monsoon	Monsoon	Post-monsoon	Pre-monsoon	Monsoon	Post-monsoon	Pre-monsoon	Monsoon	Post-monsoon
Bathinda	39.3	33.6	20.5	35.2	34.3	22	34.4	33.7	20.7
Faridkot	39.4	33.9	20.7	35.2	29.6	18.3	33	32.5	26
Mansa	39.4	33.4	20.7	35.3	34.3	21	34.6	33.6	20.5
Muktsar	39.4	33.9	20.7	35.3	34.2	20.9	34.3	35.8	20.5

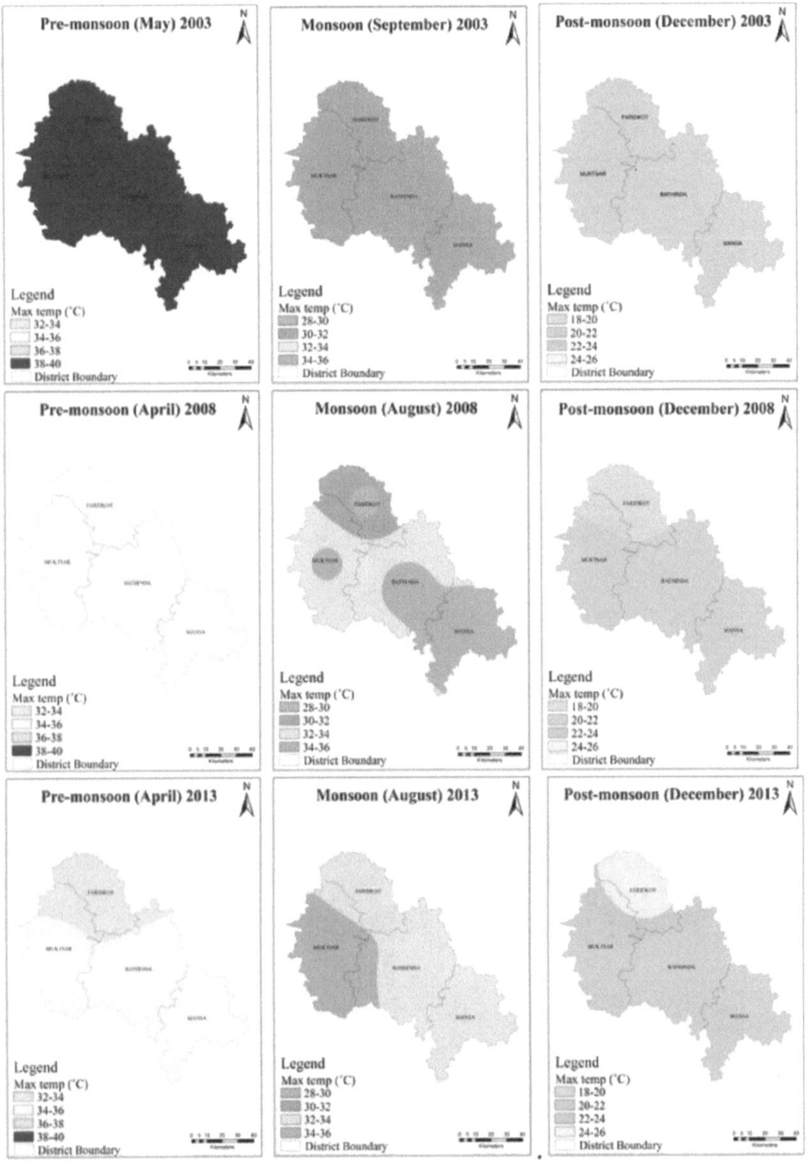

FIGURE 6.3 Spatial variation in maximum temperature during 2003, 2008 and 2013.

6.3.3 Spatio-Temporal Changes in Minimum Temperature

Among the pre-monsoon seasons of different years, minimum temperature was significantly lower during the year 2008 (Table 6.5). Minimum temperature during the monsoon season was not compared because an insignificant change in minimum temperature occurs from August to September. For the post-monsoon

TABLE 6.5
Minimum Temperature (°C) in the Study Area During 2003, 2008 and 2013

	Year								
	2003			2008			2013		
District	Pre-monsoon	Monsoon	Post-monsoon	Pre-monsoon	Monsoon	Post-monsoon	Pre-monsoon	Monsoon	Post-monsoon
Bathinda	22.6	23.3	7.3	17.2	25.5	7.8	18.2	26.5	6.9
Faridkot	22.7	23.4	7.1	21.4	26.8	8.0	21.6	24.4	10
Mansa	23.1	23.4	7.8	18.4	25.3	7.6	19.1	26.7	6.8
Muktsar	22.7	23.4	7.2	21.3	26.4	7.1	21.8	25.9	6.5

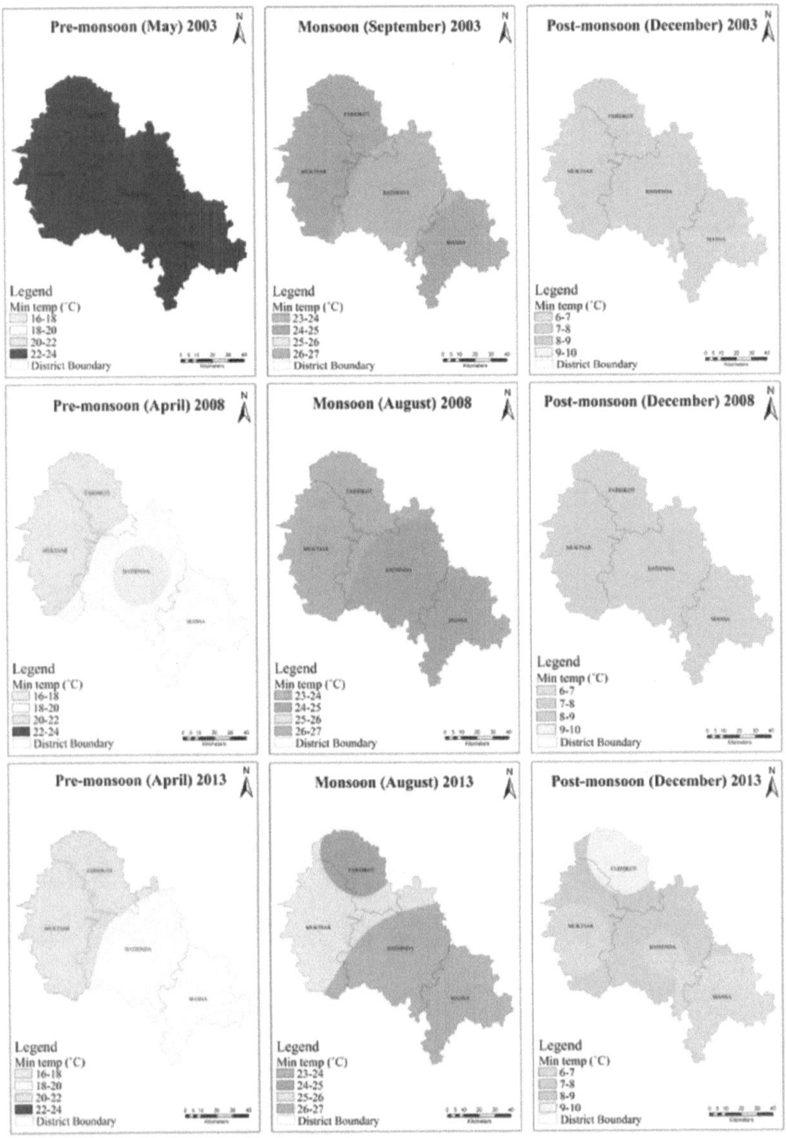

FIGURE 6.4 Spatial variability in minimum temperature during 2003, 2008 and 2013.

seasons, minimum temperature was lower than other districts in Bathinda, Mansa and Muktsar districts during the year 2013 (Figure 6.4).

6.3.4 Spatio-Temporal Changes in Average Temperature

Among the pre-monsoon seasons of the different years, average temperature was significantly lower in all districts during the year 2008 than in the other years

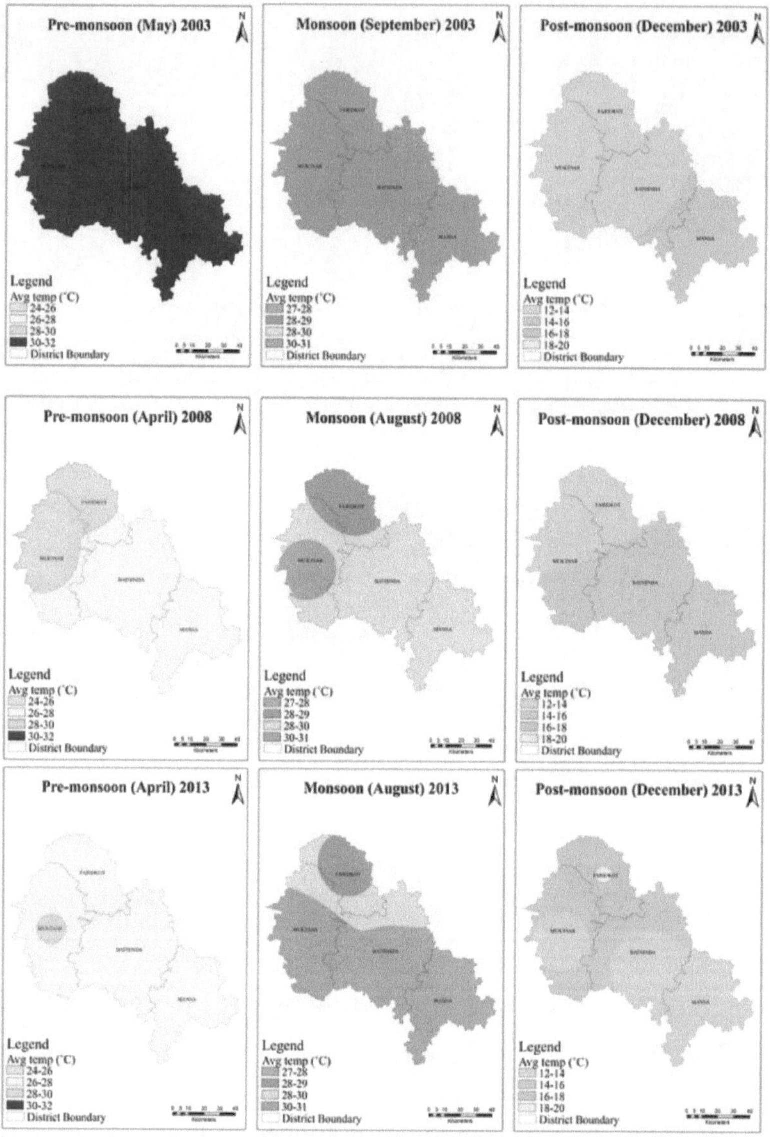

FIGURE 6.5 Spatial variability in average temperature during 2003, 2008 and 2013.

(Figure 6.5). For the post-monsoon seasons, average temperature was lower in Bathinda, Mansa and Muktsar districts during the year 2013 (Table 6.6).

6.3.5 Spatio-Temporal Changes in Soil Salinity

In 2003, the salinity was higher in the pre-monsoon season than in the monsoon and post-monsoon seasons in all four districts. Among these four districts, salinity was highest (0.30%) in Faridkot district (Figure 6.6).

TABLE 6.6

Average Temperature (°C) in Bathinda, Faridkot, Mansa and Muktsar Districts in the Pre-monsoon, Monsoon and Post-Monsoon Seasons of 2003, 2008 and 2013

	Year								
	2003			2008			2013		
District	Pre-monsoon	Monsoon	Post-monsoon	Pre-monsoon	Monsoon	Post-monsoon	Pre-monsoon	Monsoon	Post-monsoon
Bathinda	30.9	28.5	13.9	26.2	29.9	14.9	26.3	30.1	13.6
Faridkot	31.1	28.7	13.9	28.3	28.2	13.2	27.3	28.5	18.1
Mansa	31.3	28.4	14.3	26.9	29.8	14.3	26.9	30.3	13.7
Muktsar	31.1	28.7	13.9	28.3	30.3	14	28.1	30.9	13.5

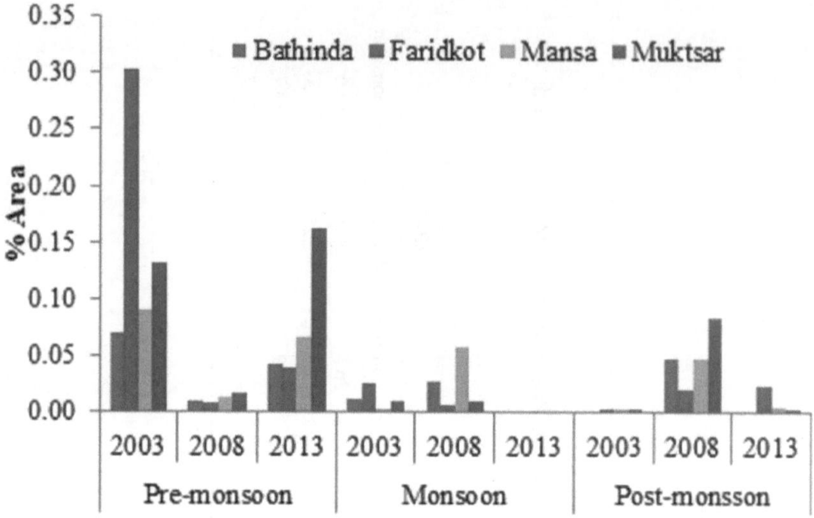

FIGURE 6.6 Percentage area under salinity during the years 2003, 2008 and 2013.

In 2008, salinity was higher from average in the post-monsoon season in all the four districts. Compared with the monsoon season, salinity was 40% higher from average in the post-monsoon season in Bathinda district, 50% higher in Faridkot district, 87.5% higher in Muktsar district but 20% lesser in Mansa district. In 2003, salinity was higher in the western direction than in the north-eastern directions of Bathinda district, in the northern direction than in the south-western areas of Faridkot district, in the southern than the north-western areas of Mansa district and in the northern than the south-eastern areas of Muktsar district. In 2008, salinity was higher in the south-western direction of Bathinda district, the north-eastern direction of Faridkot district, the south-eastern direction of Mansa district and the north-eastern direction of Muktsar district. In 2013, salinity was higher in the south-western direction of Bathinda district, north-western direction of Faridkot district, south-eastern direction of Mansa district and the north-western direction of Muktsar district (Figure 6.7).

6.3.6 Relationship between Salinity and Climate Parameters

NDSI was significantly negatively correlated with rainfall during all the seasons because salts are leached down to deeper layers with increasing rainfall. NDSI was positively correlated with temperature (minimum, maximum and average temperature). Among all the seasons, the value of the correlation coefficient between NDSI and maximum temperature was highest during the pre-monsoon season, irrespective of the year. The correlation coefficients between NDSI and minimum temperature and between NDSI and average temperature were highest during the post-monsoon season. Increased temperature causes evaporation of water from the soil surface, resulting in the capillary rise of water, which causes the upward movement of salts (Table 6.7).

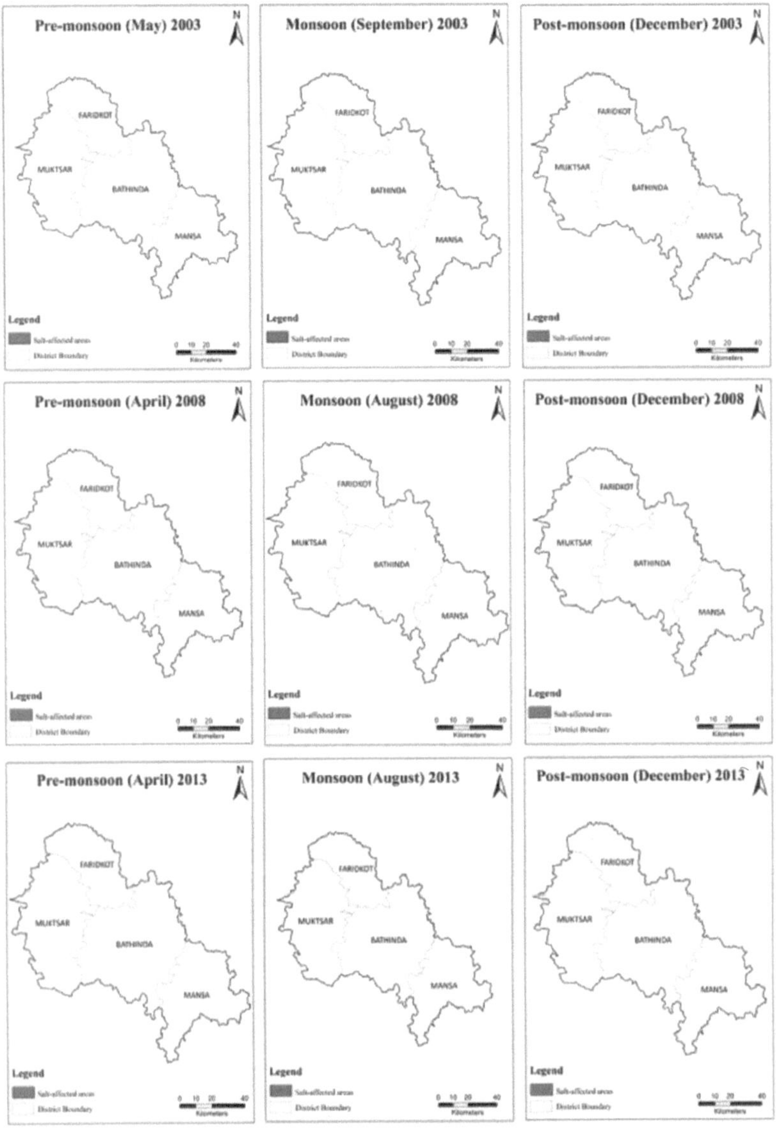

FIGURE 6.7 Spatial distribution of salinity during 2003, 2008 and 2013 (Note: Salinity was not detectable in the monsoon season of 2013).

6.3.6.1 Integration of Salinity with Rainfall and Temperature

The area under salinity was highest during the pre-monsoon season of 2003. Therefore, the NDSI map of May 2003 was integrated with rainfall and maximum temperature maps of the pre-monsoon season (May) of 2003 in GIS. It was found that salinity was higher in the direction where maximum temperature was higher, but it was lower in the direction where rainfall was higher (Figure 6.8).

TABLE 6.7
Correlation Coefficients among NDSI and Climate Parameters

NDSI/Month	Rainfall	Maximum Temperature	Minimum Temperature	Mean Temperature
		Year 2003		
Pre-monsoon	−0.84	0.94	0.68	0.72
Monsoon	−0.75	0.88	0.88	0.89
Post-monsoon	−0.63	0.86	0.92	0.97
		Year 2008		
Pre-monsoon	−0.91	0.92	0.60	0.80
Monsoon	−0.83	0.90	0.71	0.89
Post-monsoon	−0.93	0.82	0.93	0.94
		Year 2013		
Pre-monsoon	−0.79	0.95	0.58	0.95
Post-monsoon	−0.86	0.73	0.98	0.97

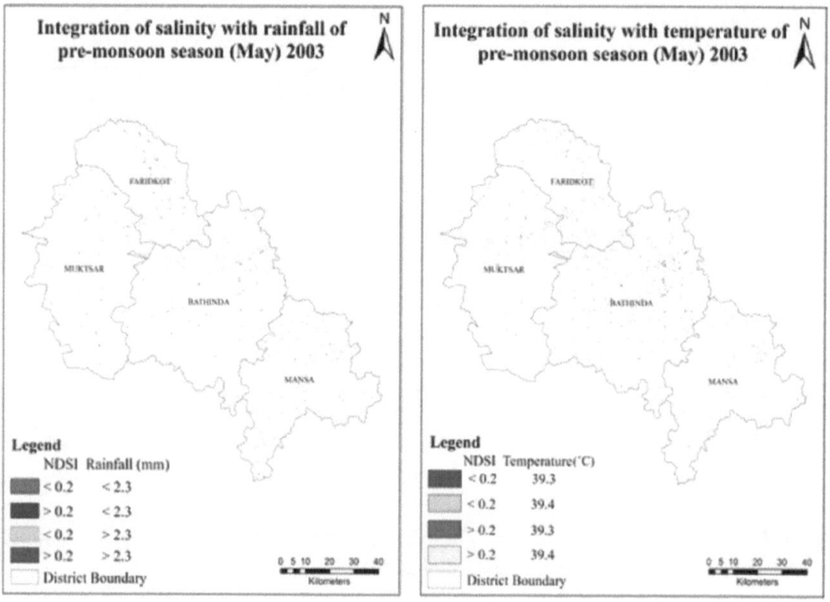

FIGURE 6.8 Integration of salinity with rainfall and temperature for the pre-monsoon season of 2003.

6.3.6.2 Accuracy Assessment

By evaluating the accuracy of the NDSI image of May 2003, the electrical conductivity (EC) of field-collected soil samples was measured. The EC of saturated paste of soil samples was >4 dS m^{-1} and NDSI identified the salt-affected areas with an overall accuracy of 85% and a Kappa coefficient of 0.35.

6.4 CONCLUSIONS

The NDSI can be used to map the spatio-temporal variability and distribution of salt-affected areas, using Landsat satellite data. Salinity in south-western Punjab is mainly due to climate, topography, depth to ground water level and soil characteristics. Among these factors, rainfall and temperature have significant impact on salinity. The integration of climate variables with topography and ground water level in GIS can help to develop the decision support system which can enable the researchers and other end-users to better understand the complex processes of salinity dynamics. It may be useful for policy makers and scientists to help plan the strategies needed to manage the salt-affected areas and the soil resources.

REFERENCES

Bhatt, C. M., R. K. Singh, P. K. Litoria and P. K. Sharma. 2004. Use of remotely sensed data and GIS techniques for assessment of waterlogged and salt-affected area tehsil-wise in Muktsar district of Punjab. http://www.gsdidocs.org/gsdiconf/GSDI-7/papers/Pcmb.Pdf (accessed on 26-03-2013).

Chatterjee, C., R. Kumar, B. Chakravorty, A. K. Lohani and S. Kumar. 2005. Integrating remote sensing and GIS techniques with groundwater flow modeling for assessment of waterlogged areas. *Water Resour Mgmt* 19:539–54.

Chatterjee, C., R. Kumar and P. Mani. 2003 Delineation of surface waterlogged areas in parts of Bihar using IRS-1C LISS-III data. *J Indian Soc Remote Sens* 31:57–65.

Hingane, L. S., K. K. Rupa and R. B. V. Murty. 1985. Long term trends of surface air temperature in India. *J Climatol* 5:521–28.

Hundal, S. S. and Y. P. Abrol. 1991. Perspectives of green house gases in climatic change and plant productivity, In *Proc. Indo-US Workshop on "Impact of Global Climatic Changes on Photosynthesis and Plant Productivity"*, New Delhi, January 8–12, pp. 767–79.

Hundal, S. S. and P. Kaur. 1996. Climatic change and its impact on crop productivity in Punjab, India. In *Climatic Variability and Agriculture*, ed. Y. P. Abrol et al., 377–93. Narosa Publishing House, New Delhi.

Kaul, H. A. and S. T. Ingle. 2011. Severity classification of waterlogged areas in irrigation projects of jalgaon district, Maharashtra. *J Appl Technol Environ Sanitation* 1:221–32.

Kingra, P. K., R. Setia and S. Singh et al. 2017. Climatic variability and its characterisation over Punjab, India. *J Agromet* 19:246–50.

Kingra, P. K., R. Setia and S. Kaur et al. 2018. Analysis and mapping of Spatio-temporal climate variability in Punjab using classical statistics and geostatistics. *Mausam* 69:147–55.

Kulkarni, H. and M. Shah. 2013. Punjab water syndrome. *Econ and Polit Weekly* 48:65.

Rangi, P. S. and M. S. Sidhu. 1998. *Growth of Punjab Agriculture and Sustainable Development*, Vol. 2, Ecological Society, Ludhiana and Centre for Research in Rural and Industrial Development Chandigarh.

Setia R., M. Lewis, P. Marschner, R. Raja Segaran, D. Summers and D. Chittleborough 2013. Severity of salinity accurately detected and classified on a paddock scale with high resolution multispectral satellite imagery. *Land Degrad Dev.* 24:375–84.

Singh, S. 2013. Water logging and its effect on cropping pattern and crop productivity in south-west Punjab: A case study of Muktsar district. *J Eco Soc Dev* 9:71–80.

Sontakke, N. A. 1990. *Indian Summer Monsoon Rainfall Variability During the Longest Instrumental Period 1813–1988*. M.Sc thesis, University of Poona.

World Development Report. 2008. *Agriculture for Development*, Oxford University Press, Washington, DC, p. 26.

Xu, H. 2007. Extraction of urban built-up land features from Landsat imagery using a thematic-oriented index combination technique. *Photogramm Eng Remote Sens* 73:1381–91.

7 Monitoring Changes in Land Cover in the Indian Sundarbans Region, Using Geospatial Sciences

Deepak Kumar and Tirthankar Ghosh

CONTENTS

7.1 INTRODUCTION

The Sundarbans is the world's largest contiguous mangrove forest (Iftekhar and Saenger, 2008; Datta and Deb, 2017), stretched over two countries, namely Bangladesh and India. It is also designated as one of UNESCO's World Heritage Site. The Indian Sundarbans region, which is located in the southern part of the Indian state of West Bengal, is home to almost 4.5 million people, residing on different islands in the Bay of Bengal.

Climate change is the greatest challenge facing mankind today, and the Sundarbans region is particularly vulnerable. Because of its tropical location and its location in the delta region of the Bay of Bengal, it is one of the most fragile and susceptible ecosystems in the world. As the global temperature rises, the polar ice caps and glaciers melt, adding to the mean sea levels. This process is now accelerating due to excessive heating of the planet, causing an increase in the relative mean sea level. In one study (Hazra et al., 2010), it was found that the rise in relative mean sea level at Sagar Island in the Sundarbans region, was 17.88 mm/year during the period 2000–2009. This significant rise in sea level caused enormous destruction to the local environment – loss of land, coastal erosion, and loss of property, like houses, schools, etc. Increased global temperature affects not only the atmospheric temperature but also the sea surface temperature (SST). This rise in SST results in greater evaporation, which, in turn, leads to increased annual rainfall and to changes in the weather patterns. The region now faces increased frequencies of cyclones and hurricanes. All of these changes affect the local environment, flora, and fauna. Coastal erosion is another major threat. The southern-most islands, such as Sagar, Ghoramara, Bulchery, Jambudwip, etc., are facing rapid coastal erosion and subsequent submergence of the islands. It was found that the land loss due to erosion in the Indian Sundarbans was about 5.50 km^2/year during the period 2001–2009, with the most-affected region being the south-western part (Hazra et al., 2010). Any major impact from a tsunami or cyclone will destroy a few of them and submerge the impacted part of the island. People are leaving these islands and migrating to the nearby cities or towns inland as they are losing their homes, and their agricultural lands are also becoming badly affected by salinity caused by ocean inundation. These people are becoming a new set of "climate refugees" in this country.

The Indian Sundarbans region is famous for its dense mangrove forest. The region is classified as moist tropical forest. Mangroves are usually shrubs or small trees that are salt tolerant, and mangrove forests usually form in the fragile coastal ecosystems, where there is an intersection between land and water. They are usually found within the tropical and sub-tropical regions, with around 15 countries containing about 75% of the total global area of mangrove forests. South Asia has 1,187,476 ha of mangrove forest, which represents approximately 7.00% of the global area (Giri et al., 2015). Mangroves help to protect the coastal area from natural disasters like tsunami, cyclones, etc. (Dahdouh-Guebas et al., 2005; Danielsen et al., 2005; Williams, 2005; Forbes and Broadhead, 2007). Mangrove forests are also a source of firewood, whereas some species of mangrove are also known for their medicinal value (Pham and Yoshino, 2015). There are a number of species of mangrove found in the Indian Sundarbans, including *Avicennia officinalis* (Indian mangrove), *Rhizophora mucronate* (Asiatic mangrove), etc.

Due to the extreme changes in the environment as a consequence of climate change, the mangrove ecosystems are under threat. Because of coastal erosion and increased numbers of tidal creeks (part of a stream affected by the ebb and flow of ocean tides), the Sundarbans is losing its mangrove forest. In the past 20 years, approximately 36% of the total global mangrove area has been lost, as a result of natural calamities and phenomena, such as cyclones and tsunamis, etc. On a smaller scale, Asia lost about 12% of its mangrove area between 1975 and 2005 (Giri et al., 2015), due mainly to coastal erosion and to a significant rise in seawater level. During the present study, it was identified that, in some places, the mangrove forest was degraded, then converted to saline blanks, where high accumulation of salinity prevents vegetation establishment. Hazra et al. (2010) noted that drivers of such changes are mainly surge inundation during high tides, water-logging, and precipitation of salt. However, due to climatic and anthropogenic influence, the mangrove ecosystems across the globe are under serious threat, and, with the present rate of loss, it is very possible that there will be no mangrove forests in the region by the end of the 21st century, if the present scenario continues (Valiela et al., 2001; Ellison, 2002; Duke et al., 2007).

In this context, satellite-based remote sensing is now by far the most-used technology with which to study and monitor changes in land cover over time. The present study focuses on the monitoring and detection of changes in forested areas and on land area changes on islands throughout the past 20 years (2000–2020). Multi-temporal Landsat (Landsat 5 Thematic Mapper (TM) and Landsat 8 Operational Land Imager [OLI]) images, with a spatial resolution of 30 m, were downloaded from the United States Geological Survey (USGS) and used in the current study. The main objectives of this investigation are to gather factual information and data to reflect the effects of the stresses operating in the region, using Landsat satellite images. The methods for detecting change on Land Use and Land Cover (LULC) maps are used to compare the past and present conditions of the study area. The information from this investigation will also help the authorities concerned to take the measures necessary to help protect the Sundarbans from any further degradation and to look forward to a sustainable future.

7.2 MATERIALS AND METHODS

7.2.1 Study Area

The study area is located in the southernmost part of the Indian state of West Bengal. Situated along the coast of the Bay of Bengal, it is close to the world's broadest delta, the Ganga-Brahmaputra Delta. Sundarbans is located over a location of 21° 32′–22° 40′ N latitude and 88° 05′–89° 51′ E longitude, stretching across a region of approximately 10,000 km²; of this total area, 62% lies in Bangladesh, with the remaining 38% area being in India (Spalding et al., 2010). Administrative blocks in the Indian Sundarbans comprise 13 blocks in the district of South 24 Parganas and six blocks in the district of North 24 Parganas. The South 24 Parganas district consists of islands like Sagar, Namkhana, Kakdwip, etc. On the other hand, the six blocks in the North 24 Parganas district consists of Haroa, Hasnabad, Minakhan, etc. The elevation is between 0.9 and 2.1 m above Mean Sea Level (MSL) (Iftekhar and Islam, 2004). The temperature here ranges from 20°C to 48°C. It is located in the tropical climatic zone, with the monsoon season extending from May to October (Ghosh et al.2016). In monsoon, the rainfall is heavy. The overall annual precipitation is in the range 1,500 mm–2,000 mm. The seven main rivers and various other channels and estuaries make a dense network and end in the delta.

The name "Sundarbans" was derived from the name of a native plant called the Sundari tree (*Heritiera fomes*), a local species of mangrove, which forms pneumatophores, aerial roots which aid with gaseous exchange, facilitating respiration and photosynthesis in the waterlogged soils. In the Sundarbans, about 245 genera and 334 species of plants were reported in 1903 (Prain, 1903), with 24 true mangrove taxa, associated with nine different families, being found within the Indian Sundarbans (Barik and Chowdhury, 2014). Some species, like *Aegialitis rotundifolia*, *Heritiera fomes*, *Sonneratia apetala*, or *Sonneratia griffithii*, are endemic. In addition to the mangroves, the Sundarbans are rich in biodiversity, with more than 200 other plant species, 400 species of fish, around 300 species of birds, 35 species of reptiles, 42 species of mammals, and countless other invertebrates, bacteria, and fungi, etc. (Gopal and Chauhan, 2006). The forest of the Sundarbans is home to almost 100 tigers, with the Royal Bengal Tiger being a native species. Since the Sundarbans is a forest reserve, tigers are protected here. The Sundarbans was announced as one of the World Heritage sites in the year 1987 and has also been included as a Ramsar wetland site since January 2019, designated to be of international importance under the Ramsar Convention. The area is connected by rail, road, and water ferries. Many islands come under the Sundarbans Forest Reserve. The main occupations here are related to apiculture, agriculture, aquaculture, and fishing. Approximately 2,000 people in the region are involved in beekeeping activities, producing around 90% of the natural honey generated in India (Spalding et al., 2010). The nearest large city is Kolkata, the capital city of West Bengal Province (Figure 7.1).

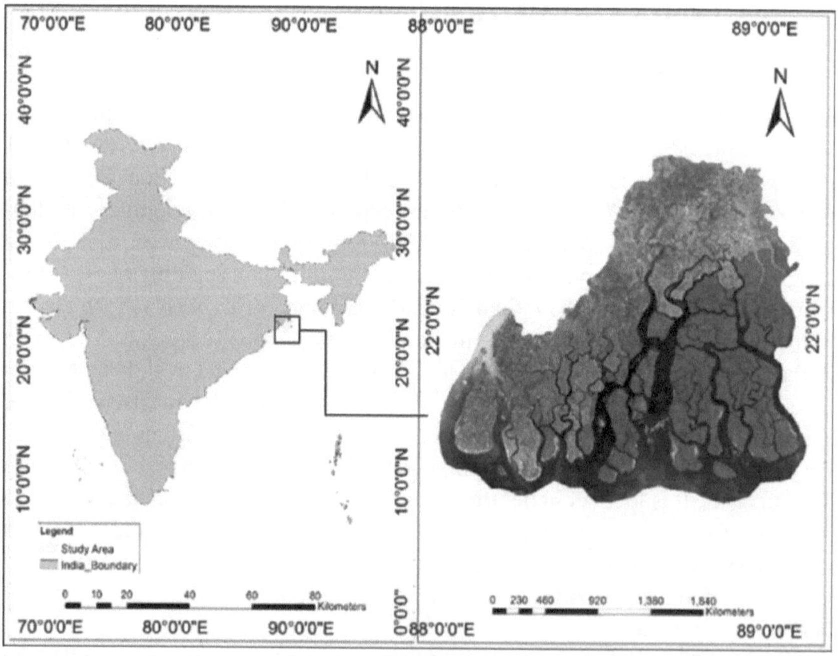

FIGURE 7.1 Location of the study areas, as viewed through satellite images (In false colour composites of red (R)=4, green (G)=3, and blue (B)=2 channels).

7.2.2 Methodology Used and Data Obtained

In this research, Landsat 5 Thematic Mapper (TM) satellite images, with optical bands of spatial resolution 30 m, acquired on 26 January 2000 (ID-LT51380452000026BKT00; Path/Row:138/45), and 21 January 2010 (ID-LT51380452010021KHC0; Path/Row:138/45), and Landsat 8 Optical Land Imager (OLI) satellite images, with optical bands of spatial resolution 30 m, acquired on 1 January 2020 (ID-LC81380452020001LGN00; Path/Row:138/45) were analyzed. All the satellite datasets were obtained from the repository of the United States Geological Survey (USGS). Images were selected which were 90% cloud-free and from the same month, to avoid errors due to seasonal differences. All the datasets were geometrically corrected using the Universal Transverse Mercator (UTM) coordinate system, 45th North Zone, and WGS 84 datum system (Table 7.1).

7.2.3 Software Used

In this study, various geoprocessing software were used. ERDAS IMAGINE 201x was utilized to subset the area of interest (AOI) from the raw satellite image data and make some geometrical corrections required for further processing. A supervised classification process was also conducted using this software. ArcMap 10.x, from the ArcGIS desktop package, was used to visualize the data and create LULC maps. Furthermore, the area was calculated for the respective spectral classes from the Table of parameters. Manual digitization was also carried out to create layers, using ArcMap 10.x software.

7.2.4 Working Methodology

A multi-temporal approach, based on remote sensing datasets, was used to monitor and observe the changes in the mangrove forested area and islands. Conventional satellite datasets, like Landsats, allowed observation of land cover changes from the 1970s (Lillesand et al., 2004). Landsat satellite images were used to prepare LULC maps of all three years (namely 2000, 2010, and 2020) by using the Supervised Classification method with the "Maximum Likelihood" technique on ERDAS IMAGINE 201x software. In the past two decades, the method of supervised classification has been exploited to monitor changes in mangrove forest cover (Hazra et al., 2010; Giri et al., 2015; Pham and Yoshino, 2015; Ghosh et al., 2016; Jones et al., 2016). Training signatures were developed with a minimum of 30 per class from all around the image to maintain accuracy. Accuracy assessment for all three years, 2000, 2010, and 2020, was performed by a method using an error matrix generating Producer's, User's, and Overall accuracy values with Kappa coefficients. The changes detected in the area of different land cover classes were calculated and compared in post-classification analysis. The shorelines of the most vulnerable islands in the southern region of the study area were manually digitized for all three years under examination to monitor

TABLE 7.1
Meta-information of Datasets Being Used

Sl. No	Satellite	Sensor	Resolution	Path/Row	Date of Acquisition	Source	Scene ID
1.	Landsat 5	TM	30 m	138/45	26.01.2000	USGS	LT51380452000026BKT00
					21.01.2010		LT51380452010021KHC0
2.	Landsat 8	OLI_TIRS			01.01.2020		LC81380452020001LGN00

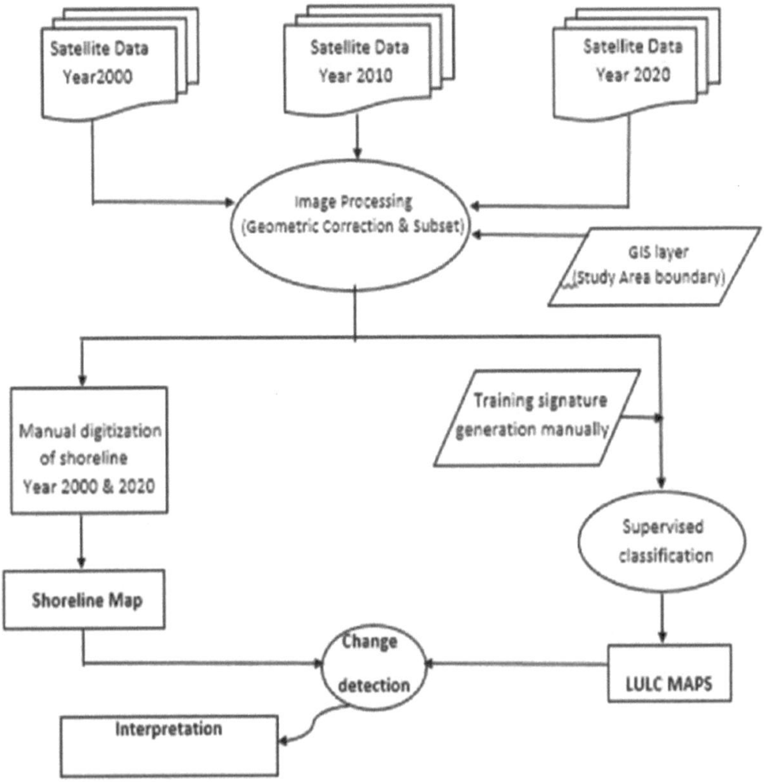

FIGURE 7.2 A comprehensive methodology for data processing and analysis.

the changes and the recession or shifting of the shoreline due to submergence or increased MSL, using ArcMap 10.5 software (Figure 7.2).

7.3 RESULTS AND DISCUSSION

LULC maps of the years 2000, 2010, and 2020 were generated with four land use classes – waterbody, mangrove forest, agricultural land, and other vegetation. A quantitative analysis of the spatio-temporal changes of the LULC of the three different years, from 2000 to 2010 and then from 2010 to 2020, was performed. Because our main focus was on the change in the area of mangrove forest areas over the test period, the area of mangrove forest in the region was calculated as 1921.07 km^2, 1813.99 km^2, and 1886.35 km^2 for the years 2000, 2010, and 2020, respectively (Table 7.2). It was noted that there was a marked fall of 5.57% in the area covered by mangrove forest from the year 2000 to 2010. The study also showed an increase of 3.98% in the area covered by mangrove forest from the year 2010 to the current situation in 2020. Therefore, the overall change from the

TABLE 7.2
Calculation of the LULC Area in 2000, 2010 and 2020

	Area (km^2)			Change in Area (%)		
LULC classes	**2000**	**2010**	**2020**	**2000–2010**	**2010–2020**	**2000–2020**
Water body	2799.3	2805.33	2718.66	0.2154	−3.0894	*−2.8807
Mangrove forest	1921.07	1813.99	1886.35	−5.5739	3.9889	−1.8073
Agricultural land	1376.24	1524.6	1667.22	10.78	9.3545	21.1431
Other vegetation	2059.88	2012.47	1883.309	−2.3015	−6.418	−8.5719

*−ve sign signifies negative change in land area

year 2000 was a fall of 1.86% from 2000 to 2020. Due to land inundation caused by high tidal waves, some parts of the mangrove forest have been permanently converted to degraded forest and ultimately into salt blanks. The increase in area from the year 2010 to 2020 is due mainly to the addition of sediment due to the natural process of coastal accretion and also to the regeneration of some of the degraded forest areas. Though this is relatively small, it still indicates that the conditions are favorable for the growth of mangrove in the region.

Large-scale changes in the mangrove ecosystem structure and coastal degradation usually happen as a result of occasional natural disasters, like cyclones, tsunamis, etc. One such incident was the "Aila" cyclone. On 25 May 2009, cyclone Aila struck the coasts of the Sundarbans in India and Bangladesh, causing havoc and mass destruction. Over 8,000 people were lost, with many more left homeless. Along the coastline of the Bay of Bengal, cyclonic activities increased by up by 26% from 1881 to 2001 (Singh, 2002). A statistical model predicts that there will be a further escalation in the recurrence of cyclonic activities, specifically in the late monsoon period (Unnikrishnan et al., 2011), with the intensity of storms increasing in the period May to June between 2070 and 2100 (Parth Sarthi et al., 2015), resulting in changes on the coastline and in the structure of the mangrove ecosystem. Thus, artificial mangrove plantations are required to avoid such permanent damage to the ecosystem of the Indian Sundarbans.

The accuracy assessment of LULC-classified maps was carried out with the help of the unclassified images, as there were no better multi-spectral satellite imageries available than the input images for the years under investigation, except for 2020. High-quality satellite images from Google Earth Pro were used as the reference image for the test pixels in the case of year 2020, with the remainder using visual interpretation of the unclassified reference image and another literature review (Figure 7.3).

7.3.1 Accuracy Assessment

A total of 100 test pixels was generated using a random sampling method in ERDAS Imagine software. Producer's accuracy, user's accuracy, and overall accuracy

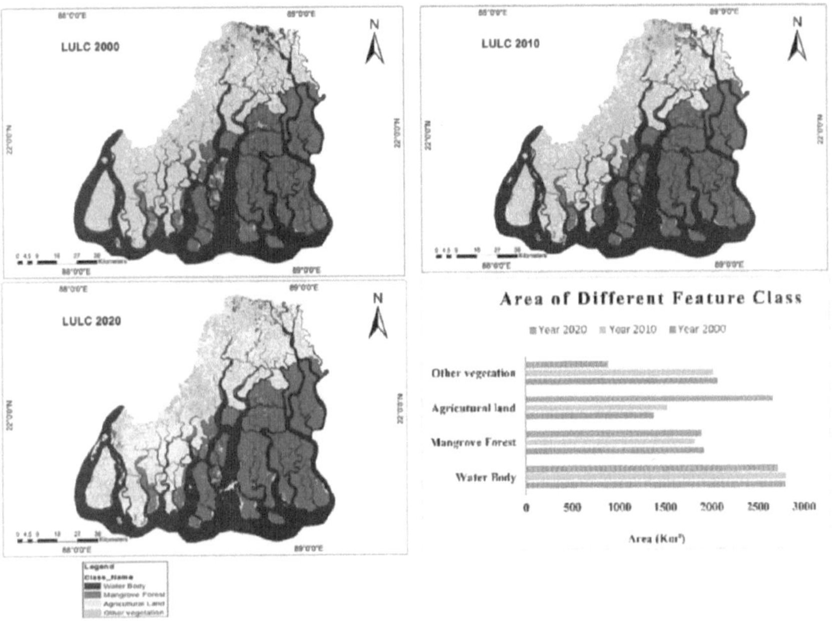

FIGURE 7.3 LULC maps for the years 2000, 2010, and 2020, along with graphical representation.

values were calculated along with Kappa coefficient statistics for the three-year LULC maps 2000 (Table 7.3 (a)), 2010 (Table 7.3(b)), and 2020 (Table 7.3 (c)); the overall accuracy of the classification was observed as 93.33%, 91.14%, and 90.63%, respectively, and the Kappa coefficient was calculated as 0.905, 0.876, and 0.870 for the years 2000, 2010, and 2020, respectively.

On the other hand, the shoreline was compared on some of the most vulnerable islands by manually digitizing the coastline for the years 2000 and 2020. Once the different layers are overlaid, it shows any changes in the shoreline. In some places, due to accretion, land had been added, but mostly there was a decrease in the land cover area of the islands. A marked change in area was noted in the islands that are located in the south-western and southern regions of the Indian Sundarbans, namely Ghoramara, Sagar, Mousuni, Jambudwip Bulchery, Bhangaduni, Dalhousie, Lothian, Dhanchi, Dakshin Surendranagar, and the Namkhana Islands.

Of all these islands, the most affected is Jambudwip Island (–51.80%), whereas, in real terms, Bhangaduni Island lost the most land area (9.56 km^2). The island Ghoramara is now almost submerged, with only 3.76 km^2 area of land left above water. It is believed that, should a large storm or cyclone hit the island, it will be completely submerged under water, which will be an irreversible change. Such storms, etc. represent one of the fundamental causes of the destruction of natural resources. Climate change is ultimately responsible for

TABLE 7.3(A)
Error Matrix For the Year 2000

Class Name	Reference Total	Classified Total	Corrected Number	Producer's Accuracy (%)	User's Accuracy (%)
Water body	29	29	29	100.00	100.00
Mangrove forest	17	16	16	94.12	100.00
Agricultural land	12	12	10	83.33	83.33
Other vegetation	17	18	15	88.24	83.33
Unclassified	25	NA	NA	NA	NA

Overall accuracy: 93.33%
Kappa coefficient: 0.907
NA: not available

TABLE 7.3(B)
Error Matrix For the year 2010

Class Name	Reference Total	Classified Total	Corrected Number	Producer's Accuracy (%)	User's Accuracy (%)
Water body	30	30	29	96.67	96.67
Mangrove forest	22	20	19	86.36	95.00
Agricultural land	11	10	9	81.82	90.00
Other vegetation	16	19	15	93.75	78.95
Unclassified	21	NA	NA	NA	NA

Overall accuracy: 91.14%
Kappa coefficient: 0.876
NA: not available

this degradation of the environment. The agricultural lands become inundated, leaving no other options to the residents other than to migrate to another place which is environmentally more stable. The area of each island (in 2000 and 2020) and the changes in percentage land area between these two years are given in Table 7.4 (Figures 7.4 and 7.5).

7.4 CONCLUSION

This multi-temporal study of the Indian Sundarbans region, using remote sensing data and geographic information system (GIS) techniques, has revealed much information. As the region faces major challenges due to climate change and global warming, it is necessary to monitor environmental degradation and to take necessary preventive measures. The main goal of this study was to monitor and

TABLE 7.3(C)
Error Matrix For the Year 2020

Class Name	Reference Total	Classified Total	Corrected Number	Producer's Accuracy (%)	User's accuracy (%)
Water Body	20	20	20	100.00	100.00
Mangrove Forest	19	18	18	94.74	100.00
Agricultural Land	16	22	16	100.00	72.73
Other vegetation	9	4	4	44.44	100.00
Unclassified	36	NA	NA	NA	NA

Overall accuracy: 90.63%
Kappa coefficient: 0.870
NA: Not available

TABLE 7.4
Changes in Areas Of Islands Located in the Southern and South-western Region Between 2000 and 2020

Islands	Area in 2020 (km²) A_{2020}	Area in 2000 (km²) A_{2000}	Area Change (km²) $\Delta_A = A_{2020} - A_{2000}$	% Change
Ghoramara	3.76967	5.40601	−1.63634	*−30.26890442
Sagar	233.575	236.39	−2.815	−1.190828715
Mousuni	26.7624	28.9071	−2.1447	−7.419284536
Jambudwip	3.83821	7.96468	−4.12647	−51.80961445
Bulchery	57.0962	65.2759	−8.1797	−12.53096472
Bhangaduni	21.709	31.2718	−9.5628	−30.57962765
Dalhousie	57.6056	65.7572	−8.1516	−12.39651323
Lothian	35.9982	33.3821	2.6161	7.836834711
Dhanchi	33.3291	34.5887	−1.2596	−3.641651753
Dakshin Surendranagar	41.3155	42.2411	-0.9256	-2.191230815
Namkhana	146.085	143.667	2.418	1.68305874

*−ve sign denotes the decline in the area from the preceding test year

evaluate the changes in the mangrove forested area and the changes in the land cover area of the vulnerable islands in the region over 20 years (2000–2020). Different techniques and software were used to produce and quantify land use maps. The Landsat images were classified, using the maximum likelihood method, to generate various spectral classes. This helped to analyze the coastal changes and the loss of mangrove forested area in the region. From the study, it is evident that the area has experienced some loss of mangrove forested area due to the conversion of some area into salt blanks, as a result of inundation. On

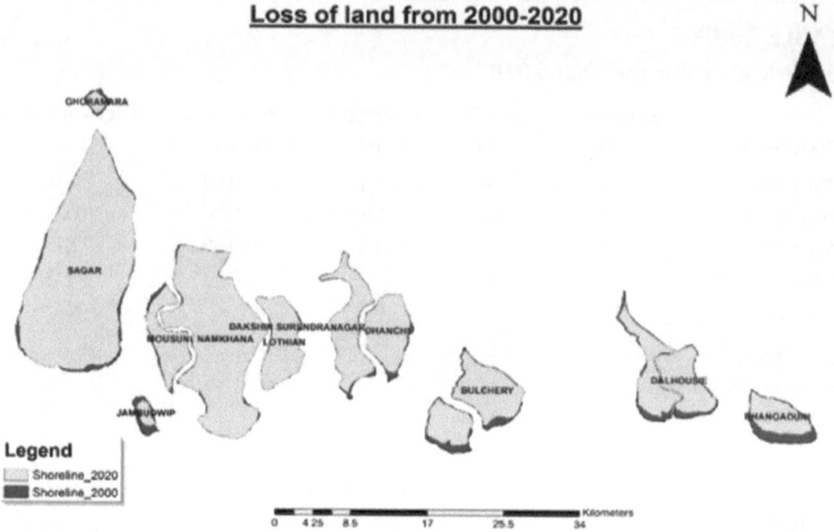

FIGURE 7.4 The loss of land area from 2000 to 2020 in the Indian Sundarbans region.

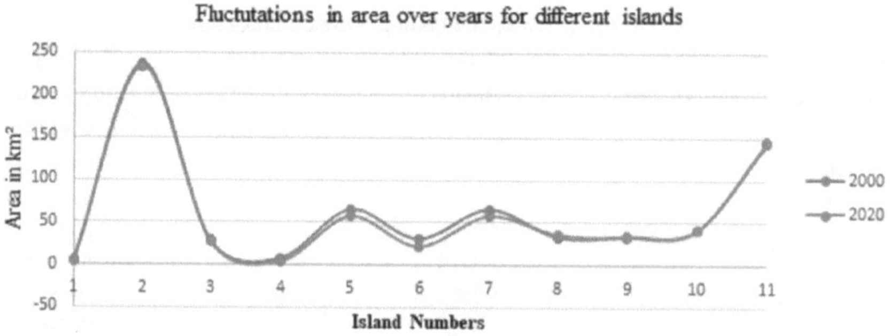

FIGURE 7.5 Variations in island areas over two decades.

a positive note, there were also some areas where some growth of mangrove forest was observed, due to accretion or the addition of sediment. But overall, it was found that there was a fall in the forested area from the year 2000 to 2020. Also, the study reveals that there was a significant loss of land due to soil erosion and a rise in the relative mean sea level. This has led to the loss of property, with people living on those islands tending to leave their homes and move to towns or cities for shelter. Government initiatives have been taken to construct embankments at the land–water interface but no concrete solutions have been generated for the improvement of the lifestyles of the people living on these islands. Day by day, the scenario is getting worse. Stronger embankments and natural barriers need to be constructed to protect the coastal islands, and planting of mangroves is also required to balance the loss due to the adverse effects discussed above. Based on the findings and the results obtained, the results from this research project may be useful for developing a proper "sustainable mitigation plan" and may lead to proper management of the mangrove forests in the Indian Sundarbans region.

Field photographs from the area by Ritaban Ghosh ©

REFERENCES

Barik, J. and Chowdhury, S. (2014) "True mangrove species of Sundarbans delta, West Bengal, Eastern India," *Check List*. doi:10.15560/10.2.329.

Dahdouh-Guebas, F. *et al.* (2005) "How effective were mangroves as a defence against the recent tsunami?," *Current Biology*. doi:10.1016/j.cub.2005.06.008.

Danielsen, F. *et al.* (2005) "The Asian tsunami: A protective role for coastal vegetation," *Science*. doi:10.1126/science.1118387.

Datta, D. and Deb, S. (2017) "Forest structure and soil properties of mangrove ecosystems under different management scenarios: Experiences from the intensely humanized landscape of Indian Sunderbans," *Ocean and Coastal Management*. doi:10.1016/j.ocecoaman.2017.02.022.

Duke, N. C. *et al.* (2007) "A world without mangroves?," *Science*. doi:10.1126/science.317.5834.41b.

Ellison, A. M. (2002) "Macroecology of mangroves: Large-scale patterns and processes in tropical coastal forests," *Trees - Structure and Function*. doi:10.1007/s00468-001-0133-7.

Forbes, K. and Broadhead, J. (2007) *The role of coastal forests in the mitigation of tsunami impacts, Fao.*

Ghosh, M. K., Kumar, L. and Roy, C. (2016) "Mapping long-term changes in mangrove species composition and distribution in the Sundarbans," *Forests*. doi:10.3390/f7120305.

Giri, C. *et al.* (2015) "Distribution and dynamics of mangrove forests of South Asia," *Journal of Environmental Management*. doi:10.1016/j.jenvman.2014.01.020.

Gopal, B. and Chauhan, M. (2006) "Biodiversity and its conservation in the Sundarban mangrove ecosystem," *Aquatic Sciences*. doi:10.1007/s00027-006-0868-8.

Hazra, S. *et al.* (2010) "Temporal change detection (2001–2008) study of Sundarban," *Change*, p. 127.

Iftekhar, M. S. and Islam, M. R. (2004) "Managing mangroves in Bangladesh: A strategy analysis," *Journal of Coastal Conservation*. doi:10.1652/1400-0350(2004)010[0139:MMIBAS]2.0.CO;2.

Iftekhar, M. S. and Saenger, P. (2008) "Vegetation dynamics in the Bangladesh Sundarbans mangroves: A review of forest inventories," *Wetlands Ecology and Management*. doi:10.1007/s11273-007-9063-5.

Jones, T. G. *et al.* (2016) "Madagascar's mangroves: Quantifying nation-wide and ecosystem specific dynamics, and detailed contemporary mapping of distinct ecosystems," *Remote Sensing*. doi:10.3390/rs8020106.

Lillesand, T. M., Kiefer, R. W. and Chipman, J. W. (2004) *Remote sensing and image interpretation, Nev York Chichester Brisbane Toronto 6IS s.*

Parth Sarthi, P., Agrawal, A. and Rana, A. (2015) "Possible future changes in cyclonic storms in the Bay of Bengal, India under warmer climate," *International Journal of Climatology*. doi:10.1002/joc.4053.

Pham, T. D. and Yoshino, K. (2015) "Mangrove mapping and change detection using multi-temporal landsat imagery in Hai Phong city, Vietnam," in *The international symposium on cartography in internet and ubiquitous environments*. 17th–19th March, Tokyo.

Prain, D. (1903) *Bengal Plants*. Vol. 2, Botanical Survey of India, Calcutta. p. 1184.

Singh, O. P. (2002) "Interannual variability and predictability of sea level along the Indian coast," *Theoretical and Applied Climatology*. doi:10.1007/s007040200009.

Spalding, M., Kainuma, M. and Collins, L. (2010) "Book review: World atlas of mangroves," *Wetlands*. doi:10.1007/s13157-011-0224-1.

Unnikrishnan, A. S., Ramesh Kumar, M. R. and Sindhu, B. (2011) "Tropical cyclones in the Bay of Bengal and extreme sea-level projections along the east coast of India in a future climate scenario," *Current Science*.

Valiela, I., Bowen, J. L. and York, J. K. (2001) "Mangrove forests: One of the world's threatened major tropical environments," *BioScience*. doi:10.1641/0006-3568(2001)051[0807:mfootw]2.0.co;2.

Williams, N. (2005) "Tsunami insight to mangrove value," *Current Biology: CB*. doi:10.1016/j.cub.2005.01.015.

8 Remote Sensing of Photosynthesis, Vegetation Productivity and Climate Variability in Bangladesh

Nur Hussain, Farhana Firdaus and Muhammad Rizwan

CONTENTS

8.1 INTRODUCTION

Remote sensing provides a synoptic view of the Earth observatory features of the land surface, which is valuable for exploring vegetation status over time. It has been used to observe biomass, including vascular plant diversity, as well as spatial and temporal vegetation productivity over an array of ecosystems (Nagendra, 2001). Satellite remote-sensing-based vegetation monitoring is an indispensable approach to sustainable natural resource management schemes, including the need to manage natural resources as well as to achieve ecological evaluation at the global and regional scale (Turner, 2011). However, satellite-based remote sensing has the capability of evaluating vegetation productivity status through many geospatial ecological investigations (Cohen and Goward, 2004). Remote sensing of vegetation function has advanced prominently over the past half-century in the ability to obtain phenological characteristics and physiological and structural quantities by measuring electromagnetic reflectance radiance over a range of global and regional scales (Dubovyk et al., 2015; Houborg et al., 2015). Phenological characteristics depend on the vegetation type, solar radiation, climatic conditions, environmental parameters and land use (David and Phillip, 2004). Furthermore, Gross Primary Production (GPP) delivers vegetation productivity status, whereas satellite-obtained solar-induced chlorophyll fluorescence (SIF) provides phenological function between the plant canopy and the atmospheric variables as well as the efficiency with which this energy is used for photosynthesis (Gitelson et al., 2015); Vegetation Indices (VIs) provide a measure of vegetation health (Dubovyk et al., 2015).

Remote-sensing-based vegetation parameter retrieval depends on the spectral properties of the vegetation; vegetation absorbs visible light strongly, including the energy needed for photosynthesis, and strongly reflects near-infrared radiation. Satellite-based Normalized Difference Vegetation Index (NDVI) and Enhanced Vegetation Index (EVI) datasets are used to explore vegetation parameters (Dech et al., 1998; Guo et al., 2017; Labus et al., 2002; Los et al., 2000; Wang et al., 2017). Remotely sensed VIs are effective for understanding the distribution and dynamics of vegetation changes in response to varying environmental conditions and have been used increasingly by ecologists to better predict the effects of global warming, seasonal dynamics trends and anthropogenic activity on terrestrial ecosystem function (Han et al., 2019). NDVI and EVI are the most widely used indices to detect vegetation characteristics, including forestry and agricultural monitoring, in the past three decades, especially with increased concern of the impacts of global warming (Carlson and Arthur, 2000; Gao, 1996). Frequent synoptic data from Earth observation satellites is gradually used to evaluate spatial and temporal scale of vegetation cover (Brown et al., 2013). The Moderate Resolution Imaging Spectroradiometer (MODIS) MYD13A1 Version 6 product provides VIs of 1 km × 1 km spatial resolution, with daily global coverage of NDVI and EVI data (Brown et al., 2013; Didan et al., 2015; Fensholt, 2004; Fensholt and Sandholt, 2005). These datasets are generally used to estimate vegetation phenology and vegetation distribution (Duchemin et al., 2006; Tarpley et al., 1984), so that VI datasets are able to explore the transition of plant health (Dobrowski et al., 2005; Sun et al., 2015).

Monitoring vegetation productivity status is important for obtaining a better understanding of how the Earth's environment responds to climatic variability and anthropogenic activity (Dubovyk et al., 2015). The satellite-obtained GPP provides vegetation productivity variations, resulting from changing climatic characteristics, where GPP depends on phenological indicators (Richardson et al., 2009). The GPP product has an 8-day temporal resolution and is intended for monitoring seasonal and spatial patterns in photosynthetic activity. However, remote-sensing data have outlier and atmospheric noise effects when trying to obtain information on climatic and vegetation status (Kogan, 1990). Despite outlier and atmospheric noise effects, satellite-derived vegetation indices provide a key opportunity to obtain the phenological characteristics. In addition, SIF has a strong relationship with photosynthesis activity and can rapidly identify the response when plants are under extreme climatic conditions (Chappelle et al., 1985, 1984; Moya et al., 2004; Oceanic, 1995; Song et al., 2018).

Satellite-derived SIF acts to reflect physiological canopy status and novel measurements to monitor vegetation conditions which may be able to convey climatic phenological responses (Xu et al., 2018). The solar-induced chlorophyll fluorescence SIF is an electromagnetic spectrum signal emitted by chlorophyll molecules after absorbing solar radiation (Porcar-Castell et al., 2014). Furthermore, SIF can provide as an early indicator of exploring vegetation condition, due to its direct relationship with the photosynthetic system (Zhao et al., 2018). Satellite-based SIF data provide remarkable accessibility for monitoring terrestrial photosynthesis (Frankenberg et al., 2011, 2014; Joiner et al., 2011, 2013; Li et al., 2018). In recent developments, the GOSIF, Global, OCO-2 based SIF product, opened up new opportunities for accessing data from terrestrial photosynthesis. The SIF may be used as a proxy scheme for assessing photosynthesis (Guanter et al., 2015; Köehler et al., 2018; Veefkind et al., 2012), which depends on the energy efficiency of utilization of photosynthetically active solar radiation and the impacts of climatic conditions.

Satellite determinations of land surface temperature (LST) offers significant information for the assessment of surface energy equilibrium mechanisms on large geospatial scales. Earth surface energy and photosynthetically active solar radiation are contingent on seasonal variation and the length of daytime. As a result, LST observation may be able to provide climatic and atmospheric energy response data for vegetation response. The MODIS MYD11A1 Version 6 product provides 1 km × 1 km spatial resolution, with daily global coverage of LST data (Nina et al., 2011; Shen et al., 2016; Wan, 1999). In this way, LST data may be used to understand climatic effects on vegetation status. The present study focused on monitoring remote-sensing-based vegetation productivity and photosynthesis activity in Bangladesh, employing solar-induced chlorophyll fluorescence observations consistent with climatic vegetation datasets. The aims of this study are: i) to explore the seasonal vegetation health status, based on VIs; ii) to obtain photosynthetic activity and vegetation productivity response data through GPP and SIF over the study domain, using GOSIF datasets; and iii) to understand the climatic response of vegetation appearance.

8.2 MATERIALS AND TECHNIQUES

8.2.1 Study Domain

Bangladesh is located between 20°34″and 26°38″ N latitude and between 88°01″ and 92°41″ E longitude. In terms of its location, it is situated at the north-eastern part of South Asia, almost wholly bordered by Indian states except for the south and the south-east. The Bay of Bengal covers the southern coastal areas of the entire country (Figure 8.1). Bangladesh contains 144,000 km^2 of a total of 147,570 km^2 of low-lying floodplain of the large rivers system from the Himalayas. Given its

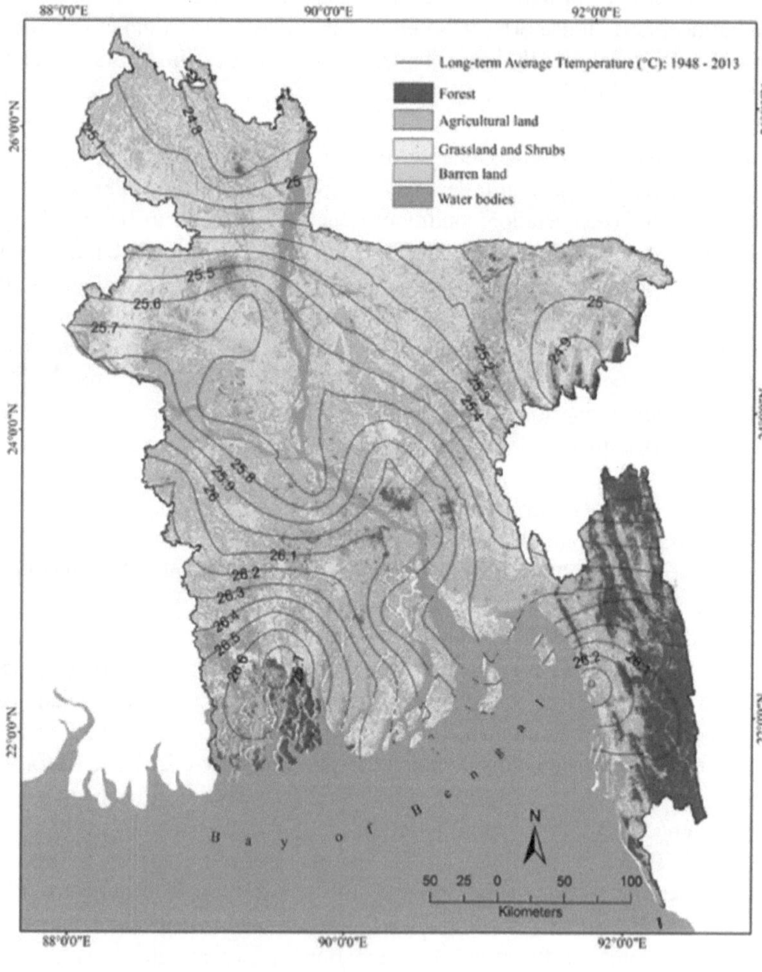

FIGURE 8.1 Long-term average temperature and vegetation cover details over the country. The average temperature was calculated using historical temperature data from 1948 to 2013 from the Bangladesh Meteorological Department (BMD). The vegetation density status was obtained, using MODIS-NDVI data.

geographic location, Bangladesh is deemed one of the most climate-change-vulnerable countries (ADB, 2012). Bangladesh has a tropical humid climate, characterized by moderately high temperatures and high humidity (Alamgir et al., 2015; Brammer, 2014). The climate of Bangladesh is characterized by strong variability with respect to air pressure, wind direction, rainfall and temperature. Based on long-term climate variability, Bangladesh has mainly four seasons: a) a hot, pre-monsoon summer season (March, April and May); b) a rainy summer monsoon season (June, July and August); c) a brief autumn season (September, October and November); and d) a cool dry winter season (December, January and February). There is no clear difference between any two successive seasons, and they may overlap one another by a few days to a week or so. The agriculture and vegetation growth of Bangladesh are characterized by the seasonal conditions. However, changes in vegetation monitoring and photosynthetic activity, reflecting prevailing climatic conditions, are significant for the development and sustainability of the country (Rahman et al. 2016).

8.2.2 Monitoring Vegetation Greenness

The MODIS-Terra satellite overpasses the equator at ~10:30 a.m. (local time) in descending node, and the MODIS-Aqua satellite overpasses the equator at ~1:30 p.m. (local time) in ascending node, which both of the satellite providing 500 m spatial resolution at 8-day intervals and 1-km spatial resolution at 16-day intervals for global coverage as well as NDVI and EVI datasets (Wan, 1999). We used MODIS-Aqua MYD11A1 Version 6 NDVI and EV 1-km spatial resolution at 16-day interval products from 2005 to 2018, to explore vegetation productivity status according to the availability and retrieval time of other datasets. The NDVI and EVI datasets were obtained from the download URL (https://earthexplorer.usgs.gov/).

8.2.3 GPP Datasets

The MODIS gross primary productivity (GPP) product generates widely used globally accepted vegetative productivity data, critically used for measuring plant productivity, carbon balance and carbon-climate reactions. The Global Ecology Group from the Earth System Research Center, University of New Hampshire, USA, develops 0.05° fine spatial resolution and 8-day, 30-day and annual global coverage GPP data for the period from 2000 to 2018. These GPP data are very useful because of their strong correlations with a total of 91 Eddy Covariance (EC) flux site-based ground measurements from all over the world. These GPP data are able to explore spatial and seasonal plant productivity variations, using coarse resolution (1°) for global estimation, and fine resolution (0.05°) for regional short-term (8-day, monthly) and long-term (annual) GPP estimates (Li et al., 2018; Li and Xiao, 2019). In this study, we used 0.05° fine spatial resolution and 8-day GPP from the Global Ecology Group for the period of 2005–2018. We obtained GPP data from the download URL (http://globalecology.unh.edu).

8.2.4 Satellite-Obtained SIF Datasets

The Global Ecology Group from the Earth System Research Center recently developed GOSIF solar-induced chlorophyll fluorescence data. GOSIF data were estimated from the Orbiting Carbon Observatory-2 (OCO-2) satellite. GOSIF was directly aggregated from the discrete OCO-2 observatory and the EC flux tower-based GPP measurement. The GOSIF data provide 1° spatial coarse resolution and 0.05° spatial fine resolution global SIF coverage over 8-day, monthly and annual temporal intervals. The GOSIF data have high accuracy on the basis of highly significant correlation coefficients ($r = 0.73$, $p < 0.001$) with a total of 91 ground-measured EC flux sites (Li et al., 2018; Li and Xiao, 2019). In this study, we used 0.05° fine spatial resolution and 8-day SIF from the Global Ecology Group for the period 2005–2018. The SIF datasets were collected from the Global Ecology Group, *via* the download URL (http://globalecology.unh.edu).

8.2.5 Climate Datasets

To explore the climatic phenomena, we used satellite-based LST datasets. The MODIS-Aqua MYD11A2 Version 6 data provide 1 km × 1 km spatial resolution with 8-day global coverage LST, as well as temperature product (Nina et al., 2011; Shen et al., 2016; Wan, 1999). With respect to climatic change, night temperatures decrease at a quicker rate than day temperatures because of less radiant warmth change, considering the increased cloudiness. LST informational indices are registered from the satellite-based remotely detected pictures with high resolution at a moderate resolution scale, allowing the day and late evening observations of the Earth's surface. LST is an important factor in environmental evaluation. Global MODIS LST items are accessible at different levels. The LST datasets were used to explore the response of vegetation status and photosynthesis activity to climate change. The MODIS LST datasets were collected from the download URL (https://earthexplorer.usgs.gov/).

8.3 RESULTS

8.3.1 Seasonal Vegetation Greenness Response

The vegetation greenness response was determined from MODIS vegetation index datasets. NDVI and EVI showed similar spatial and temporal seasonal patterns (Figure 8.2 and 8.3). We employed 16-day smoothed VI (EVI and NDVI) measurements to understand vegetation greenness responses from 2005 to 2018, relative to seasonal characteristics. For both EVI and NDVI, the lowest vegetation greenness response occurred in the winter season. With respect to seasonal characteristics, the winter season of Bangladesh has a comparatively short daylength and lower soil moisture than do the other seasons, as a result of which winters are the less-productive seasons, vegetatively speaking. Furthermore, maximum greenness was observed in the autumn season, at which point the NDVI response was better than EVI. Due to seasonal behavior in the monsoon season, heavy

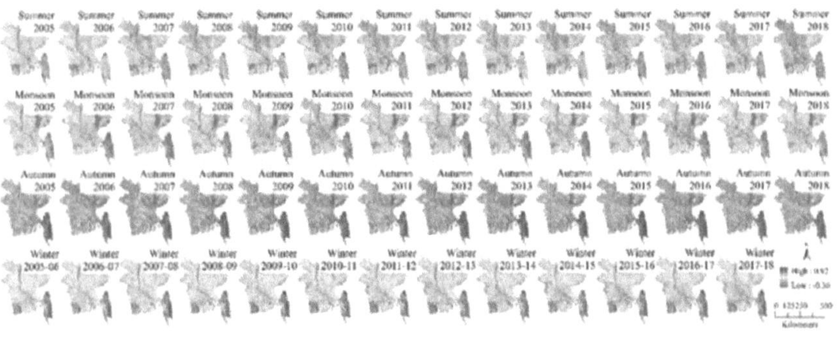

FIGURE 8.2 The geo-spatial and historical distribution of NDVI over Bangladesh. The spatial and temporal distributions of NDVI were employed using 1-km spatial resolution of MODIS-Aqua MYD13A1-V6 vegetation index datasets.

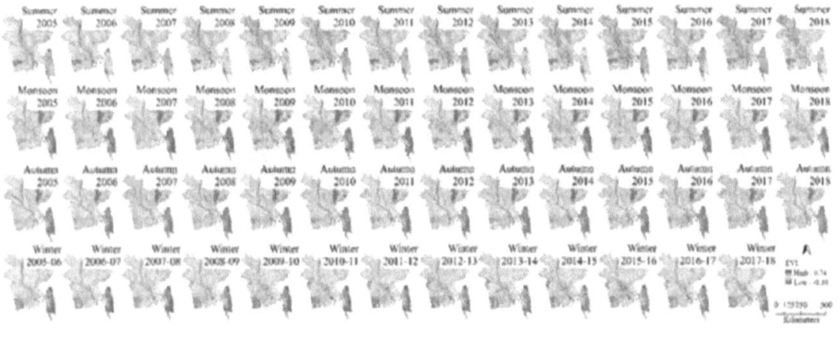

FIGURE 8.3 The spatial and temporal distribution of EVI of the study area during 2005 to 2018. The spatial and temporal distributions of EVI were conducted using 1-km spatial resolution of MODIS-Aqua MYD13A1-V6 vegetation index datasets.

rainfall occurs, resulting in greater plant production after the monsoon in the autumn. The spatial and temporal patterns of EVI and NDVI were strongly correlated, with a correlation coefficient r of 0.75.

In Figure 8.2, the autumn season is the season with the highest plant productivity, from 2005 to 2018. The average NDVI responses were 0.50, 0.43, 0.55 and 0.41 in the summer, monsoon, autumn and winter seasons, respectively. Similarly, Figure 8.3 represents the seasonal spatial distribution of EVI in the study area during the study period from 2005 to 2018. In Figure 8.3, the autumn season was also the more-productive season in terms of vegetation. The average EVI response were 0.31, 0.26, 0.35, and 0.23 in the summer, monsoon, autumn and winter seasons, respectively. With respect to spatial distribution, the south-eastern region of the study area was mostly vegetative, due to evergreen hilly forest. On the other hand, the north-eastern area was the least-vegetated region, in terms of permanent and seasonal wetlands.

8.3.2 Seasonal Vegetation Productivity and Photosynthesis Response

Vegetation productivity was monitored from GPP measurements (Figure 8.4). We employed 16-day smoothed GPP values for obtaining measures of vegetation productivity from 2005 to 2018, following the seasonal aspect. We also explored photosynthetic activity responses from GOSIF 16-day smoothed SIF datasets (Figure 8.5). Observing both GPP and SIF responses, the lowest primary production and photosynthesis responses were in the winter season. In the winter season, the photosynthetically active solar radiation and energy emissivity are lower than in the other seasons, so that vegetation production and photosynthetic activity are low in the winter season. Furthermore, the maximum primary production and photosynthetic activity were observed in the summer season. The days of the summer season are long, and plant receive more photosynthetically active solar radiation and energy, helping plant biomass production and photosynthetic activity. The spatial and temporal patterns of GPP and SIF were strongly correlated, with a significant correlation coefficient r of 0.95.

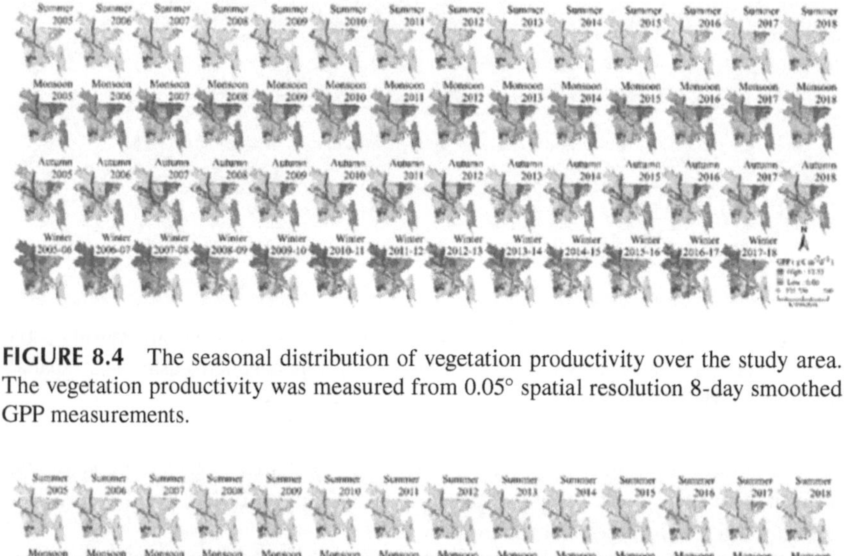

FIGURE 8.4 The seasonal distribution of vegetation productivity over the study area. The vegetation productivity was measured from 0.05° spatial resolution 8-day smoothed GPP measurements.

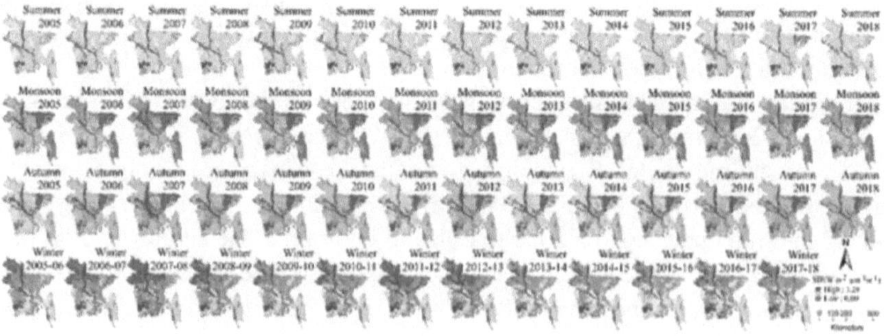

FIGURE 8.5 The spatial and temporal distribution of photosynthesis responses. Photosynthesis response was measured from 0.05° spatial resolution 8-day smoothed GOSIF measurements.

The spatial and temporal distributions of GPP and SIF showed a pattern similar to vegetation greenness response. The average vegetation production values were 7.91 (g C (carbon) m^{-2} d^{-1}), 9.08 (g C m^{-2} d^{-1}), 8.39 (g C m^{-2} d^{-1}) and 5.46 (g C m^{-2} d^{-1}) for the summer, monsoon, autumn and winter seasons, respectively, during the period 2005–2018. Similarly, the average photosynthetic activity was 4.04 (W m^{-2} μm^{-1} sr^{-1}, where sr is 1 steradian), 3.58 (W m^{-2} μm^{-1} sr^{-1}), 3.77 (W m^{-2} μm^{-1} sr^{-1}) and 2.47 (W m^{-2} μm^{-1} sr^{-1}) for the summer, monsoon, autumn and winter seasons, respectively, from 2005 to 2018. With respect to spatial distribution, the biomass production and photosynthetic activity followed a distribution similar to the greenness status, with the south-eastern region being mostly vegetative, due to evergreen hilly forest, whereas the north-eastern area was less responsive, being mostly wetland ecosystems.

8.3.3 Climatic Response of Photosynthetic Activity and Vegetation Greenness

We explored climatic conditions, using MODIS-Aqua MYD11A2-V6–LST and MODIS-Aqua MYD15A2H-V6–FPAR datasets to understand climatic responses for vegetation greenness and photosynthetic activity (Figure 8.6). The LST-based average temperatures were 31.52 °C, 26.96 °C, 28.76 °C and 21.85 °C for the

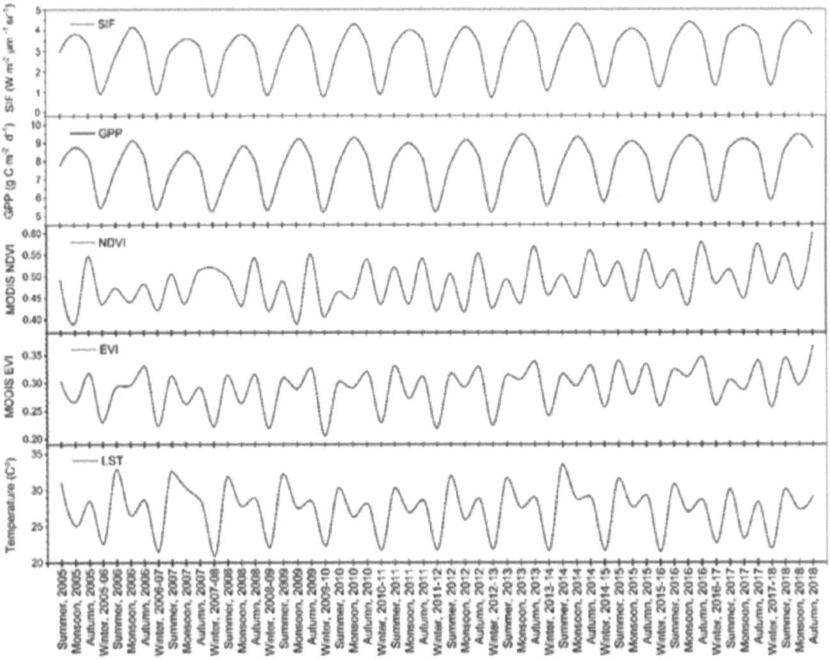

FIGURE 8.6 The seasonal trends of climatic effects on photosynthetic activity and vegetation greenness during 2005–2018.

TABLE 8.1
The Correlation Coefficient Matrix of Climatic Phenology Response

	SIF	GPP	NDVI	EVI	LST
SIF	1				
GPP	0.91	1			
NDVI	0.71	0.57	1		
EVI	0.77	0.74	0.75	1	
LST	0.58	0.63	0.48	0.79	1

summer, monsoon, autumn and winter seasons, respectively. Monitoring the average climatic features, the LST followed a similar temporal trend to SIF and GPP from 2005 to 2018, where the SIF and GPP also followed similar patterns. The long-term climatic variability showed a hot summer (March, April and May), mild monsoon (June, July and August), a brief autumn (September, October and November) and a cool dry winter (December, January and February) climatic behavior. The vegetation health, productivity and photosynthesis responses were dominated by these seasonal characteristics.

Reviewing the important vegetation productivity and phenological functions, in terms of perception or estimation, the SIF and GPP mostly followed similar seasonal responses. The variation in temperature variability and phenological condition were significantly correlated (Table 8.1). The relationships among temperature intensity, vegetation greenness status, vegetation productivity and photosynthetic activity responded dynamically to seasonal variation.

8.4 DISCUSSION

8.4.1 Vegetation Productivity

The significant positive correlation between EVI and NDVI (r = 0.75) demonstrated the vegetation greenness acceptance. The MODIS-NDVI is able to most closely reflect the actual vegetative conditions, whereas the SPOT-VEGETATION (VGT) sensor is based on bidirectional compositing information (Chen et al., 2006). The NDVI values are strongly correlated to the photosynthetically active radiation intercepted, which is absorbed by the plant canopy (Gallo et al., 1985). Few studies have methodically examined whether satellite greenness data appropriately reflect vegetation growth information in the fields. Huete found that NDVI was sensitive to canopy contextual transformation signals for high-biomass conditions (Huete et al., 2002). The other vegetation index, EVI, recommended by Huete et al. (1994), was enhanced by augmented sensitivity for biomass detection in dense vegetation canopies through a decoupling of the canopy background response and a reduction in the influence of atmospheric and soil reflectance (Chen et al., 2006; Huete et al., 1994, 2002). In this present study, the greenness response, as well as the vegetation productivity status of EVI, was better than NDVI due to stronger correlation with the SIF anomaly. Moreover, EVI and NDVI were both able to provide greenness responses, reflecting vegetation productivity, in consistent spatial and temporal information, with regard to global vegetation.

8.4.2 Photosynthetic Activity and Vegetation Productivity

In this present study, it was essential to understand the relationship between photosynthetic activity and vegetation productivity, in terms of phenological function. Taking account of the SIF anomaly, we found clear relationships between the SIF anomaly and VIs. The SIF was able to provide a measure of the physiological canopy electromagnetic reflectance emitted by chlorophyll molecules after the absorption of solar radiation signal, which could be considered to be a proxy for terrestrial photosynthesis (Porcar-Castell et al., 2014; Song et al., 2018; Xu et al., 2018). In addition, SIF has a direct connection with plant photosynthetic activity and can rapidly reflect the changing values for vegetation growth and respiration. Furthermore, the GPP-based measure of vegetation productivity followed the spatial and temporal trends in vegetation greenness response and SIF. Sun et al. (2018) showed that the association between SIF and existing modeled GPP products diverged somewhat across biomes at the regional environment (Sun et al., 2018); similarly, we found a stronger correlation between GPP and SIF ($r = 0.91$). However, the pattern of terrestrial photosynthesis fitted temporally and spatially with that of vegetation productivity, where we found strong significant correlations (Figure 8.7).

The coefficient of determination (COD) r_2 was 0.91 ($p<0.0001$) between SIF and GPP where the r_2 was 0.84 ($p<0.0001$) between SIF and EVI. We also found that, the r_2 value was 0.71 ($p<0.0001$) between SIF and NDVI. Observing the VIs and SIF patterns, we traced that the photosynthetic activity and vegetation productivity values gradually decreased from December to February (winter season), with the lowest values being obtained in January, helping to understand the climatic environmental response.

8.4.3 Climatic Response on Photosynthetic Activity and Vegetation Productivity

The consistent temporal SIF anomaly and VI variations suggested that photosynthetic activity and vegetation productivity were also sensitive to temporal climatic variables. The temperature variability and SIF intensity both correlated with vegetation condition and photosynthetic response. The SIF is produced by the photosynthetic tissue in a leaf canopy, which is also a key variable in the assessment of vegetation productivity. The leaf canopy absorbs photosynthetically active radiation, with which it can conduct one of three functions: take part in the photochemical reaction, diffuse energy through non-photochemical quenching (NPQ) or emit longer wavelengths as fluorescence (Baker, 2008). The fluorescence reflects climatic sensitivity, e.g., when plants suffer temperature stress under extreme

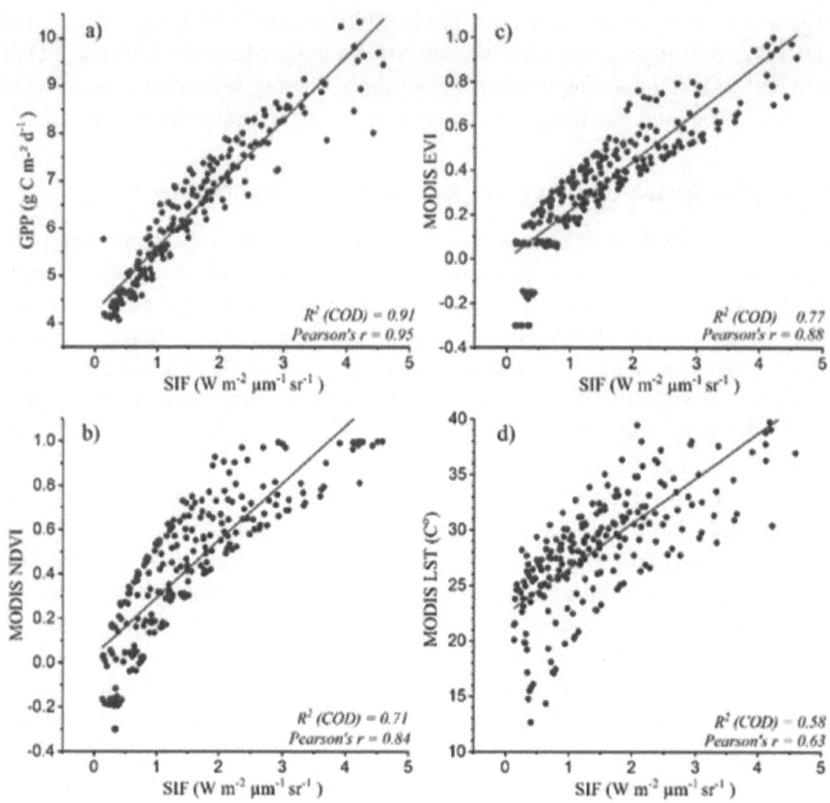

FIGURE 8.7 a) The relationship between SIF and GPP; b) the relationship between SIF and NDVI; c) the relationship between SIF and EVI; and d) the relationship between SIF and LST- based temperature.

environments (Chappelle et al., 1984, 1985; Moya et al., 2004; Oceanic, 1995; Song et al., 2018). When plants are exposed to abiotic or biotic stresses, the maximal fluorescence is frequently decreased (Baker, 2008). However, within the current study area, the photosynthetic response and vegetation productivity decreased in the winter season and increased in summer to autumn, when the temperature also followed a similar trend.

8.5 CONCLUSION

In spite of outliers and atmospheric noise effects, the remote-sensing-based vegetation productivity and photosynthesis observations are significant for a better understanding of how plants respond to climatic change. The relationships between GPP, VIs and SIF demonstrated that vegetation productivity as well as temperature variability reflect the response of vegetation greenness to climatic variation. Furthermore, terrestrial photosynthetic CO_2 assimilation is regulated by temporal climatic variability and dynamic seasonal response (Lee et al., 2013). The GOSIF was able to achieve a better understanding of territorial photosynthetic activity due to its greater spatial resolution than others, with satellites monitoring SIF and MODIS-LST reflecting the climatic condition response on vegetation. Using GPP and VI data, it was possible to obtain the vegetation productivity status in terms of spatial, temporal and seasonal aspects. However, it obtained the marked variations in vegetation productivity status and photosynthesis anomaly gathered temperature intensity. Study results indicated a strong synchronization of vegetation productivity with photosynthetic activity, in response to the climatic environment. The results suggest that the GOSIF was able to reflect terrestrial photosynthesis measurements similarly the LST-based temperature provide climatic response including extreme environment, that may provide guidelines for future management research into vegetation productivity.

ACKNOWLEDGMENTS

We would like to thank Dr. Jingfeng Xiao and the Global Ecology Group from the Earth System Research Center, University of New Hampshire, USA for SIF and GPP datasets. We are also thankful to S. Dara Shamsuddin, former professor, Geography and Environment Department, Jahangirnagar University, Bangladesh for his academic direction in seasonal climatic variability of Bangladesh.

REFERENCES

ADB, 2012. *Addressing Climate Change and Migration in Asia and the Pacific*. Manila. https://doi.org/10.1126/science.323.5916.876b

Alamgir, M., Shahid, S., Hazarika, M.K., Nashrrullah, S., Harun, S.B., Shamsudin, S., 2015. Analysis of meteorological drought pattern during different climatic and cropping seasons in Bangladesh. *J. Am. Water Resour. Assoc.* 51, 794–806. https://doi.org/10.1111/jawr.12276

Baker, N.R., 2008. Chlorophyll fluorescence : A probe of photosynthesis in vivo. *Annu. Rev. Plant Biol.* 59, 89–113. https://doi.org/10.1146/annurev.arplant.59.032607.092759

Brammer, H., 2014. Climate risk management Bangladesh's dynamic coastal regions and sea-level rise. *Clim. Risk Manag.* 1, 51–62. https://doi.org/10.1016/j.crm.2013.10.001

Brown, J.C., Kastens, J.H., Coutinho, A.C., Victoria, D.de C., Bishop, C.R., 2013. Classifying multiyear agricultural land use data from Mato Grosso using time-series MODIS vegetation index data. *Remote Sens. Environ.* 130, 39–50. https://doi.org/10.1016/j.rse.2012.11.009

Carlson, T.N., Arthur, S.T., 2000. The impact of land use - land cover changes due to urbanization on surface microclimate and hydrology: A satellite perspective. *Glob. Planet. Change* 25, 49–65. https://doi.org/10.1016/S0921-8181(00)00021-7

Chappelle, E.W., Wood, F.M., Mcmurtrey, J.E., Newcomb, W.W., 1984. Laser-induced fluorescence of green plants. 1: A technique for the remote detection of plant stress and species differentiation. *Appl. Opt.* 23, 134–138.

Chappelle, E.W., Wood, F.M., Newcomb, W.W., Iii, J.E.M., 1985. Laser-induced fluorescence of green plants. 3: LIF spectral signatures of five major plant types. *Appl. Opt.* 24, 74–80.

Chen, P.-Y., Fedosejevs, G., Tiscareno-Lopez, M., Arnold, A.J.G., 2006. Composite data using agricultural measurements : An. *Environ. Monit. Assess.* 119, 69–82. https://doi.org/10.1007/sl0661-005-9006-7

Cohen, W.B., Goward, S.N., 2004. Landsat's role in ecological applications of remote sensing. *Bioscience* 54, 535–545.

David, H., Phillip, F., 2004. Phenological description of natural vegetation in southern Africa using remotely-sensed vegetation data. *Appl. Veg. Sci.* 7, 19–28.

Dech, S.W., Tungalagsaikhan, P., Preusser, C., Meisner, R.E., 1998. Operational value-adding to AVHRR data over Europe: Methods, results, and prospects. *Aerosp. Sci. Technol.* 2, 335–346. https://doi.org/10.1016/S1270-9638(98)80009-6

Didan, K., Munoz, A.B., Huete, A., 2015. *MODIS Vegetation Index User's Guide (MOD13 Series). Vegetation Index and Phenology Lab*, The University of Arizona.

Dobrowski, S.Z., Pushnik, J.C., Zarco-Tejada, P.J., Ustin, S.L., 2005. Simple reflectance indices track heat and water stress-induced changes in steady-state chlorophyll fluorescence at the canopy scale. *Remote Sens. Environ.* 97, 403–414. https://doi.org/10.1016/j.rse.2005.05.006

Dubovyk, O., Landmann, T., Erasmus, B.F.N., Tewes, A., Schellberg, J., 2015. Monitoring vegetation dynamics with medium resolution MODIS-EVI time series at sub-regional scale in southern Africa. *Int. J. Appl. Earth Obs. Geoinf.* 38, 175–183. https://doi.org/10.1016/j.jag.2015.01.002

Duchemin, B., Hadria, R., Erraki, S., Boulet, G., Maisongrande, P., Chehbouni, A., Escadafal, R., Ezzahar, J., Hoedjes, J.C.B., Kharrou, M.H., Khabba, S., Mougenot, B., Olioso, A., Rodriguez, J.C., Simonneaux, V., 2006. Monitoring wheat phenology and irrigation in Central Morocco: On the use of relationships between evapotranspiration, crops coefficients, leaf area index and remotely-sensed vegetation indices. *Agric. Water Manag.* 79, 1–27. https://doi.org/10.1016/j.agwat.2005.02.013

Fensholt, R., 2004. Earth observation of vegetation status in the Sahelian and Sudanian West Africa: Comparison of Terra MODIS and NOAA AVHRR satellite data. *Int. J. Remote Sens.* 25, 1641–1659. https://doi.org/10.1080/01431160310001598999

Fensholt, R., Sandholt, I., 2005. Evaluation of MODIS and NOAA AVHRR vegetation indices with in situ measurements in a semi-arid environment. *Int. J. Remote Sens.* https://doi.org/10.1080/01431160500033724

Frankenberg, C., Fisher, J.B., Worden, J., Badgley, G., Saatchi, S.S., Lee, J.-E., Toon, G.C., Butz, A., Jung, M., Kuze, A., Yokota, T., 2011. New global observations of the terrestrial carbon cycle from GOSAT: Patterns of plant fluorescence with gross primary productivity. *Geophys. Res. Lett.* 38, https://doi.org/10.1029/2011GL048738

Frankenberg, C., O'Dell, C., Berry, J., Guanter, L., Joiner, J., Köhler, P., Pollock, R., Taylor, T.E., 2014. Prospects for chlorophyll fluorescence remote sensing from the Orbiting Carbon Observatory-2. *Remote Sens. Environ.* 147, 1–12. https://doi.org/10.1016/j.rse.2014.02.007

Gallo, K.P., Daughtry, C.S.T., Bauer, M.E., 1985. Spectral estimation of absorbed photosynthetically active radiation in corn canopies. *Remote Sens. Environ.* 17, 221–232.

Gao, B., 1996. NDWI a normalized difference water index for remote sensing of vegetation liquid water from space. *Remote Sens. Environ.* 58, 257–266.

Gitelson, A.A., Peng, Y., Arkebauer, T.J., Suyker, A.E., 2015. Productivity, absorbed photosynthetically active radiation, and light use efficiency in crops: Implications for remote sensing of crop primary production. *J. Plant Physiol.* 177, 100–109. https://doi.org/10.1016/j.jplph.2014.12.015

Guanter, L., Aben, I., Tol, P., Krijger, J.M., Hollstein, A., Köhler, P., Damm, A., Joiner, J., Frankenberg, C., Landgraf, J., 2015. Potential of the TROPOspheric monitoring instrument (TROPOMI) onboard the Sentinel-5 precursor for the monitoring of terrestrial chlorophyll fluorescence. *Atmos. Meas. Tech.* 8, 1337–1352. https://doi.org/10.5194/amt-8-1337-2015

Guo, X., Zhang, H., Wu, Z., Zhao, J., Zhang, Z., 2017. Comparison and evaluation of annual NDVI time series in China derived from the NOAA AVHRR LTDR and terra MODIS MOD13C1 products. *Sensors (Switzerland)* 17, 1–18. https://doi.org/10.3390/s17061298

Han, J.-C., Huang, Y., Zhang, H., Wu, X., 2019. Characterization of elevation and land cover dependent trends of NDVI variations in the Hexi region, northwest China. *J. Environ. Manage.* 232, 1037–1048.

Houborg, R., Fisher, J.B., Skidmore, A.K., 2015. Advances in remote sensing of vegetation function and traits a. *Int. J. Appl. Earth Obs. Geoinf.* 43, 1–6. https://doi.org/10.1016/j.jag.2015.06.001

Huete, A., Didan, K., Miura, T., Rodriguez, E.P., Gao, X., Ferreira, L.G., 2002. Overview of the radiometric and biophysical performance of the MODIS vegetation indices. *Remote Sens. Environ.* 83, 195–213.

Huete, A., Justice, C., Liu, H., 1994. Develpmeclassification and soil indices for MODIS-EOS. *Remote Sens. Environ.* 49, 224–234. https://doi.org/Doi 10.1016/0034-4257(94)90018-3

Joiner, J., Guanter, L., Lindstrot, R., Voigt, M., Vasilkov, A.P., Middleton, E.M., Huemmrich, K.F., Yoshida, Y., Frankenberg, C., 2013. Global monitoring of terrestrial chlorophyll fluorescence from moderate-spectral-resolution near-infrared satellite measurements: Methodology, simulations, and application to GOME-2. *Atmos. Meas. Tech.* 6, 2803–2823. https://doi.org/10.5194/amt-6-2803-2013

Joiner, J., Yoshida, Y., Vasilkov, A.P., Yoshida, Y., Corp, L.A., Middleton, E.M., 2011. First observations of global and seasonal terrestrial chlorophyll fluorescence from space. *Biogeosciences* 8, 637–651. https://doi.org/10.5194/bg-8-637-2011

Köehler, P., Frankenberg, C., Magney, T.S., Guanter, L., Joiner, J., Landgraf, J., 2018. Global retrievals of solar induced chlorophyll fluorescence with TROPOMI: First results and inter-sensor comparison to OCO-2. *Geophys. Res. Lett.* 45, 456–463. https://doi.org/10.1029/2018GL079031

Kogan, F.N., 1990. Remote sensing of weather impacts on vegetation in non-homogeneous areas. *Int. J. Remote Sens.* 11, 1405–1419. https://doi.org/10.1080/01431169008955102

Labus, M.P., Nielsen, G.A., Lawrence, R.L., Engel, R., Long, D.S., 2002. Wheat yield estimates using multi-temporal NDVI satellite imagery. *Int. J. Remote Sens.* 23, 4169–4180. https://doi.org/10.1080/01431160110107653

Lee, J.-E., Frankenberg, C., Tol, C. van der, Berry, J.A., Guanter, L., Boyce, C.K., Fisher, J.B., Morrow, E., Worden, J.R., Asefi, S., Badgley, G., Saatchi, S., 2013. Forest productivity and water stress in Amazonia: Forest productivity and water stress in Amazonia: Observations from GOSAT chlorophyll fluorescence. *Proc. R. Soc. B* 280. https://doi.org/10.1098/rspb.2013.0171

Li, X., Xiao, J., 2019. Mapping photosynthesis solely from solar-induced chlorophyll fluorescence: A global, fine-resolution dataset of gross primary production derived from OCO-2. *Remote Sens.* 11. https://doi.org/10.3390/rs11212563

Li, X., Xiao, J., He, B., Altaf Arain, M., Beringer, J., Desai, A.R., Emmel, C., Hollinger, D.Y., Krasnova, A., Mammarella, I., Noe, S.M., Ortiz, P.S., Rey-Sanchez, A.C., Rocha, A. V., Varlagin, A., 2018. Solar-induced chlorophyll fluorescence is strongly correlated with terrestrial photosynthesis for a wide variety of biomes: First global analysis based on OCO-2 and flux tower observations. *Glob. Chang. Biol.* 24, 3990–4008. https://doi.org/10.1111/gcb.14297

Los, S.O., Collatz, G.J., Sellers, P.J., Malmstr, C.M., Pollack, N.H., DeFries, R.S., Bounoua, L., Parris, M.T., Tucker, C.J., Dazlich, D.A., 2000. A global 9-yr biophysical land surface dataset from NOAA AVHRR data. *J. Hydrometeorol.* 1, 183–199. https://doi.org/10.1175/1525-7541(2000)001<0183:AGYBLS>2.0.CO;2

Moya, I., Camenen, L., Evain, S., Goulas, Y., Cerovic, Z.G., 2004. A new instrument for passive remote sensing 1 . Measurements of sunlight-induced chlorophyll fluorescence. *Remote Sens. Environ.* 91, 186–197. https://doi.org/10.1016/j.rse.2004.02.012

Nagendra, H., 2001. Using remote sensing to assess biodiversity. *Int. J. Remote Sens.* 22, 2377–2400. https://doi.org/10.1080/01431160117096

Nina, S., Lautenbach, S., Seppelt, R., 2011. Exploring indicators for quantifying surface urban heat islands of European cities with MODIS land surface temperatures. *Remote Sens. Environ.* 115, 3175–3186. https://doi.org/10.1016/j.rse.2011.07.003

Oceanic, N., 1995. Application of vegetation index and brightness temperature for drought detection. *Adv. Sp. Res.* 15, 91–100.

Porcar-Castell, A., Tyystjärvi, E., Atherton, J., Tol, C. van der, Flexas, J., Pfündel, E.E., Moreno, J., Frankenberg, C., Berry, J.A., 2014. Linking chlorophyll a fluorescence to photosynthesis for remote sensing applications: Mechanisms and challenges. *J. Exp. Bot.* 65, 4065–4095. https://doi.org/10.1093/jxb/eru191

Rahaman, K.M., Ahmed, F.R.S., Nazrul Islam, M., 2016. Modeling on climate induced drought of north-western region, Bangladesh. *Model. Earth Syst. Environ.* 2, 45. https://doi.org/10.1007/s40808-016-0089-7

Richardson, A.D., Hollinger, D.Y., Dail, D.B., Lee, J.T., Munger, J.W., O'Keefe, J., 2009. Influence of spring phenology on seasonal and annual carbon balance in two contrasting New England forests. *Tree Physiol.* 29, 321–331. https://doi.org/10.1093/treephys/tpn040

Shen, H., Huang, L., Zhang, L., Wu, P., Zeng, C., 2016. Long-term and fine-scale satellite monitoring of the urban heat island effect by the fusion of multi-temporal and multi-sensor remote sensed data: A 26-year case study of the city of Wuhan in China. *Remote Sens. Environ.* 172, 109–125. https://doi.org/10.1016/j.rse.2015.11.005

Song, L., Guanter, L., Guan, K., You, L., Huete, A., Ju, W., Zhang, Y., 2018. Satellite sun-induced chlorophyll fluorescence detects early response of winter wheat to heat stress in the Indian Indo-Gangetic plains. *Glob. Chang. Biol.* 24, 4023–4037. https://doi.org/10.1111/gcb.14302

Sun, Y., Frankenberg, C., Jung, M., Joiner, J., Guanter, L., Köhler, P., Magney, T., 2018. Overview of solar-induced chlorophyll fluorescence (SIF) from the Orbiting Carbon Observatory-2: Retrieval, cross-mission comparison, and global monitoring for GPP. *Remote Sens. Environ.* 209, 808–823. https://doi.org/10.1016/j.rse.2018.02.016

Sun, Y., Fu, R., Dickinson, R., Joiner, J., Frankenberg, C., Gu, L., Xia, Y., Fernando, N., 2015. Drought onset mechanisms revealed by satellite solar-induced chlorophyll fluorescence: Insights from two contrasting extreme events. *J. Geophys. Res. G Biogeosci.* 120, 2427–2440. https://doi.org/10.1002/2015JG003150

Tarpley J.D., Schnieder S.R., Monet R.L., 1984. Global vegetation indices from NOAA-7 meteorological satellite. *J. Clim. Appl. Meteorol.* 23, 491–503.

Turner, D.P., 2011. Global vegetation monitoring : Toward a sustainable technobiosphere. *Front. Ecol. Environ.* 9, 111–116. https://doi.org/10.1890/090171

Veefkind, J.P., Aben, I., McMullan, K., Förster, H., de Vries, J., Otter, G., Claas, J., Eskes, H.J., de Haan, J.F., Kleipool, Q., van Weele, M., Hasekamp, O., Hoogeveen, R., Landgraf, J., Snel, R., Tol, P., Ingmann, P., Voors, R., Kruizinga, B., Vink, R., Visser, H., Levelt, P.F., 2012. TROPOMI on the ESA Sentinel-5 Precursor: A GMES mission for global observations of the atmospheric composition for climate, air quality and ozone layer applications. *Remote Sens. Environ.* https://doi.org/10.1016/j.rse.2011.09.027

Wan, Z., 1999. *MODIS Land-Surface Temperature Algorithm Theoretical Basis Document (LST ATBD).* Institute for Computational Earth System Science University of California.

Wang, C., Chen, J., Wu, J., Tang, Y., Shi, P., Black, T.A., Zhu, K., 2017. A snow-free vegetation index for improved monitoring of vegetation spring green-up date in deciduous ecosystems. *Remote Sens. Environ.* 196, 1–12.

Xu, S., Liu, Z., Zhao, L., Zhao, H., Ren, S., 2018. Diurnal response of sun-induced fluorescence and PRI to water stress in maize using a near-surface remote sensing platform. *Remote Sens.* 10, 1510–1527. https://doi.org/10.3390/rs10101510

Zhao, F., Li, R., Verhoef, W., Cogliati, S., Liu, X., Huang, Y., Guo, Y., Huang, J., 2018. Reconstruction of the full spectrum of solar-induced chlorophyll fluorescence: Intercomparison study for a novel method. *Remote Sens. Environ.* 219, 233–246. https://doi.org/10.1016/j.rse.2018.10.021

Section III

Land Use/Land Cover Analysis Applications

9 Rural Land Transformation in Chandigarh Periphery
A Spatio-Temporal Analysis

Ravinder Singh and Ravinder Kaur

CONTENTS

9.1 INTRODUCTION

A rapid and unplanned expansion of a city into its periphery is a typical phenomenon of an urban landscape. It involves numerous transformation processes operating on the edges of large urban centers, such as transformation of existing rural settlements into urban settlements, thereby changing their land use, architecture, demography, land use management and even social structure. As the city expands its commercial, industrial and residential areas into the periphery, the land use transforms and becomes more complex over time. Agricultural land becoming converted into urban land uses is a common phenomenon, the rate of expansion of which is generally guided by centrifugal forces or the outward-expanding forces from the city and its impact/influence on the periphery. Land that is adjacent to the city boundary (along the highways and near the areas that are already developed or developing) is more susceptible to conversion.

Chandigarh, being the world's first planned city, was planned to be a self-sufficient city, with the rural areas at its periphery intended for agricultural functions of dairy/poultry farming and agriculture/horticulture, so that the urban core could get a steady supply of vegetables, milk, food grains, eggs and meat products from its surroundings. Its architect, Le Corbusier, pointed out the importance of the periphery of Chandigarh, and that "the functions of the city and the periphery must not be interchanged, otherwise confusion and anarchy are sure to follow" (The new capital Periphery Control Act, 1952). In the original Periphery Control Act of 1952, the start of the periphery was marked as being 8 km distance from the city boundary but it was later extended to 16 km in 1962. The periphery of Chandigarh is a controlled area of 16 km in radius from the outer boundary of the Chandigarh administrative boundary, as enacted by the Punjab New Periphery Control Act. It is almost circular in shape and is part of two states (Punjab and Haryana). Most of this area is in Punjab (73%), with minor shares in Haryana (24%) and Chandigarh Union Territory (UT) (3%).

The periphery of Chandigarh extends from 76° 33' 2.84" E in the West to 77° 1' 4.708" E in the East and from 30° 32' 5.906" N to 30° 55' 25.004" N in the North. The total geographical area of the periphery is 1266.58 km^2. excluding Chandigarh in its center (Figure 9.1). The northern parts of the periphery are hilly and they run from west to east in a continuous belt separated by Ghaggar River in two parts. In the extreme North lies the inter-montane plain running parallel to the lower hilly regions, and in the South, lies the much higher Shiwalik Hills towards North. The flood plains of Punjab and Haryana are almost flat, covering almost three-quarters of the area of the periphery. The periphery is drained by many seasonal *choes* or streams which originate from the northern hilly areas and flow toward the south-west.

A paradox has emerged in the periphery. The periphery was to protect and guard against the haphazard urban growth, as described in "The Punjab New Capital Periphery Control Act, 1952". According to this Act, no alteration in the land use patterns was allowed in the area covered by the radius of sixteen kilometers from the outer boundary of Chandigarh, with only agricultural activities being permitted within the periphery. But the Government itself flouted the law and carved out major townships (Sahibzada Ajit Singh Nagar by Punjab and Panchkula Urban Complex by Haryana) in the periphery after the division of the periphery into three subparts i.e. Punjab, Haryana and the UT of Chandigarh.

According to the original master plan of Chandigarh, the city and the periphery were both proposed to have clearly distinct functions, under the Statute of the Land spelled out by Le Corbusier. In his views, the surrounding periphery areas of the city were meant to be the provider and the urban core area was to act as an absorber.

The inherent message of Le Corbusier appears to have been unheeded as the periphery has continuously succumbed to pressure from the city. At present, the city is unable to withstand the pressure of demand from the rapidly growing population within its administrative boundaries. All this is evident from the transformed character of the periphery. The size of the agricultural/green areas in the periphery is declining, with the rural land use transforming into peri-urban areas. Parts of the periphery in physical proximity to the city are entirely urbanized and

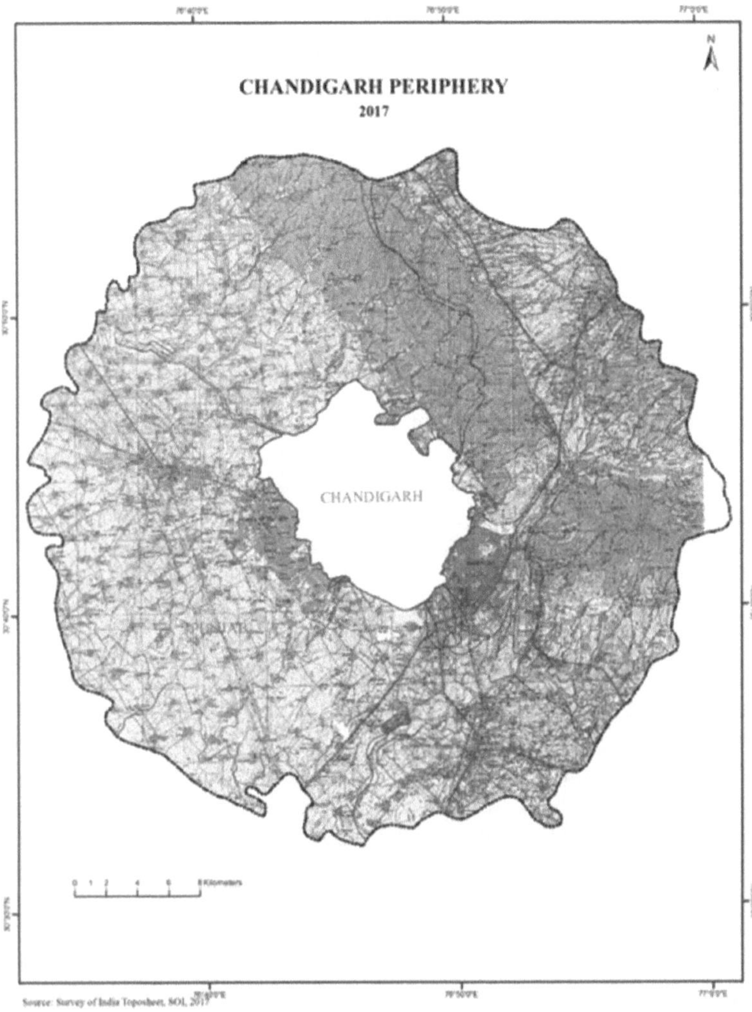

FIGURE 9.1 Administrative divisions of Chandigarh periphery.

remaining areas are also sprinkled with transformational pockets of spill-over of urban land uses. Some parts of the periphery have witnessed unplanned/disorganized growth, especially due to overspill into the rural built-up (*Abadi/deh*) areas. Creation of the satellite towns S.A.S Nagar (Mohali) and Panchkula by Punjab and Haryana, respectively, was the first step toward the violation of this periphery control act. Afterward, many small rural villages acquired a semi-urban or urban character.

The extension of urban land uses into the agricultural land intended in the Chandigarh periphery over the past two decades has cultivated strong public interest in the usage of the land in the periphery. The rapid growth of the population in the city, from 6.42 lakh (642,000) in 1991 to 9 lakh in 2001, and reaching up to 10.5 lakh in 2011, has had expected and undesirable consequences of rapidly changing the character of the regional land use (Chandigarh Master Plan 2031).

This paper tries to measure the magnitude, intensity and direction of land use/ cover transformation in the Chandigarh periphery. The paper also intends to identify the determinants of land transformation.

Defining Land Transformation: Rural land transformation has been defined in various ways, but mainly as a process of change, which bring comprehensive societal and economic change, driven by the global and homogenizing forces, that interact with the local infrastructure, institutions and various actors (government or private) to produce uneven patterns of land use and land cover. Some describe it as a process where sharp social, cultural and economic variations between the rural and urban areas gradually blend and merge into one another along a continuous gradient. Land transformation is an outcome of land use and land cover change over time and is defined as the transformation of land use and land cover from one land use category to another. It is a very dynamic concept, as land use is under a constant state of change. Both natural and anthropogenic factors control the pattern and rate of change in land use and land cover, ultimately resulting in transformation of land.

9.2 DATA SOURCES AND METHODOLOGY

The present work was carried out using satellite data sources, with analysis of the satellite data having been done using the remote sensing and geographic information system (GIS) techniques. The Land Use and Land Cover (LULC) patterns in the periphery of Chandigarh have been studied since 1972 (Table 9.1). A period of almost four decades has been selected for this study, after due consideration of the availability of satellite data. The satellite data is obtained from the United States Geological Survey (USGS) official website and data-sharing portal (Glovis/Earth Explorer https://earthexplorer.usgs.gov/). The collected data were rectified and corrected for errors. The hybrid classification technique was applied, using the ERDAS IMAGINE software package to generate maps showing present as well as prior LULC patterns. Transformations in the patterns of LULC were observed and analyzed using the ArcGIS software package. Results were verified by accuracy assessment and ground truth (Figure 9.2).

9.2.1 Classification Scheme for Land Use and Land Cover Analysis

The hybrid classification approach is applied in the current study to recognize various land use and land cover classes, depending on their spectral reflectance properties. It is followed by field verification of the classes. After the necessary corrections and modifications, the final output maps were prepared, showing the land use and land cover patterns for the period under investigation.

Multi-temporal satellite data is a significant data base of land use and land cover information analysis of certain areas. The prime aim for constructing

TABLE 9.1
Area (km²) of Chandigarh Periphery, Land Use and Land Cover, 197 –2017

S. No	Category	1972	1980	1991	2000	2005	2009	2014	2017
1	Agricultural land	300.67	322.19	272.80	361.47	452.13	390.68	265.56	242.50
2	Built-up land	55.84	98.09	131.41	152.12	164.05	182.80	237.08	273.57
3	Vacant land	119.77	124.84	130.29	138.39	202.56	331.60	321.24	294.73
4	Tree cover	783.76	717.76	730.21	612.55	444.84	357.51	436.39	451.89
5	Water body	6.53	3.70	1.86	2.05	3.00	3.99	6.32	3.90
Total		1266.58	1266.58	1266.58	1266.58	1266.58	1266.58	1266.58	1266.58

Source: The Digital Image Classification of Landsat Satellite Images, USGS,1972–2017.

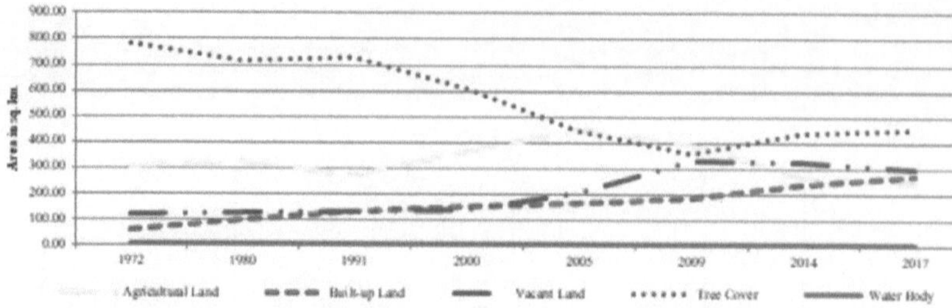

Source: Calculated from LULC Classified Images of Landsat, USGS, 1972 – 2017

FIGURE 9.2 Chandigarh periphery, land use and land cover, 1972–2017.

LULC maps for various temporal intervals is to estimate the degree of change in the different LULC classes and to evaluate the growth pattern of the built-up area into the peripheral zone.

For the present study, the first order of Anderson's digital image classification system is utilized to classify the land use in the periphery into built-up land, area under tree cover, water bodies, vacant land and agricultural land.

As already mentioned, an attempt was made to group similar LULC categories using a pixel value-based hybrid classification approach by selecting appropriate training sites. Images from Landsat, Indian Remote Sensing satellites (IRS) and Google Earth, along with the topographical maps of the study area have been used to classify the digital satellite images. Ground truth and field visits for the image classification and selection of training sites was accompanied by the accuracy assessment of the digitally classified image.

9.2.2 Land Use and Land Cover Change Detection Analysis

In the current study, after classifying the LULC classes in the area under study into discrete categories such as built-up land, tree cover, vacant land, agricultural land and water bodies, the corresponding land cover statistics were also obtained. In addition to the information that can be read from the maps, the statistics were used to calculate land cover area under each category in square kilometers, determine the percentage increase and decrease in land cover area between two consecutive periods, the rate of change and hence the land cover change, and so forth. To analyze LULC dynamics of the Chandigarh periphery, the data and the methodology of the classification scheme employed were explained in the previous paragraph.

The present paper intends to analyze the different measures obtained from the classified satellite images so that the information obtained can be used for evaluating LULC distribution for the study area in the past and present, the trend of urban built-up area expansion and the pattern of land use change in the Chandigarh periphery.

9.3 RESULTS AND DISCUSSION

9.3.1 Land Use and Land Cover: A Spatial View

The municipal limits in the case of the Chandigarh periphery have undergone changes with the passage of time. The municipal boundaries of towns like S.A.S. Nagar, Panchkula, Kharar, Zirakpur, Pinjore, Kalka and Banur, have been extended to incorporate the rapidly growing population. Consequently all these have acquired urban status. With this expansion and the upgrading of civil authorities, the boundaries have expanded and surrounding rural area was brought within the municipal limits. The patterns of use of land in the periphery will be discussed under this section. Figures 9.3 and 9.4 show the continuously changing patterns of land use in the periphery. The total geographical area of the periphery is approximately 1,266.58 km^2. The area under the periphery remained almost the same during the study period, as the periphery is identified by "The Punjab New Capital (Periphery) Control Act, 2005", under which no further expansion can be made by the state governments.

In 1972, the area under tree cover was the dominant class. It occupied an area of 783.76 km^2, representing 61.88% of the total geographical area in the periphery, followed by the area under agricultural land use (300.67 km^2) and vacant land (119.77 km^2). At this time, the built-up land and water bodies accounted for only 55.84 km^2 and 6.53 $km^2$of the geographical area respectively. In 1980, the dominant class remained to be tree cover as it covered 717.76 km^2 of the land area in the periphery. Agricultural land increased to 322.19 km^2 (25.44% of the total land area), followed by vacant land (124.84 km^2, almost 10 percent of the geographical area). The built-up land almost doubled to cover 98.09 km^2 (7.74%) of the land. The next two decades showed the same trend of decline in area under tree cover and expansion of the area under agricultural and built-up land use in the periphery. The proportion of the area under tree cover is reported to have declined considerably after 1991.

In 2000, the proportion of area under tree cover was reported to have declined considerably to 612.55 km^2, occupying less than 49 percent of the total area. On the other hand, the land under built-up and agricultural uses enlarged to 152.13 km^2 (12.01%) and 361.48 km^2 (28.54%) respectively. The vacant land also recorded an increase, being spread over 202.49 km^2 of land area.

After a decade of continuous ongoing development, the area of built-up land was recorded in 2009 at182.96 km^2. The area under tree cover decreased to only 357.82 km^2 (28.23% of the total area in the periphery), followed by agricultural land and vacant land (391.09 km^2 and 331.88 km^2 respectively). The rapid decline of the area under tree cover brought some policy changes, such as a strengthening of Joint Forest Management Programme and National Forest Action Programme and better implementation of policies to save the forest and to reclaim the area under forest. As a side-effect of the efforts made by state governments to reclaim forest areas in Shiwalik Hills and the creation of rainwater-retaining dams (Siswan, Jayanti Devi, Kaushalya Dam, etc.), area under tree cover increased to

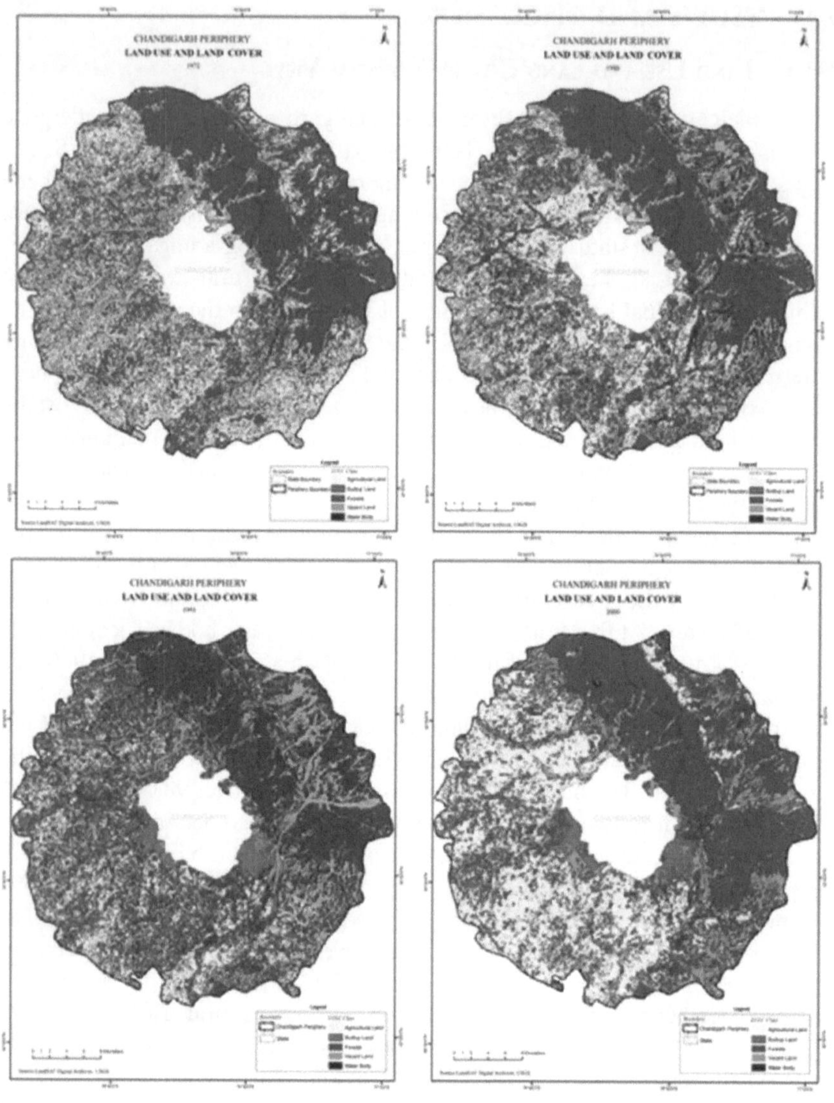

FIGURE 9.3 Land use and land cover in Chandigarh periphery, 1972–2005.

451.88 km^2 in 2017. At this time, in 2017, the agricultural land (242.50 km^2), vacant land (294.72 km^2) and built-up land (273.56 km^2) occupied almost identical proportions of the periphery land area.

As discussed above, the type of use of land and natural land cover of the Chandigarh periphery has been understood in terms of the segments of diverse classes of land use in the current geographical area. However, in an attempt to understand the actual change in the use of land and natural land cover, a discussion based on common areal limit is essential. The following section will discuss

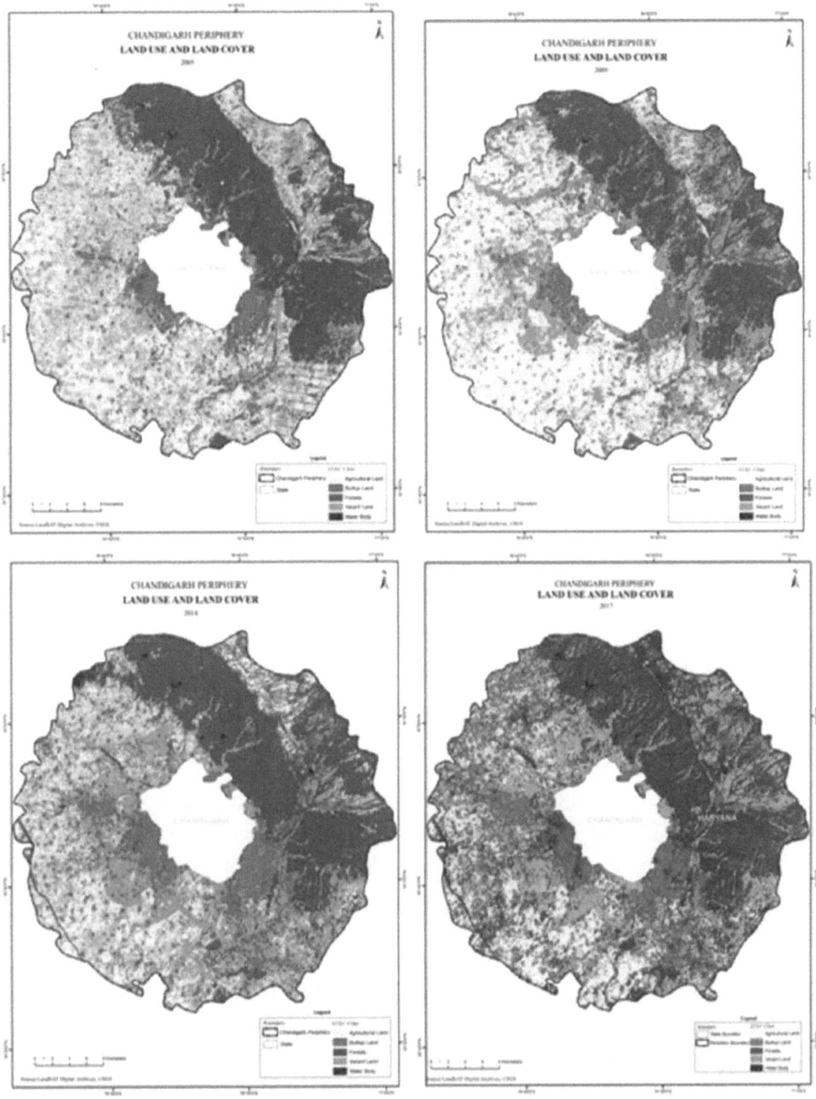

FIGURE 9.4 Land use and land cover in Chandigarh periphery, 2005–2017.

the changing use of land and land cover patterns, considering the periphery of Chandigarh as a single unit during the past four decades.

9.3.2 Land Use and Land Cover Change: A Temporal View

There was a continuous change in individual land use class and land cover categories in the period 1972 through 1980, 1991, 2000, 2005, 2009 and 2014 to 2017.

In addition to the land use and land cover classes mapping for each respective period, it is possible to compute the changes between any two periods. The present section discusses the changes in the LULC patterns in the periphery of the Chandigarh from 1972 to 2017.

The periphery was created to provide agricultural and other resources, such as milk, vegetable, fruits and grains, to the growing city and to restrict development of other urban centers close to the city. Agriculture was the major occupation of the local people during that period. In 1972, agricultural land occupied almost one-quarter of the area of the total periphery. Agricultural land increased continuously until 2005, when it occupied 452.13 km^2 of the area of land in the periphery. After 2005, there was a sharp decline of almost 250 km^2 in the land under agricultural use in the following decade.

The area of built-up land in the periphery grew to almost 273 km^2 from 55.84 km^2 over the previous four decades. The growth of built-up land was constant and gradual for the first two decades, after which the speed of conversion of non-built-up land to built-up land accelerated. Most of the periphery in 1972 was under tree cover, occupying 783.76 km^2, but the continuous population growth in the periphery, the growth of existing urban centers, the demand for agricultural land and the lack of policies to conserve the forests contributed to the rapid decline in the area under tree cover. The area under tree cover declined from 783.76 km^2 in 1972 to a minimum of 357.82 km^2 in 2009, after which the change in the approach of government and the strict actions taken by government officials reclaimed some of the area under tree cover and restored it to the current figure of 451.88 km^2 of land in the periphery.

The total area in the periphery under tree cover showed a continuous decline from 717.76 km^2 to 451.88 km^2 between 1980–2017, with an absolute decline of almost 266 km^2 (Table 9.1). This growth occurred because of strict legalization and restrictions imposed by government policies, under Joint Forest Management (JFM), National Forestry Action Programme (NFAP) and "The Forest Control Act, 1980", to conserve and protect the remaining forest cover in the hilly areas.

Remarkably, the proportion of the periphery area corresponding to built-up areas showed a corresponding increase during the study period. Built-up land, covering only 55.84 km^2 (4.41%) of the total land area in 1972, increased to 273.57 km^2 (21.60%) in 2017, registering nearly a five-fold increase in the span of just half a century (Table 9.2). The analysis shows continuous and steady growth of built-up land area in the periphery.

Vacant land, which occupied only 119.77 km^2 of land in the periphery in 1972, increased at a slow rate until 2000, representing only 138.39 km^2 of the land area. After 2000, the vacant land covered more than one-quarter (26.18%) of the total area, with an expansion to 331.88 km^2 in the periphery. At present, almost 295 km^2 of land in the periphery is classified as vacant, after being acquired from the local residents for future development of the city and urban areas. Water bodies, occupying only 6.53 km^2 in 1972, fell to only 1.86 km^2 in 1991. After 1991, the need to develop water resources was felt by officials, especially in the hilly areas of the northern periphery. This focused the attention of the government

TABLE 9.2
Chandigarh Periphery, Land Use and Land Cover Change (Area in km^2)

S. No	*Category*	**1972–1980**	**1980–1991**	**1991–2000**	**2000–2005**	**2005–2009**	**2009–2014**	**2014–2017**
1	Agricultural land	21.52	–49.16	88.45	90.51	–60.97	–125.46	–64.49
2	Built-up land	42.25	33.43	20.6	11.87	18.96	54.12	35.16
3	Vacant land	5.07	5.56	7.99	64.11	129.39	–10.65	–140.04
4	Tree cover	–66.00	13.08	–118.29	-167.9	–86.86	78.56	165.42
5	Water body	–2.84	–1.83	0.18	0.95	1.00	2.33	1.33
Total					**1266.58**			

Source: The Satellite Data, Landsat Digital Archives, USGS, 1972–2017.

on these areas. Many small-scale check dams (Siswan Dam, Jayanti Devi Dam, Kaushalya Dam, etc.) were built in order to store water during the rainy season. Under the World Bank project, 29 dams were proposed in the Shiwalik Hills in the Punjab, 12 of which lay in the Chandigarh periphery (World Bank, 2015). Nine of these dams have already been completed and are working. Continuous attention and efforts resulted in the growth of water bodies in the periphery, to cover almost 6 km^2 in 2014 (Table 9.2). At present, there are approximately 4 km^2 of water bodies. This fluctuates according to the availability of rainwater and its utilization by the local people.

9.3.3 Land Transformation

Land transformation although localized has widespread consequences, and its impact can be seen over the entire region. The Chandigarh periphery, being rural in nature, as stated in "The New Capital (Periphery) Control Act, 1952", was supposed to remain rural after its creation. But the urban centers within the periphery grew at a rapid pace and transformed the rural landscape into a peri-urban landscape. During the period 1972 to 2017, massive transformation was witnessed in every category of land use. The area under tree cover in the Chandigarh periphery was the worst hit (Figure 9.5).

It is clear from Table 9.3 that, from 1972 to 2017, 110.20 km^2 of tree cover was transformed into built-up land. Furthermore, another 116.09 km^2 area, which was previously under tree cover, was converted to agricultural land and 216.90 km^2 of land area was converted into vacant land. Only 338.83 km^2 of area under tree cover remained unchanged. Likewise, 1.74 km^2 of waste land was transformed into water body.

Next to the area under tree cover was agricultural land, where the transformation from 1972 to 2017 was very significant. Most of the agricultural land (96.04 km^2) was converted into built-up area. About 49 km^2 of agricultural land was acquired and transformed into vacant land for further transformation into built-up land in future. Almost 62 km^2 of the agricultural land in restricted forest areas was recovered and transformed into tree cover. Approximately 0.5 km^2 of the agricultural land was brought under the classification of water bodies during this period in the form of check dams, created under the World Bank project. Efforts to improve the Sukhna Catchment area and to build the Kaushalya Dam have also yielded results. During the same reference period (1972–2017), 29.91 $km^2$of vacant land was converted into built-up land. Furthermore, 30.23 km^2 of vacant land was converted to agricultural land to fulfill the continuous demand for more and more arable land. Similarly, 43.83 km^2 of land that was lying vacant was brought under tree cover by development of water bodies in the Shiwalik Hills.

During this period, 11.39 km^2 of built-up land was cleared to either construct new buildings or to legalize unauthorized construction in the periphery. The area associated with water bodies fluctuated as in the first two decades from the 1972 to 2017 period; the water bodies in villages in the forms of ponds were significantly reduced but, after the intervention of local governing authorities,

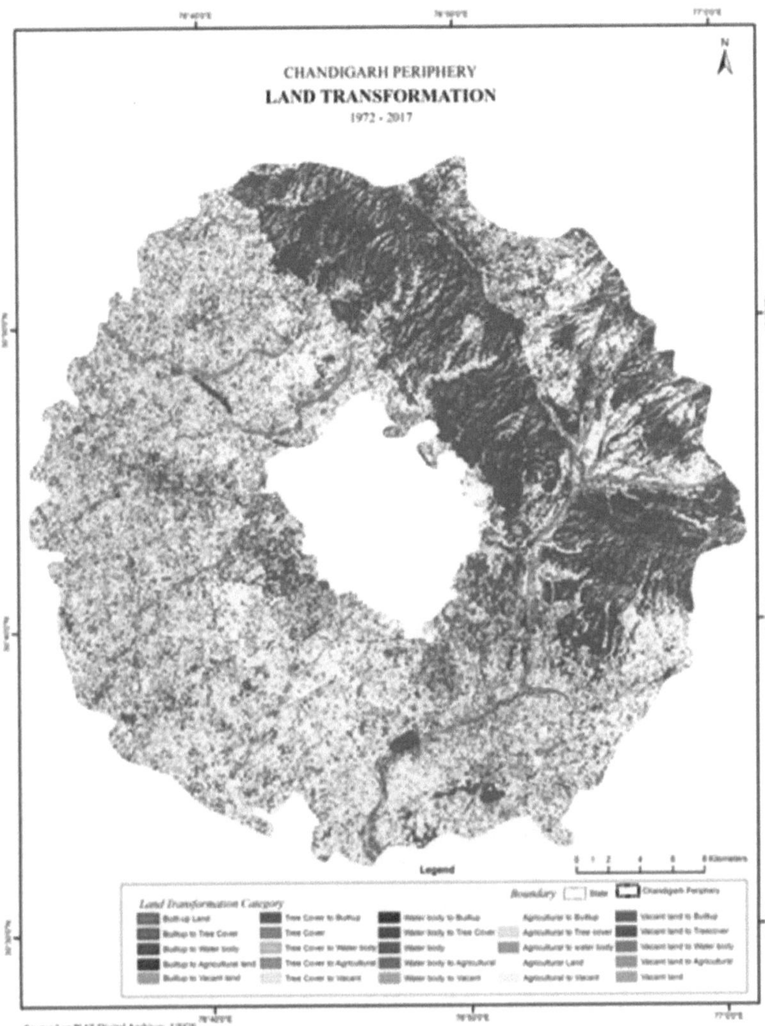

FIGURE 9.5 Rural land transformation in Chandigarh periphery, 1972–2017.

construction of check dams and the restoration of ponds brought the water bodies to the present level.

9.3.3.1 Determinants of Land Transformation

The present section attempts to understand the factors responsible for land transformation in the Chandigarh periphery. The causes or determinants of LULC change can be termed as 'actors': "various organizations, including people, which regulate several forces that cause rural land transformation" (Bryant and Bailey, 1997). The patterns and trends of land transformation is a result of certain 'alliances' and 'conflicts' between these actors.

TABLE 9.3

Land Transformation in Chandigarh Periphery, 1972–2017 (area in km^2)

Year		1972					
	Category	Built-up Land	Tree Cover	Water Bodies	Agricultural Land	Vacant Land	Total
	Built-up land	36.05	110.2	1.38	96.04	29.91	273.57
	Tree cover	5.63	338.83	1.78	61.83	43.83	451.89
2017	Water bodies	0.29	1.74	1.08	0.52	0.29	3.9
	Agricultural land	2.49	116.09	0.35	93.33	30.23	242.5
	Vacant land	11.39	216.9	1.96	48.96	15.51	294.72
	Total	55.84	783.76	6.53	300.68	119.77	1266.58

Source: The Satellite Data Analysis,1972–2017.

Based on the above discussion, different actors operating in Chandigarh periphery will be discussed here. These actors can be classified as 'Institutional' or 'Private' actors. Institutional actors can be government agencies, whereas, on the other side, private actors consist of property dealers, farmers, residents, entrepreneurs and developers, who are continually transforming the land to different classifications.

***Institutional Actors*:** These largely include state governments, who are responsible for the development and management of urban areas. Being the capital of two states, Chandigarh preserves a special status. Its periphery is controlled by "The New Capital (Periphery) Control Act, 1952". So, the governments of Punjab and Haryana are responsible for the urban development in their respective subzones of the periphery. The periphery of Chandigarh was delineated and extended with respect to its spatial extent in subsequent amendments by their state governments. Furthermore, multitudes of agencies of the two said governments have been largely responsible for the land transformations in the Chandigarh periphery. These agencies have played a significant role in transformation of the rural landscape in the periphery. The important agencies operating in the Chandigarh periphery are: Chandigarh Municipal Corporation, Municipal Corporation of S.A.S. Nagar, Municipal Corporation of Panchkula, Municipal council of Zirakpur, Municipal Corporation Kurali, Greater Mohali Area Development Authority, Punjab Urban Development Authority, Haryana Urban Development Authority, Municipal Corporation of Pinjore and the Town Planning Departments of all three states. These government agencies construct infrastructure services to fulfill the growing demand of the city and its residents. The multiplicity of such agencies in the periphery has introduced competition among them to reap the advantages of the proximity to Chandigarh.

The infrastructural development projects, such as the construction of new roads (200-feet wide airport road by the Greater Mohali Area Development Authority (GMADA), a 200-feet wide road between Chandigarh and Kurali through Mullanpur, and road widening projects, such as on the Chandigarh–Kurali road through Kharar, the Chandigarh–Ambala road through Zirkpur, etc.), the construction of IT parks, construction of rehabilitation colonies, etc., have also resulted in the conversion of land under tree cover, agricultural land and vacant land to built-up land. This transformation of the land has inflated the rent of the land of transformed locations to bring about LULC changes in the neighboring areas.

Government policies, on the other hand, have played an ancillary role in this process of land use transformation. Numerous policies, such as "The Punjab Regional and Town Planning and Development Act, 1995" (amended in 1996, 2003 and 2006), the "Haryana Development and Regulation of Urban Areas Act, 1975" and "The New Capital (Periphery) Control Act of 1952", have created a lot of confusion and competition among the Punjab, Haryana and UT governments. This conflict between the state governments has given rise to the large-scale conversion of agricultural land and area under tree cover to built-up land and vacant land to accommodate rapidly growing populations and to fulfill their demands

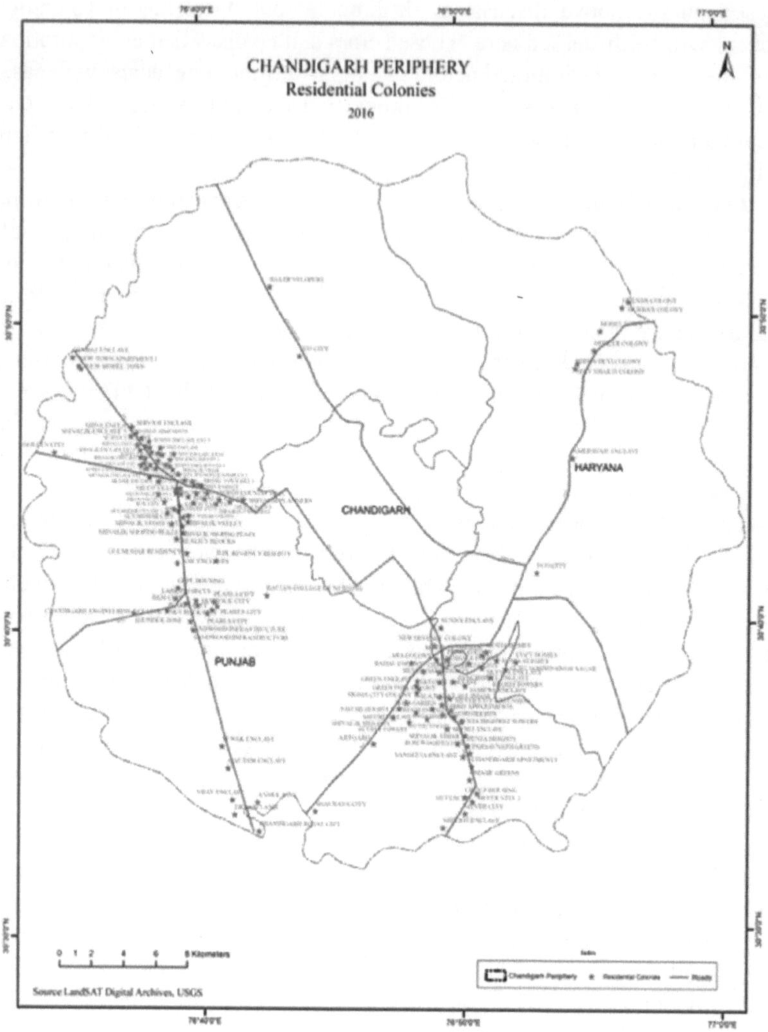

FIGURE 9.6 Spatial distribution of residential colonies in the periphery of Chandigarh.

for residential and commercial lands. Many such residential complexes, like Gilco Valley, Amravati Enclave, Sunny Enclave etc. (in the Punjab sub-region) and Kamdhenu Homes, Chandigarh Enclave, Royal Estate, etc. (in the Haryana sub-region of the periphery) have been constructed in the periphery (Figure 9.6).

This unauthorized and uncontrolled transformation of the periphery is a consequence of a lack of synchronization among the government agencies and its different departments. Presently, in the Punjab sub-region of the periphery, the Greater Mohali Area Development Authority is the sole managing government organization, whereas, in the Haryana sub-region, the Haryana Urban

Development Authority (HUDA) has similar responsibilities. The future masterplans, such as for Chandigarh (by 2030), GMADA (during 2008 to 2058) and Panchkula (by 2021), have assigned the area currently under agricultural land to built-up land in the future. The majority of the area under the masterplans has already been acquired by the government authorities and handed over to private builders for infrastructural development.

***Private Actors*:** Private actors comprise farmers, residents, property dealers, developers and entrepreneurs. Associated by their socio-economic requirements, they play a vital part in transforming the land around the city. Their contribution to the rural land transformation in the periphery of Chandigarh can be understood by classifying them into the following categories: (i) local landowners, (ii) local landless people and (iii) migrants. The periphery of Chandigarh is in its development phase and there is an ever-increasing demand for land for residential, as well as for commercial, purposes. As a consequence, almost 150 residential projects, such as Sunny Enclave, Gilco Valley, Amravati Enclave, Maya Estate, Motia Developers, etc., have been developed in the periphery during the past 15 years (Figure 9.5). The demand for land raises the value of land at certain locations, such as along the Chandigarh–Ambala road, the Chandigarh–Khara road, the New Chandigarh residential project near Mullanpur, etc., and convinces more and more local landowners to sell their land. For instance, almost eight percent of land within a radius of eight kilometers from the city boundary has already been acquired, with projects under development (GMADA, 2017). There are many factors that raise the demand, such as families growing and splitting into nuclear households, requiring additional residential space.

According to the estimates of the Agricultural Census of India 2011, twelve thousand marginal farmers in the two districts, S.A.S Nagar and Panchkula, have further divided up their land. As the land holdings become smaller, the farmer sells them off to the private developers for the handsome price offered by them. Furthermore, many residents of the city, residing in the dense parts of the town, prefer to move out due to the higher rents in the city. According to the Census of India 2011, almost five percent of the people living in the periphery migrated there from Chandigarh and three percent from Delhi. This rising demand for commodities and various services is a result of such developments in the urban fringe areas, and to realize this demand commercial developments have taken place, resulting in land transformation. Rapid population growth, creating huge demand, resulted in the construction of approximately 120 higher education institutions, twelve large shopping malls and almost one hundred marriage palaces in the periphery. Chandigarh periphery is thus developing as an educational and commercial hub.

Farmers with large land holdings, businessmen and residents in the high-income group from the tertiary sector act as entrepreneurs in the periphery. They are solely accountable for transfiguring the rural non-built-up land into peri-urban built-up land. People with high incomes purchase the land from farmers and construct farmhouses in isolated places in the periphery. For example, the entire northern periphery of the Chandigarh, starting from the city boundary to the foothills of the Shiwalik Hills, is dotted with palatial farmhouses owned by doctors, lawyers, bureaucrats and politicians. They invest heavily in the construction of commercial hubs and complexes in the proximity of the city. Large numbers of residential and commercial buildings have also been erected by entrepreneurs around Zirakpur, Kurali, Pinjore and Kharar towns. Similarly, large numbers of marriage palaces and resorts (about 100), colleges (120 colleges/institutes) and commercial complexes (12 malls) has been constructed in the periphery in past two decades. It was observed that the farmers with large land holdings are the last to transform their land from agricultural to non-agricultural land use. GMADA has acquired land from almost all the marginal and medium-sized farmers within a radius of eight kilometers of the city, but locations where the land was owned by large farmers has still not been acquired (for example along the Chandigarh–Kharar road, with respect to land belonging to Gurudwara Daun Sahib). Locational factors (proximity to a major road, type of soil, irrigation facilities, etc.) push the farmers with small land holdings to either sell their land or to transform the land for non-agricultural use.

Property dealers play the role of middlemen and are engaged in purchasing and selling of property in the periphery. According to the official records of S.A.S Nagar and Panchkula districts, there are almost five hundred property dealers operating in the periphery of Chandigarh. Farmers with small land holdings are their first preference. They purchase small properties from each of multiple landowners and sell the combined parcel to a bigger developer at a much higher price. This type of land settlement benefits the property dealers as majority of them earn their livelihood in the form of a percentage share in the conversion process, as the higher land rental leads to greater monetary benefits to them. They work with entrepreneurs and developers and help them to acquire land from multiple stakeholders, facilitating the development of large residential and commercial projects.

Furthermore, developers attempt to obtain maximum profit by expanding the residential and commercial areas. The GMADA Master Plan and the Master Plan of Panchkula, along with the Punjab Apartment and Property Regulation Act (1995), the Punjab Regional and Town Planning and Development Act (1995) and the Haryana Development and Regulation of Urban Areas Act (1975), enabled large-scale developers to come into the periphery and transform the land from non-built-up to built-up use. The Master Plans provide the guidelines to the developers and enable them to construct residential and commercial hubs around the urban centers. These can be witnessed in New Chandigarh, Panchkula Extension, Zirakpur Extension, etc., in the periphery. The developers were allowed to privatize the construction in the periphery by the respective privatization policies of the Government of India.

The developers augmented the government efforts to provide residential dwellings and commercial complexes in response to the high demand from the population. The residential colonies were constructed close to the urban centers (e.g. Sunny Enclave and Gilco Valley near Kharar), along all the major roads (North Country Mall on the Chandigarh–Kharar road, the Amravati Enclave on the

Chandigarh–Pinjor road), connecting the city with the neighboring urban centers (Map 5). Consequently, exploratory developers acquired the land market, which led to the variation in demand and supply of urban land. According to Government records, there are more than 150 residential projects in the periphery, but more than three-fifths of the flats are lying vacant. According to the law, a private developer can only purchase land from the Government but they have somehow managed to buy fertile agricultural land from the local people. A report published by The Tribune in 2013 indicated that more than five thousand 'Kanal' (625 acres) or artificial waterways through the common land of the villages Shamlat and Panchayat in the periphery have been illegally occupied by such private owners. Furthermore, the report indicated that it was all happening with the political support of local influential people (The Tribune, 2013).

These are the roles played by the actors in the nexus involving land transformation in the periphery. It seems that reaping the benefits from the periphery has become the sole motive of these governmental and private agencies, which has led to massive land use transformation on fertile agricultural land.

CONCLUSION

To sum up, it can be inferred that the periphery, which was created to provide a productive buffer around the city, has come under attack from government agencies on one hand and private property dealers and developers on the other. The spatial expansion has taken place on the prime agricultural land from where Chandigarh should get its supplies of vegetables, fruits, milk and grains. It is pertinent for us to ask where the city will draw its food supplies in the future, once the agricultural land has been converted to built-up land for residential and commercial use.

REFERENCES

Ahmed, A. (2011). *A bursting city with cage in the borders: Addis Ababa city since 1973.* Seminar paper on environmental economics, Ph.D program on environmental planning, Ethiopian Institute of Architecture, Building Construction, Ethiopia, 95

Amin, A. & Fazal, S. (2012). Quantification of land transformation using remote sensing and GIS technologies. *American Journal of Geographic Information System*, 1(2), 17–28.

Anderson, J.R., Hardy, E.E., Roach, J.T. & Witmer, R.E. (1976). *A land use and land cover classification system for use with remote sensor data.* Geological survey professional paper no. 964, U.S., Government Printing Office, Washington, DC, 28.

Brookfield, H. & Stocking, M. (1999). Agrodiversity: Definition, description and design.*Global Environmental Chang*, 9(1), 77–80.

Bryant, R.L. & Bailey, S. (1997). *Third world political ecology.* London: Routledge: 144–168.

Census of India (1991). *General population tables, census of India, New Delhi.* Retrieved from www.censusindia.gov.in

Census of India (2000). *General population tables, census of India, New Delhi.* Retrieved from www.censusindia.gov.in

Census of India (2011a). *general population tables, census of India, New Delhi.* Retrieved from www.censusindia.gov.in

Census of India (2011b). *Migration tables, census of India, New Delhi.* Retrieved from www.censusindia.gov.in

Dimple Singh, R. (2014). Land use and land cover change along Shiwaliks between river Ghagghar and Yamuna. M. Singh et al. (eds.), *Climate change and biodiversity: Proceedings of IGU Rohtak conference 1, advances in geographical and environmental sciences*, Springer, Japan, 139–150.

FAO (2000). *Land Cover Classification System (LCCS): Classification concepts and user manual*, Food and Agriculture Organization, Rome, 42–45.

FAO/UNEP (1977). Assessing soil degradation. food and agriculture organization, FAO soils bulletin, United States. *Roma*, 34, 89–96.

Fazal, S. (2000). Urban expansion and loss of agricultural land—A GIS based study of Saharanpur City, India. *Environment and Urbanization*, 12, 133–149.

Lambin, E.F. & Geist, H.J. (eds.) (2003). Causes and trajectories of land use/cover change. *Land-Use and Land-Cover Change: Local Processes and Global Impacts*, Springer, 41–54.

Leulsegged, K., Gete, Z., Dawit, A., Fitsum, H. & Andreas, H. (2012). *Impact of urbanization of Addis Ababa city on peri-urban environment and livelihoods.* Ethiopian Economic Association (EEA), Ethiopia, 65–89.

Mather, A.S. (2006). Driving forces. H.J. Giest (ed.), *Our earth's changing land: An encyclopedia of land use and land cover change*, Green Wood Press, Westport, 179–185.

Meyer, W.B. & Turner, B.L. (1996). Land-use/land-cover change: Challenges for geographers. *Geography Journal*, 39(3), 237–240.

Rohilla, S.S., Singh, R. & Batra, D. (2017). Psycho-geographical study of mental health, well-being and perceived stress among students belonging to urban and rural areas of Chandigarh. *Asian Resonance*, *6*(3), 176–181.

Singh, R. (2016). Land transformation on the northern periphery of planned city of Chandigarh: A study based on remote sensing and GIS technology. *Online International Interdisciplinary Research Journal*, 6(4), 86–100.

Singh, R. (2017). Developing village level water quality information system: A case study of Amroha Tehsil, Uttar Pradesh, India. *International Journal of Science and Research Technologies (IJSART)*, 3(12), 240–247.

Singh, R. & Preeti, K. (2017). Land transformation in the western periphery of Chandigarh: A case study using remote sensing and GIS. *International Journal of Interdisciplinary and Multidisciplinary Studies (IJIMS)*, 4(3), 170–178.

Skole, D.L. (1994). *Data on global land-cover change: Acquisition, assessment, and analysis, changes in land use and land cover: A global perspective.* Cambridge University Press, Cambridge, 155–168.

Tadesse, W., Tsegaye, T.D. & Coleman, T.L. (2001). Land use/cover change detection of the city of Addis Ababa, Ethiopia using remote sensing and geographic information system technology, *IGARSS, Scanning the Present and Resolving the Future. Proceedings. IEEE International Geoscience and Remote Sensing Symposium Sydney, NSW*, 1, 462–464.

The Hindustan Times (2015). *SC order on Sectors 76–80 leaves 350 allottees in lurch*, Report. The Hindustan Times, Chandigarh. Retrieved from http://www.Hindustantimes.com/chandigarh/sc-order-on-sectors-76-80-leaves-350-allottees-in-lurch/story-kpFSDOq5gLrz3x1AsNc2uO.html

The Tribune (2008). *GMADA dumps Sahara City rejects licence; investors in soup*, Report. The Tribune, Chandigarh. Retrieved from http://www.tribuneindia.com/2008/20081230/cth1.htm

The Tribune (2013). *How high &mighty grabbed public and in Punjab villages near Chandigarh*, Report. The Tribune, Chandigarh. Retrieved from http://www.tribuneindia.com /2013/20131110/main6.htm

The Tribune (2014). *Land acquisition: Farmers reject meagre compensation hold protest outside GMADA office, threaten to move court*, Report. The Tribune, Chandigarh. Retrieved from http://www.tribuneindia.com/2014/ 20140 111/cth1.htm

The World Bank (2015). *Projects and programmes in India*, Report. U.N. Retrieved from www.worldbank.org/countr ies/India/Projects

The World Bank (2017). *World development indicators, data*. Retrieved from www.data .worldbank.org/indicators

Turner, B.L., Skole, D., Sanderson, S., Fischer, G., Fresco, L. & Leemans, R. (1995). *Land-use and land-cover change science/research plan, joint publication of the international geosphere-biosphere programme (Report No. 35) and the human dimensions of global environmental change programme (Report No. 7)*, 255–263.

UNEP (2001). *Depleted uranium in Kosovo, post-conflict environmental assessment*. United Nation Environmental Protection, Balkans Task Force Report. Nairobi, Kenya, 75–89.

UNFPA (2007). *State of the world population, 2007, unleashing the potential of urban growth*, Report. U.N.

Verburg, P.H., DE Groot, W.T. & Veldkamp, A. (2003). Methodology for multi-scale land-use change modelling: Concepts and challenges. A.J. Dolman, A. Verhagen & C.A. Rovers (eds.), *Global environmental change and land use*. Kluwer Academic, Dordrecht, 233–254.

10 Examining the Temporal Change in Land Cover/Land Use in Five Watersheds in Goa, India

Using a Geospatial Approach

Ashwini Pai Panandiker, Lewlynn de Mello, Mahender Kotha, and A.G. Chachadi

CONTENTS

10.1 INTRODUCTION

The bio-physical cover of the Earth's surface, which includes vegetation, bare rocks, bare soils, and water areas constitutes land cover (Di Gregorio and Jansen, 1998), whereas land use is the extent of human activities that have an impact on land and its resources (Young, 1994). Some of the most significant human modifications affecting the surface of the Earth are caused by land-use and land-cover change (LULC) (Lambin et al., 2003).

Change in land use is a dynamic and intricate process that occurs due to both natural causes and human interventions. This directly influences soil, water, and the atmosphere (Turner, 2006) and hence is directly related to many global environmental issues. Land-use change is also one of the key aspects in the process of climate change and the relationship between the two is interdependent; changes in land use may affect the climate whilst climatic change also influences future land use (Dale, 1997; Watson et al., 2000). Furthermore, river-land ecosystems around the world have been markedly altered by anthropogenic activities, particularly land-use change. At watershed level, land-use changes can affect sediment transport, thereby causing siltation of rivers (Kondolf et al., 2012). Land-use changes can affect surface water runoff, flood frequency, baseflow, and annual mean discharge (Huntington, 2006; Brown, 2013; Wei et al., 2013). Hence, examining the hydrologic response of watersheds to physical (land use) and climatic change (precipitation and air temperature) becomes a crucial element of water resource planning and management (Vorosmarty et al., 2000). Many studies across the world have examined the impact of land-use/land-cover change on the streamflow (Guzha et al., 2018; Nugroho et al., 2013; Vicente et al., 2018; Welde and Gebremariam, 2017). However, there are not many studies addressing this issue in Goa, India.

Goa, a small state on the west coast of India, has been witnessing pressure owing to increasing population, tourism, mining, and urban development, which are leading to land-use/land-cover (LULC) change at varying spatial and temporal scales (Misra et al., 2014). There have been many studies, such as Kotha and Kunte (2013), Roy and Srivastava (2011), Singh et al. (2004), Sampath et al. (2014), and NNRMS, ISRO (2014), that have examined the change in the land use in Goa, whereas Yedage et al. (2015) studied the land-use change for only the South Goa district. Furthermore, Misra et al. (2014) studied the land-use change in the Mandovi-Zuari river estuarine region. With the growth in tourism and related activities over the past two decades and variations in the geological setup, changes in land-use patterns have been observed. This has resulted in altered ecosystems and degradation of forest and coastal areas (Mascarenhas et al., 1997). Yedage and Harmalkar (2017) calculated the changes in LULC of Quepem Taluka over a 10-year period, and predicted the probable LULC changes in the future. In the study by Singh et al. (2004), the use of satellite-based remote sensing for analyzing changes in mangrove forest cover along the coastline in Goa was explored. It was noted that the study on land-use change at a river watershed scale had not been done before in Goa. To address this gap, the current study was undertaken with the specific objective of improving the understanding of the land-use changes that have occurred over the years in five watersheds in Goa from a hydrological modeling perspective.

10.2 STUDY AREAS

A tourist destination of international repute, Goa is one of the smallest states on the west coast of India. With an area of 3702 km^2, it is located between the

TABLE 10.1
Description of the Watershed Areas/Study Area

River	Discharge Gauging Station	Latitude (N)	Longitude (E)	Watershed Area (km²)
Kalna	Hassapur	15°44'09"	73°54'09"	119.23
Sal	Verna	15°19'24.84"	73°56'5.82"	37.01
Talpona	Ordohond	14°59'44"	74°05'16"	149.32
Galjibag	Borus	14°57'40.8"	74°5'42.12"	54.27
Khandepar	Collem	15°20'21"	74°14'59"	113.96

geographical co-ordinates between 14° 53'54" North to 74° 20'13" East and is 1.02 m above the mean sea level. According to the 2011 census, the resident population of Goa is 1,458,545. Five watersheds in Goa were selected for this study. The details of these watersheds are provided in Table 10.1 and are shown in Figure 10.1.

The villages around the watersheds of the various rivers are as follows:

River Galjibag – the villages Tirwan, Borus, Poingunim, and Loliem fall in Goa whereas Maigini is in Karnataka

River Kalna – Hassapur, Chandel (Goa). Other villages are part of Maharashtra and include Netrade, Phondye, Kalane, Ugade, Bhike Konal, Talkat, Kolzar, Padve Majgaon, Zolambe, Fukeri, Khadpade, Kumbhavade, Isapur, Chaukul, and Adali

River Khandepar – Villages in Goa are Sonaulim, Caranzol, Ghansuli, and Collem. The ones falling in Karnataka are Kuveshi, Atle, Castlerock, and Ghat Kunang.

River Sal – the one village, Verna, falls in Goa

River Talpona – all four villages, namely Pajimol, Gaodongrem, Ardodhond, and Khotigao, are in Goa.

10.3 DATA AND METHODOLOGY

10.3.1 DATA

The gauge station (operated by the Water Resources Department, Government of Goa) was used as an outlet to delineate the watershed area. Using Indian Space Research Organization (ISRO)'s online Bhuvan platform, the watershed areas were delineated on the basis of the Shuttle Radar Topography Mission (SRTM) Digital Elevation Model (DEM) and hydrological parameters. Multi-spectral satellite image data was obtained from National Remote Sensing Centre through the Bhuvan geo-platform and from the United States Geological Survey's global data platform EarthExplorer. All satellite images were obtained for the months of February or March, according to availability, for the years 1977, 1988, 1999, 2009,

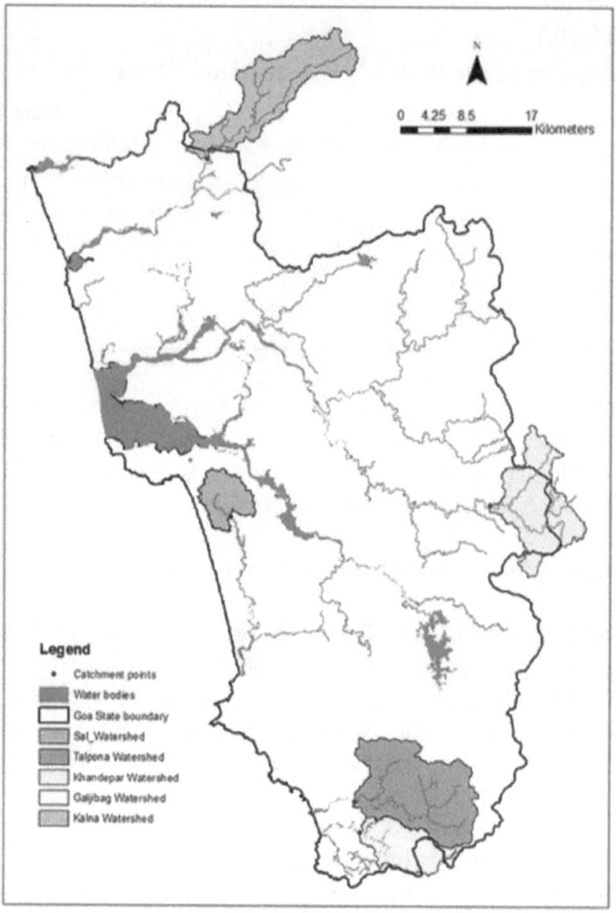

FIGURE 10.1 Study area showing five river watershed areas.

and 2018. For all of the years under study, the satellite image data for February or March were available, and hence those months were chosen. The remote-sensing data (from LISS III, Landsat 1–5, and Landsat 8) was geo-referenced and merged. The resolution for the 1977 image was 60 m, whereas all other images had a resolution of 30 m. For this study, software such as ArcGIS 10.3, TerrSet 18.31, and Microsoft Office 2010 were used.

10.3.2 Methodology

The image classification for change analysis and the various spatial analyses was performed using ERDAS IMAGINE 2014 and ArcGIS 10.3. False Colour Composite (FCC) images of the satellite images were prepared and used in the land-use classification. The satellite data were enhanced before classification to improve the image quality and to achieve greater classification accuracy. A supervised signature extraction with the maximum likelihood procedure was employed to classify the digital data of the satellite images for LULC mapping. In supervised classification, spectral signatures were developed from specified locations in the image. The representative samples for each land-cover class in the digital image, called "training sites", were selected. The image classification software used the training sites to identify the land-cover classes in the entire image. A vector layer was digitized over the raster scene. The vector layer consisted of various polygons overlaying different land-use types. The training sites helped to develop spectral signatures for the outlined areas. Maximum likelihood classification, a statistical decision criterion, was used. It assisted in the classification of overlapping signatures; pixels were assigned to the class of highest probability. A snapshot of the methodology that was followed is shown in the schematic given in Figure 10.2.

Based on the availability of the data, the FCCs for five years (1977, 1988, 1999, 2009, and 2018) were used. Based on the National Remote Sensing Agency (NRSA), India, land-use/land-cover classification system, Level 1 has been used for this study. Five categories of land use that were used for classification were agricultural land, built-up area, barren land, forests (dense vegetation, shrubs, and marshes), and water bodies (rivers, ponds, lakes, and open mine pits filled with water). The built-up category includes settlements, mining areas, and man-made features, such as roads. After classification, the area under different land-use categories was calculated.

The TerrSet Land Change Modeller was used to compare and identify the land-use maps generated using the Maximum Likelihood method. The land-use changes over the past four decades and over the past decade were computed.

10.4 RESULTS AND DISCUSSION

In the present study, the False Colour Composite (FCC) of the satellite image for five years were created prior to application of the classification procedure in order to observe the variation in land-cover categories and to obtain an approximation of the number of land-cover classes for subsequent classification. This was first carried out for the entire state of Goa. As a next step, the watershed areas of the five rivers that are described in the earlier section were clipped from the FCC images that had been prepared for Goa.

The area (km^2) under the land-use categories for the river watersheds studied is given in Table 10.2. The FCC images (1977, 1988, 1999, 2009, and 2018) and the corresponding land-use maps and their FCC satellite images for the five studied watershed areas are shown in Figures 10.3 to 10.7.

The land-use changes of the past four decades and the past decade were computed using the TerrSet Land Change Modeller. The gains and losses for each land-use category for the five watershed areas during the four-decade period (1976 to 2018) are depicted in Figure 10.8. Furthermore, the gains and losses for each land-use category for the watershed areas for the most recent decade (2009–2018) are presented in Figure 10.9.

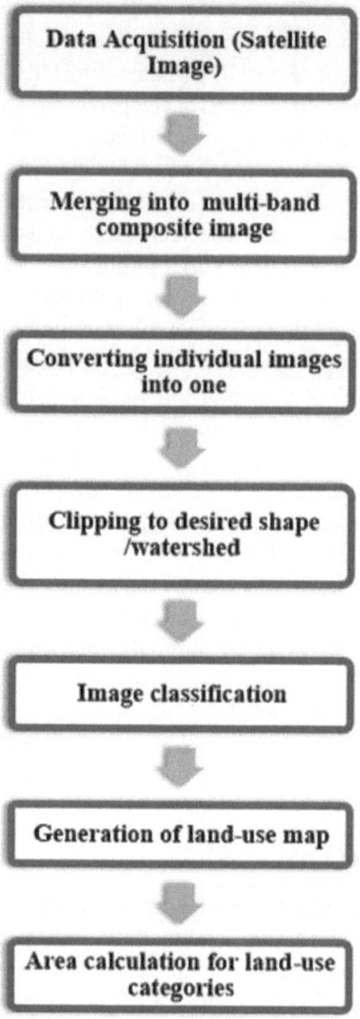

FIGURE 10.2 Stepwise methodology adopted for land-use classification.

A close look at the results from the past four decades indicates that there was a decrease in the forest cover in all the watersheds, except for the River Talpona watershed, and an increase in the built-up land in all the river watersheds. All of the river watersheds showed an increase in agricultural land, except for the River Sal watershed.

The land-use changes of the watershed areas were compared for the past four decades (1977 to 2018) and tabulated in Table 10.3. With a view to understanding the more recent status, percentage change was computed for the last decade (2009 to 2018). This has been summarized in Table 10.4.

TABLE 10.2
Land-Use /Land Cover Area under the Five River Watersheds Studied for the Four-Decade Period

		1977	1989	1998	2009	2018
	Barren lands	4.80	4.64	4.3	3.99	3.99
	Water bodies	0.65	0.57	0.54	0.43	0.43
Talpona	Agricultural lands	23.90	24.20	23.5	23.22	24.34
	Forest cover	114.42	122.70	118.9	119.24	116.33
	Built-up land	5.55	5.51	5.8	6.56	6.68
	Barren lands	2.47	2.10	2.31	2.05	1.95
	Water bodies	0.15	0.15	0.14	0.11	0.12
Galjibag	Agricultural lands	2.95	3.85	5.17	4.28	3.45
	Forest cover	57.43	52.66	51.64	50.86	51.88
	Built-up land	1.40	1.61	0.15	0.13	0.48
	Barren lands	8.44	8.07	7.63	7.50	6.99
	Water bodies	0.07	0.07	0.07	0.06	0.06
Khandepar	Agricultural lands	3.28	3.28	4.70	4.68	5.13
	Forest cover	106.52	104.25	118.76	165.23	208.51
	Built-up land	0.65	0.72	1.05	1.15	1.34
	Barren lands	9.01	7.99	5.54	4.21	5.17
	Water bodies	0.25	0.24	0.24	0.24	0.23
Kalna	Agricultural lands	13.82	14.04	21.01	18.45	14.18
	Forest cover	103.52	93.72	92.13	79.84	76.93
	Built-up land	2.63	3.47	5.49	7.72	12.94
	Barren lands	5.46	6.32	6.62	7.91	9.52
	Water bodies	6.32	7.11	6.71	5.82	5.03
Sal	Agricultural lands	22.04	20.11	18.75	17.91	16.21
	Forest cover	0.4	0.41	0.48	0.49	0.46
	Built-up land	2.78	3.05	4.44	4.87	5.78

A close look at the results indicates that forests dominate the land use in all the watersheds, except for that of the River Sal. The next-largest category of land use is agricultural land in all the watersheds except for the River Khandepar at Collem. The results for the respective watersheds are described below.

10.4.1 River Kalna Watershed

The River Kalna originates in Maharashtra in the village Kalna or Kalane (Karmali Wada). Kalna is approximately 4–5 km from the Water Resource Department, Goa (WRD) river monitoring station at Hassapur, Goa. The origin of the river is on a mountain that is approximately 15–20 km away. River Kalna joins the River Colvale (at Ozorim village in Goa) and finally the River Chapora

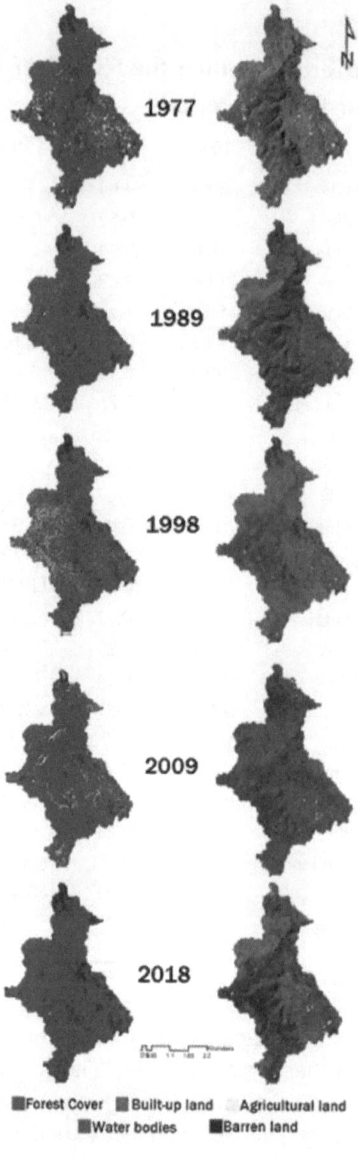

FIGURE 10.3 FCC satellite images and their corresponding land-cover/land-use maps of the River Khandepar watershed for five year points 1977 to 2018.

in Goa. There was a sizeable decrease in the forest area of 11.28 % (Table 10.3 and Figure 10.8) over the four-decade period studied in this watershed. However, the 2.37% decrease in forest cover in the last decade (Table 10.4 and Figure 10.9) is lower than the average decrease over the 40 years studied. The increase in the area of built-up land in the north-eastern region of the watershed and a shift away from agriculture as the major occupation in the region in the recent decade has

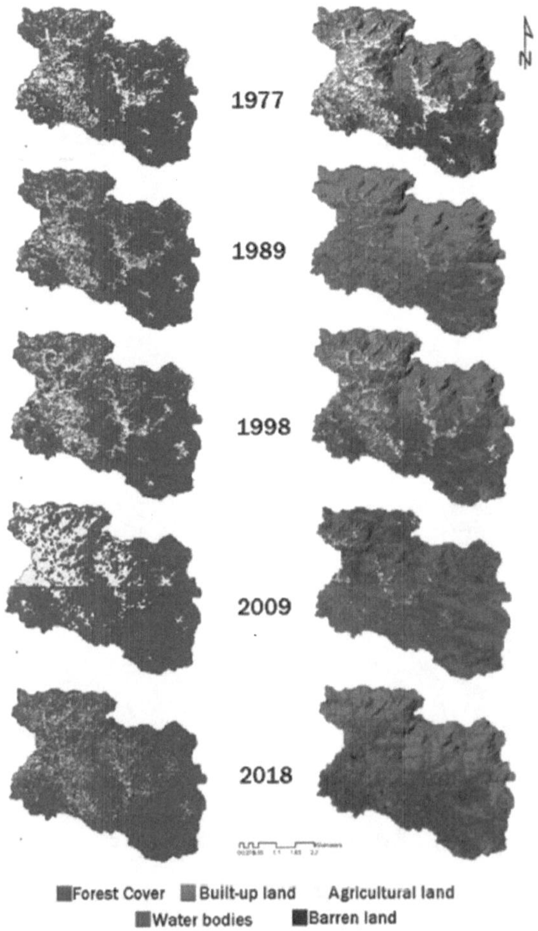

FIGURE 10.4 FCC satellite images and their corresponding land-cover/land-use maps of the River Talpona watershed for five year points 1977 to 2018.

resulted in a marked decrease in the area of cultivated land. This north-eastern region of the watershed falls in the state of Maharashtra. Basalt (a minor mineral) is found in this watershed. Furthermore, the Dodamarg *tehsil* ("township") of Maharashtra (north-eastern region in the watershed) had open-cast activities for the mining of iron/manganese (Vanashakti, 2010). This, along with quarrying for basalt and the general increase in the settlements in the region, has mainly contributed to the increase in built-up land area.

10.4.2 River Khandepar Watershed

The watershed at Collem (River Khandepar) is located in Sanguem *taluka*, or township, close to the Western Ghats and thereby is predominantly forest areas.

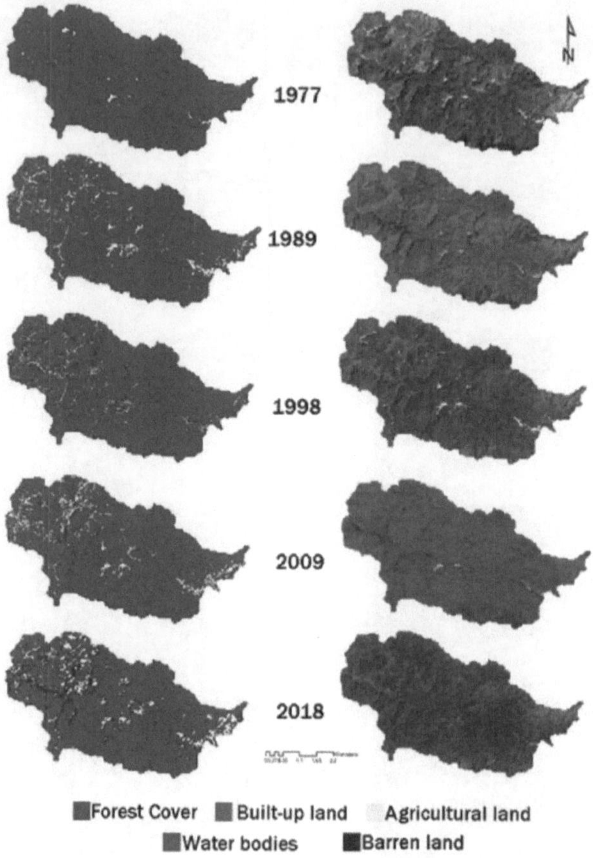

FIGURE 10.5 FCC satellite images and their corresponding land-cover/land-use maps of the River Galjibag watershed for five year points 1977 to 2018.

The origin of the river is the Dudhsagar Falls, which is also referred to as the River Dudhsagar. The watershed lies in the states of Goa and Karnataka. In the past four decades, there has been a 5.80% increase in the forest area (Table 10.3 and Figure 10.8); however, the past decade (2009–2018) has witnessed a decrease in the forest area of 3.17% (Table 10.4). More recent years (Table 10.4) indicated an increase in the agricultural area (3.58%) and the built-up area (0.26%). Similar trends were also reported by Yedage et al. (2015).

10.4.3 River Talpona Watershed

The River Talpona watershed in South Goa is the largest watershed area in this study. It has witnessed an increase in the area of agricultural land and the forest cover over the past four decades (Table 10.3). However, the last decade (Table 10.4) saw a reduction in the area under forest cover (2.95% decrease). The barren lands

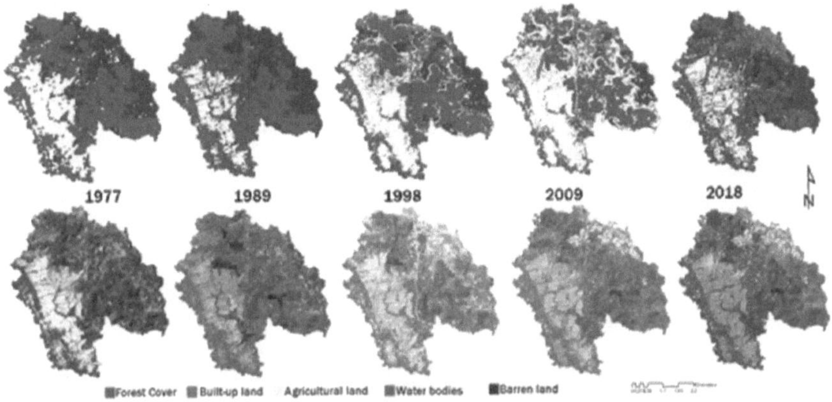

FIGURE 10.6 FCC satellite images and their corresponding land-cover/land-use maps of the River Sal river watersheds for five year points 1977 to 2018.

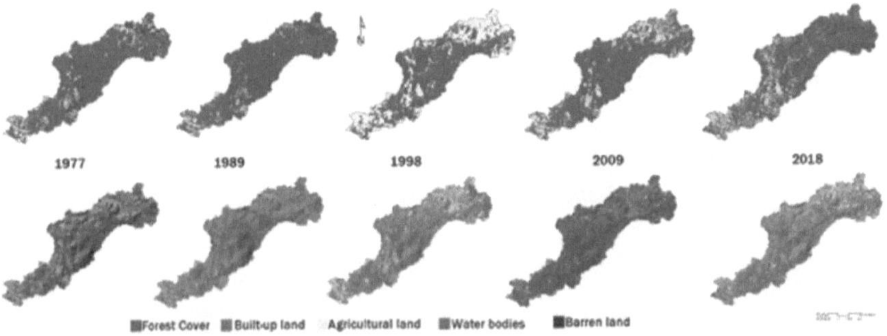

FIGURE 10.7 FCC satellite images and their corresponding land-cover/land-use maps of River Kalna watersheds for five year points 1977 to 2018.

were reclaimed by the forests in the watershed area, although some of the forest areas near the settlements were converted into agricultural lands (2.34% increase) or built-up land (0.61% increase).

10.4.4 River Galjibag Watershed

The River Galjibag watershed at Borus in the Canacona *taluka* is also dominated by forest cover, with over 90% of the area under forests. Large parts of the river-banks have been cultivated. This trend continued to increase over the past four decades. The decrease in the forest cover of 9.10% (Table 10.3) in this region is due largely due to its conversion to agricultural lands and associated rural or urban settlements. Many parts of the forest show a decline in tree cover in areas where barren lands are present. In recent years (2009 to 2018) (Table 10.4), an increase in built-up land (12.77%) has been seen.

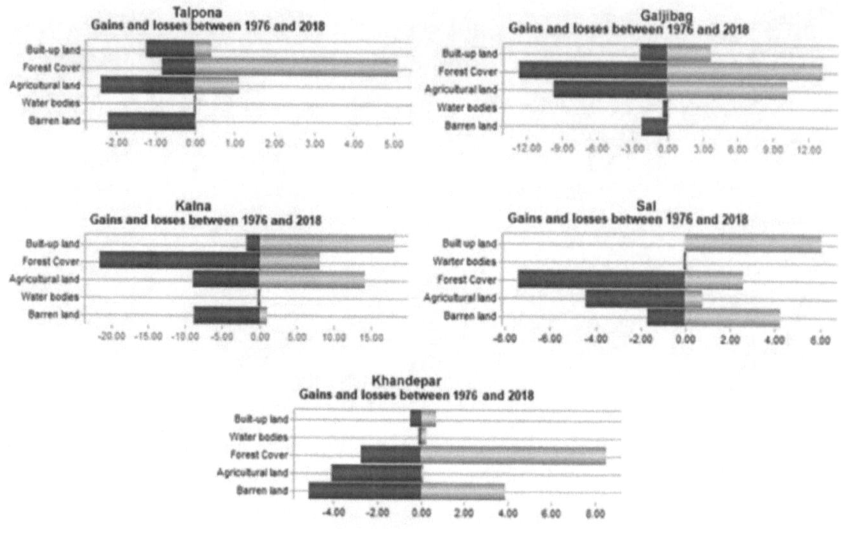

FIGURE 10.8 The gains and losses for each land-use category for the five watershed areas during the four-decade period studied.

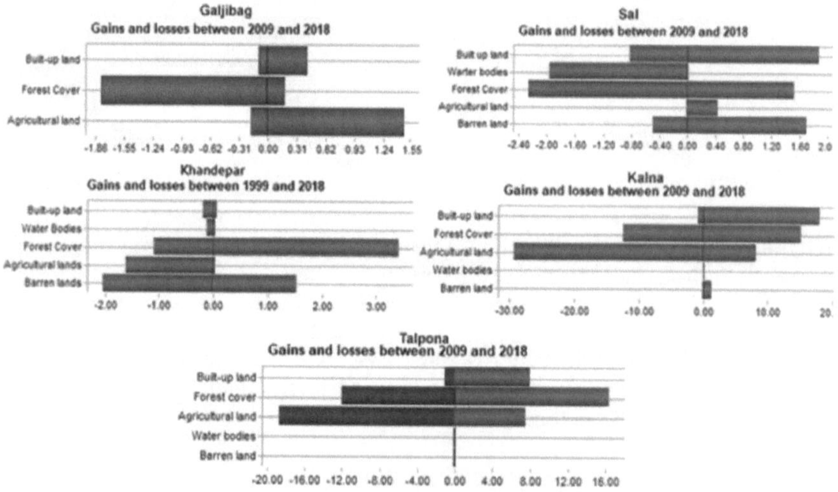

FIGURE 10.9 The gains and losses for each land-use category for the watershed areas of the last decade period.

10.4.5 River Sal Watershed

The River Sal watershed, at Verna, is the only river located near major settlements. However, 50% or more of the watershed area is agricultural land. As a result of the increase in urbanization, much farmland was left uncultivated or was converted into settlements. This can be seen from the results shown in Tables 10.3

TABLE 10.3
Percentage Change in Land-Use/Land-Cover in Five River Watersheds for the Past Four Decades

	Change From 1977 to 2018				
	Kalna	**Khandepar**	**Talpona**	**Galjibag**	**Sal**
Barren lands	−6.60	−1.22	−4.07	−5.25	1.05
Water bodies	−0.21	−0.24	−46.80	−0.17	−0.07
Agricultural lands	4.50	−4.00	45.02	3.59	−1.42
Forest cover	−11.28	5.80	5.20	−9.10	−6.27
Built-up land	13.60	−0.32	−1.74	10.93	6.72

TABLE 10.4
Percentage Change in Land-Use/Land-Cover in Five River Watersheds over the Past Decade

	% Change from 2009 to 2018 (Last Decade)				
	Kalna	**Khandepar**	**Talpona**	**Galjibag**	**Sal**
Barren lands	0.86	−0.15	0.00	−0.24	8.59
Water bodies	−0.01	0.00	0.00	0.09	0.00
Agricultural lands	−17.55	3.58	2.34	−20.43	- 6.65
Forest cover	−2.37	−3.17	−2.95	7.83	−5.75
Built-up land	14.33	0.26	0.61	12.77	3.88

and 10.4. There was a decrease in the area of agricultural land and an increase in the area of barren and built-up land over the past four decades, particularly so in the last decade. The reason for this land-use change can be attributed to clearance of farms/plantations/orchards for the construction of settlement zones. The concentration of the population is high along the National Highway 66 (previously known as NH-17), especially in and around Margao City. Rapid urbanization has been observed in this watershed.

10.5 CONCLUSION

In the most recent decade (2009–2018), there has been a decrease in the forest cover in all the watersheds, except for the River Galjibag. Built-up land area has increased in all five watersheds. This decrease was greatest for the River Sal watershed (−5.75%). Despite the increase in forest cover area in the River Galjibag watershed (7.83%) over the last decade, in the four decades studied, the watershed area has seen a 9.10% decrease in the overall forest cover. The increase in the forest cover is due largely to the various afforestation activities taken on by the state government and by the local population in the area. The built-up areas in the watersheds of the Galjibag and Talpona rivers show the greatest increases in forest cover, each greater than 10% in the past decade. The built-up areas indicate that the River Sal watershed has the greatest urbanized area and the River Kalna watershed has the least.

It was observed that, sometimes, the narrow rivers/water bodies become covered with natural vegetation. This creates an obstruction, and the water bodies are not detectable by satellite images. In many places, agriculture is practised near the river only in the post-monsoon season. During the monsoon season, these lands are covered with water as the river flow increases. The satellite images taken in February/March reflect the agricultural activity. This may distort the results to a small degree. However, these results can definitely provide useful insights for generating the future land-use scenarios, to carry out hydrological modeling exercises in these five watersheds and to predict future water availability. These results would be very useful for achieving better management of water resources in these areas.

ACKNOWLEDGEMENT

The authors gratefully acknowledge the Water Resources Department (WRD), Government of Goa for providing the relevant data and financial support to carry out this study. We are most grateful for the timely suggestions and overall guidance extended by Professor L. Nandagiri, National Institute of Technology Karnataka, Surathkal.

REFERENCES

Brown, A.E. 2013. Impact of forest cover changes on annual streamflow and flow duration curves. *Journal of Hydrology* 483:39–50.

Dale, Virginia, H. 1997. The relationship between land-use change and climate change, Ecological Applications, 7(3):753–769 https://doi.org/10.1890/1051-0761(1997)007[0753:TRBLUC]2.0.CO;2

Di, Gregorio, and Jansen, L.J.M. 1998. A new concept for a land cover classification system. *Land* 2(1):55–65.

Guzha, A.C., Rufino, M.C., Okoth, S., Jacobs, S., and Nobrega, R.L.B. 2018. Impacts of land use and land cover change on surface runoff, discharge and low flows: Evidence from East Africa. *Journal of Hydrology: Regional Studies* 15:49–67.

Hungtinton, T.G. 2006. Evidence for intensification of the global water cycle: Review and synthesis. *Journal of Hydrology* 319:83–95. doi:10.1016/j.jhydrol.2005.07.003

Kondolf, G.M., Piegay, H., and Landon, N. 2002. Channel response to increased and decreased bedload supply from land use change: Contrasts between two watersheds. *Geomorphology* 45:35–51.

Kotha, M., and Kunte, P.D. 2013. Land-cover change in Goa-An integrated RS - GIS approach. *International Journal of Geoinformatics* 9:2.

Lambin, E., Geist, H., and Lepers, E. 2003. Dynamics of land-use and land-cover change in tropical regions. *Annual Review of Environment and Resources* 28:205–241, http://dx.doi.org/10.1146/annurev.energy.28.050302.105459

Mascarenhas, A., Sawkar, K., and Chauhan, O.S. 1997. *The Coastal Zone of Goa: Then and Now.* Paper presented at Seminar on Coastal Zone Environment and Management Mangalore University, India.

Misra, Ankita, Mani Murali, R., and Vethamony, P. 2014. Assessment of the land use/land cover (LU/LC) and mangrove changes along the Mandovi-Zuari estuarine complex of Goa, India. *Arabian Journal of Geosciences* https://www.researchgate.net/publication/263580411_Assessment_of_the_land_useland_cover_LULC_and_mangrove_changes_along_the_MandoviZuari_estuarine_complex_of_Goa_India

Murali, Mani, Misra, Ankita, and Vethamony, P. 2014. Assessment of the land use/land cover (LU/LC) and mangrove changes along the Mandovi–Zuari estuarine complex of Goa, India. *Arabian Journal of Geosciences* 8(1):267–279.

National Natural Resources Management System (NNRMS), ISRO. 2014. Regional Land-use Land Cover map using IRS LISS-III data, https://bhuvan-appl.nrsc.gov.in/2dresources/thematic/LULC502/MAP/GA.jpg.

Nugroho, Prasetyo, Djoko, Marsono, Putu, Sudira, Hatma, and Suryatmojo. 2013. Impact of land-use changes on water balance. *Procedia Environmental Sciences* 17:256–262.

de Paulo Rodrigues da Silva, Vicente, Silva, Madson Tavares, Singh, Vijay P., Pereira de Souza, Enio, Campos Braga, Celia, de Holanda, RomildoMorant, Almeida, Rafaela Silveria R., de AssisSalviano de Souza, Francisco, and Braga, Armando Cesar Rodrigues. 2018. Simulation of streamflow and hydrological response to land cover changes in a tropical river basin. *Catena* 162: 166–176.

Roy, A., and Srivastava, V.K. 2011. Geospatial approach to identification of potential hotspots of land-use and land-cover change for biodiversity conservation. *Current Science* 102:8.

Roy, P.S., and Roy, Arijit. 2010. Land use and land cover change in India: A remote sensing & GIS perspective. *Journal of the Indian Institute of Science* 90:4.

Sampath, Kumar P., Mahtab, Anjum, Roy, Arijit, Srivastava, V., and Roy, Parth. 2014. *Impact of Drivers on the Land Use/Land Cover Change in Goa, India*, Paper presented at International Symposium, India Geospatial Forum, International Convention Centre, Hyderabad, India.

Singh, S,K., and Kushwaha, S.P.S. 2004. Assessment and monitoring of estuarine mangrove forests of Goa using satellite remote sensing, Paper presented at International Symposium, India Geospatial Forum, *International Journal of the Indian Society of Remote Sensing*, https://www.researchgate.net/publication/226323595

TERI. 2000. *Developing Scenarios for Tourism in Goa: A Multi-Stakeholder Workshop.* http://www.teriin.org/teri-wr/coastin/scenarios

Turner, B.L. 2006. Land change as a forcing function in global environmental change. In: Geist, H.J. (Eds.), *Our Earth's Changing Land: An Encyclopaedia of Land-Use and Land–Cover Change*, vol. 1 (A-K). Greenwood Press, London, pp. 25–32.

Vanashakti. 2010. Dossier of the proposed iron ore mining in the Sawantwadi-Dodamarg stretch of the Sindhudurg District, Maharashtra, India. Mining projects in the northern Western Ghats (*Sahyadris*). http://www.indiaenvironmentportal.org.in/files/file/Sindhudurg.pdf

Vicente-Settano, S.M., Peña-Gallardo, M., Hannaford, J., Murphy, C., Lorenzo-Lacruz, J.,and Dominguez-Castro, F., 2019. Climate, irrigation, and land cover change explain streamflow trends in countries bordering the Northeast Atlantic. *Geophysical Research Letters*, 46, 10,821–10,833. https://doi.org/10.1029/2019GL084084

Vorosmarty, C.J., Green, P., Salisbury, J., and Lammers, R.B. 2000. Global water resources: Vulnerability from climate change and population growth. *Science* 289: 284–288.

Watson, R.T., Noble, I.R., Bolin, B., Ravindranath, N.H., Verardo, D.J., and Dokken, D.J. 2000. Land Use, Land-Use Change, and Forestry: A Special Report of Intergovernmental Panel on Climate Change. https://www.cabdirect.org/cabdirect/abstract/20083294691

Wei, X.H., Liu, W.F., and Zhou, P.C. 2013. Quantifying the relative contributions of forest change and climatic variability to hydrology in large watershed: A critical review of research methods. *Water* 5:728–746, doi:10.3390/w5020728

Welde, Kidane, and Bogale, Gebremariam. 2017. Effect of land use land cover dynamics on hydrological response of watershed: Case study of Tekeze Dam watershed, northern Ethiopia. *International Soil and Water Conservation Research* 5:1–16.

Yedage, Anil, Nandkumar, Sawant, and Vishal, Malave. 2015. Change detection analysis using geo-spatial technique: A case study of South Goa. *International Journal of Scientific & Engineering Research* 6(6). ISSN 2229-5518

Yedage, A., and Harmalkar, S. 2017. Analysis and simulation of landuse / land cover change in the Quepem Tehsil Goa, India. *International Journal of Advances in Remote Sensing and GIS* 5:1.

Young, A. 1994. *Towards an International Classification of Land Use.* UNEP/FAO Consultancy Report. FAO Remote Sensing Centre, Rome, 43p.

11 Impacts of Land-Use and Land-Cover Change on Land Degradation

A Case Study in Dolakha District, Nepal

Pawan Thapa

CONTENTS

11.1 INTRODUCTION

Land-use and land-cover change are significant in Nepal's rural areas, as they can damages the land and the environment. It is important to find out how these changes arise and to map these changes for policymaking, monitoring, and managing resources (Aspinalls and Hill, 2008; Foody and Atkinson, 2002). Land-cover patterns are affected by both human activities and biophysical processes (Li & Shao, 2014). Therefore, proper study and mapping are essential (Thapa, 2020a) to identify land degradation. Land degradation has been one of the environmental issues that most affects humans and nature (Uddin et al., 2018).

Land degradation is mainly caused by overgrazing, climate, vegetation cover, and soil (Wang et al., 2006). It is a significant issue for a developing nation like Nepal, which constitutes a vulnerable and threatened ecosystem, where the cultivation of marginal lands and livestock overgrazing is directly accelerating land degradation problems (Paudel et al., 2009). Nepal has unique land-use patterns, reflecting a very steep topography and a monsoon climate, with a mostly rural population living on steep terrain (Jaquet et al., 2012).

Few studies on the land use/land cover (LULC) changes in Nepal have been conducted. The aim of the current study is to visualize these changes in the Dolakha District. The study gathered information on the impact of these change on land degradation in the years 2001, 2010, and 2018 through Landsat images, Open Data Nepal, and International Centre for Integrated Mountain Development (Nepal) (ICIMOD, 2010) data, which would form the framework for further study, as well as being useful for different stakeholders for decision making and for formulating plans and policies for sustainable development. The findings could provide useful information about land degradation in Nepal, which would be valuable for implementing appropriate land policy and activities.

11.2 STUDY AREA, DATA, AND METHODOLOGIES

The Dolakha District is located in the Janakpur zone, north-east of Kathmandu, and extends from an altitude of 2088 m (OpenNepal, 2015). In recent years, large-scale human activities, such as deforestation, farming on steep slopes, and rapid population growth, resulted in land degradation in this district (Shengji and Sharma, 1998). Other factors impacting land are changes in land use and cover (Pielke et al., 2002, 2011), such as the increasing population and changes in the LULC effects. Landsat satellite images were utilized for the current LULC study (Ding and Shi, 2013; Xiao and Weng, 2007). Landsat 8 has two sensors, with eleven bands of 30 m spatial resolution for the satellite image (Figure 11.1).

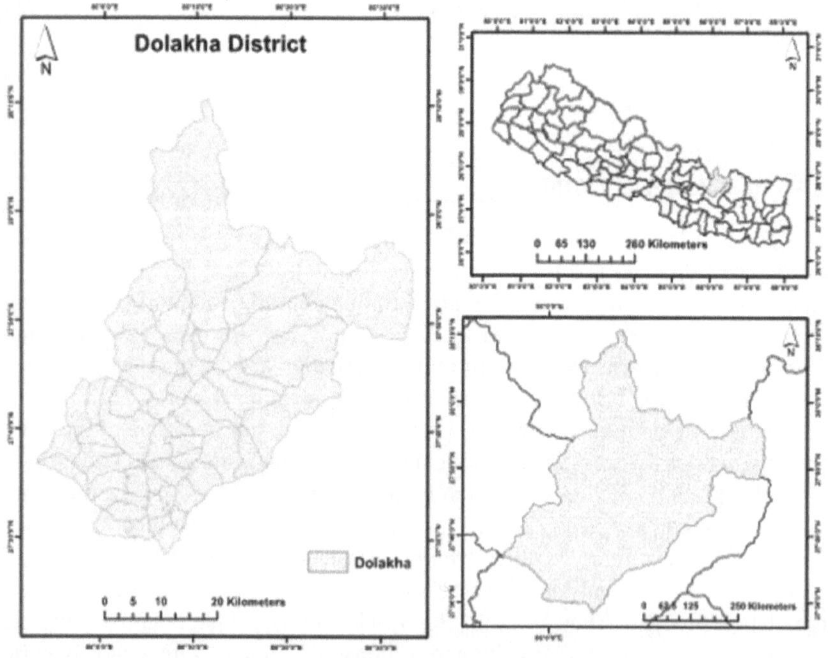

FIGURE 11.1 Map of the study area, the city of Dolakha.

11.2.1 Methodologies Used

During image acquisition, the no-cloud option was selected, to avoid the need for atmospheric correction. Both multispectral images (from the two sensors) were acquired on one day, 1 December 2018. Using ArcGIS 10.5 software, the Maximum Likelihood supervised classification was extracted for LULC.

11.2.1.1 Image Classification

Over the past 17 years, LULC change in Dolakha was observed by image classification and segmentation, with a total of 61,154 point samples being randomly selected from 1 December 2018. The ground truth samples were extracted from Google Earth. For this study, six types of land use were used: forest, river, pond and lake, barren land, cultivated land, and snow and glacier, shown in Table 11.1.

11.2.1.2 Accuracy Assessment

Accuracy is checked on image data by creating confusion matrices, using the test samples, as shown in Table 11.2. Overall accuracy was calculated by the Kappa

TABLE 11.1
Land-Use/Land-Cover Types for Classification

Land-use/Land-cover Type	Description
Forest	Farmland trees, mountain forest, roadside trees, trees around water bodies
River	River, stream, flowline
Barren land	Uncultivated land, vacant land
Cultivated land	Cultivated land
Pond and lake area	Lake, pond, reservoir, wetland
Snow and glaciers	Sites with snow and glaciers

TABLE 11.2
LULC Classification and Confusion Matrices, 2018

	Reference Data					
	Barren Land	Forest	River	Cultivated Land	Pond and Lake Area	Snow and Glacier
Barren land	28	0	3	0	3	6
Forest	0	34	0	0	0	0
River	2	2	12	1	1	0
Cultivated land	0	2	1	51	0	0
Pond and lake area	0	1	1	3	6	3
Snow and glacier	3	2	1	2	2	7
Producer's accuracy	84.85%	82.93%	66.67%	89.47%	50.00%	43.75%

coefficient and the confusion matrices (Lillesand et al., 2004). The two equations defined below are used for calculation of the Kappa coefficient and overall accuracy calculation:

$$\text{Overall Accuracy} \sum_{i=1}^{r} \frac{\text{Xii}}{x} \tag{11.1}$$

where X_{ii} is the diagonal elements in the error matrix, x is the total number of samples in the error matrix.

$$\text{Kappa coefficient } (\hat{K}) \frac{\text{n} \sum_{i=1}^{r} Xii - \sum_{i=1}^{r} Xi + X + i}{\text{n2} - \sum_{i=1}^{r} Xi + X + i} \tag{11.2}$$

where r is the number of rows in the matrix, X_{ii} is the number of observations in row i and column i, X_{i+} and X_{+i} are marginal totals for row i and column i, respectively, and n is the number of observations (pixels).

Table 11.2 shows that the overall accuracy for 2018 was 77.97%, and the Kappa coefficient for the corresponding maps was 0.7795.

11.3 RESULTS/DISCUSSIONS

The pie chart (Figure 11.2) shows the results of LULC for 2001, 2010, and 2018. In 2001, forest area, cultivated land, and barren land were the predominant land-use types, with percentages of 37%, 26%, and 19%, respectively. In contrast, in 2010, forest, cultivated land, and barren land were the dominant land-use types, at 30%, 24%, and 27%, respectively. In 2018, the proportions were 30%, 27%, and 24% for forest, barren land, and cultivated land, respectively.

Comparison of 2001, 2010, and 2018 land use demonstrates the significant decrease in cultivated land within the study area, with increases in the areas of barren land and forest. The barren land area significantly increased from 8% in 2001 to 27% in 2018. Over the past 17 years, a significant decrease in the cultivated land area has taken place, resulting in reduced production of crops in the study area. This land-use change leads to poverty, and the

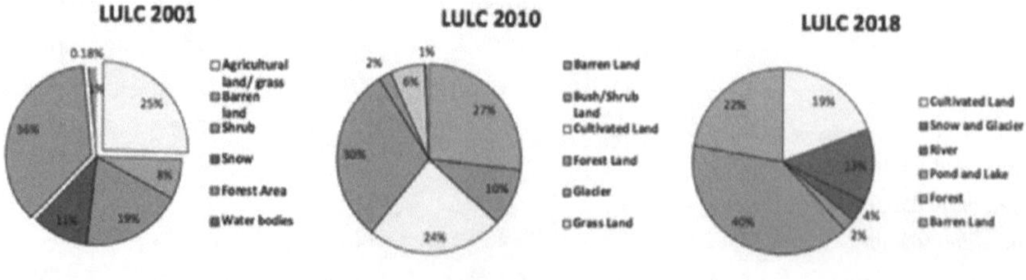

FIGURE 11.2 LULC pie chart of the study area.

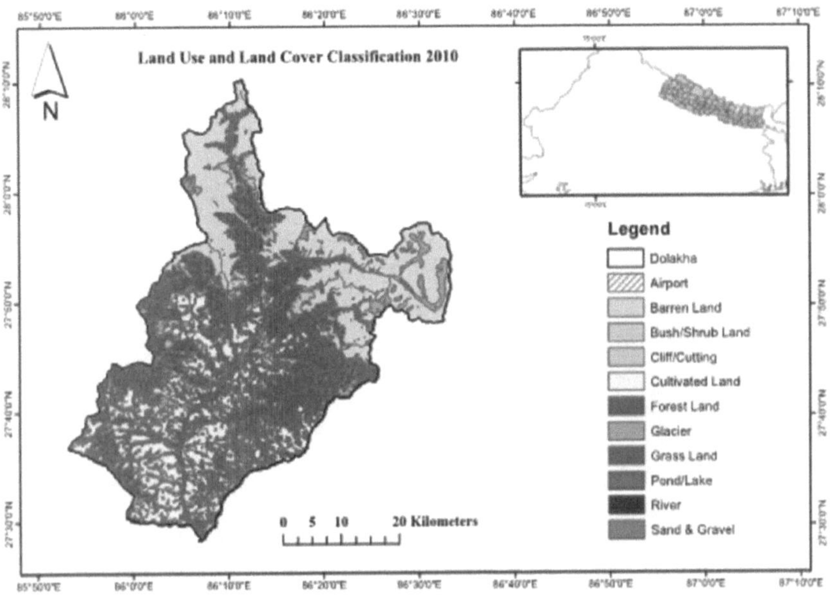

FIGURE 11.3 Study area map showing LULC (2010).

conversion of cultivated land to barren land. Consequently, people of this area started to clear forests and cultivate crops on steeper slopes, without consideration of soil conservation practices, leading to soil degradation (Acharya, 2012; Taib, 2015).

The LULC maps for 2010 and 2018 are shown in Figures 11.3 and 11.4, respectively. The area of cultivated land decreased significantly, whereas the area of barren land increased very markedly in 2010 and 2018, relative to 2001. These land-use changes are associated with increases in size of urban areas and increased population growth, as well as altered climatic conditions, with drying-up of water resources, lack of human resources, and low food production. Land degradation in Asia is due to a growing population, increasing pressure on the land, decline in soil and water resources, changing climate, and deterioration in the living conditions for many people (Sivakumar and Ndiang'Ui, 2007). Increasing land degradation, from inappropriate land use and management practices, leads to increased poverty and hunger (Taib, 2015).

11.4 CONCLUSIONS

This study of LULC changes in 2001, 2010, and 2018 shows that there was a significant effect on land degradation in the study area. Cultivated land is decreasing in such a way that local people face a shortage of food for the entire year. This type of research indicates that remotely sensed images can

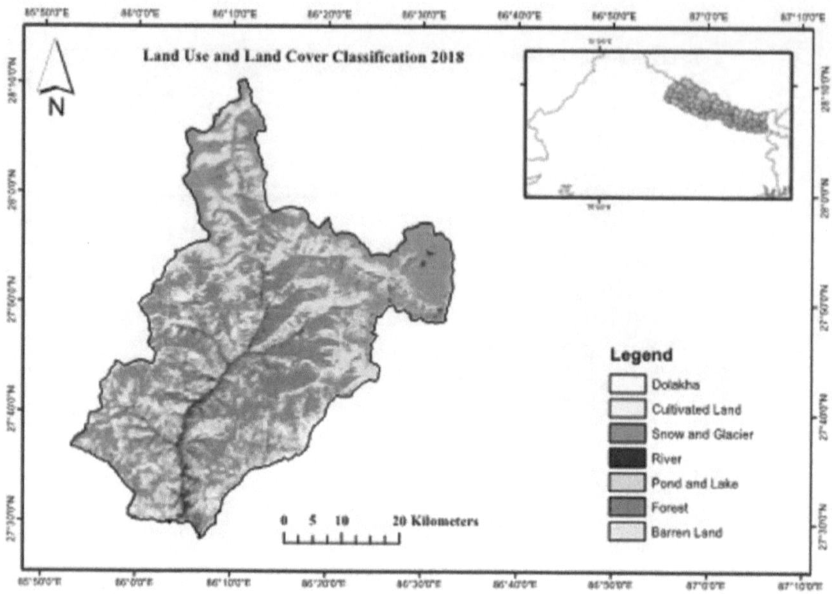

FIGURE 11.4 Study area map showing LULC (2018).

reveal changes in LULC and its impact on land, which are valuable for decision makers to develop and implement plans for sustainable development for this study area. However, the remote-sensing method has limitations, such as cloud cover and limited revisit duration of a satellite over a particular location. There should be research into the technique of integrating data from multi-sensor satellites with climatic data from ground-based stations, which may increase information accuracy.

REFERENCES

Aspinalls, R. J. and Hill, M. 2008. Land use change, science policy and magamegent. New York CRC Press, Taylor and Francis Group.

Ding, H. & Shi, W. 2013. Land-use/land-cover change and its influence on surface temperature: A case study in Beijing City. *International Journal of Remote Sensing* 34(15): 5503–5517.

Foody, G. M. and Atkinson, P. M. 2002. Uncertainty in remote sensing and GIS, JohnWiley and Sons, London.

(ICIMOD, 2010) International Centre for Integrated Mountain Development (ICIMOD). Available online: http://rds.icimod.org/ (accessed on 30 March 2020).

International Centre for Integrated Mountain Development (ICIMOD). Available online: http://rds.icimod.org/ (accessed on 30 March 2020).

Jaboyedoff, M. 2013. Stephanie Jaquet, Karen Sudmeier-Rieux, Marc-Henri Derron and Michel Jaboyedoff. *The Role of Ecosystems in Disaster Risk Reduction*, 343.

Karki, S., Thandar, A. M., Uddin, K., Tun, S., Aye, W. M., Aryal, K., … & Chettri, N. (2018). Impact of land use land cover change on ecosystem services: a comparative analysis on observed data and people's perception in Inle Lake, Myanmar. *Environmental Systems Research*, 7(1), 25.

Li, X. and Shao, G. 2014. Object-Based Land-Cover Mapping with High-Resolution Aerial Photography at a County Scale in Midwestern USA. *Remote Sensing* 6 (11) : 11372–11390. Doi : 10. 3390/rs61111372.

Lillesand, T.M., Kiefer, R.W. & Chipman, J.W. 2004. *Remote Sensing and Image Interpretation*, Fifth Edition. WILEY, New Jersey, pp. 556–557, 586–591.

Paudel, P. P., Devkota, B. D. & Kubota, T. (2009). Land degradation in Nepal: A review on its status and consequences, *Journal of the Faculty of Agriculture, Kyushu University*, 54(2), 477–479.

Proposed Loan for Additional Financing Nepal. 2015. *Kathmandu Valley Water Supply Improvement Project.* Asian Development Bank, Manila.

Sivakumar, M.V. & Ndiang'Ui, N. 2007. *Climate Change and Land Degradation*. Springer Nature.

Wang, X., Wang, T., Dong, Z., Liu, X., & Qian, G. (2007). Nebkha development and its significance to wind erosion and land degradation in semi-arid northern China. *Journal of Arid Environments*, 65(1), 129–141

Xiao, H. & Weng, Q. 2007. The impact of land use and land cover changes on land surface temperature in a karst area of China. *Journal of Environmental Management* 85(1): 245–257.

12 Remote Sensing-Based Approach to Identify the Influence of Land Use/Land Cover Change on the Urban Thermal Environment

A Case Study in Chattogram City, Bangladesh

Abdulla-Al Kafy, Muhaiminul Islam, Md. Soumik Sikdar, Tahera Jahan Ashrafi, Abdullah-Al-Faisal, Md. Arshadul Islam, Abdullah Al Rakib, Md. Hasib Hasan Khan, Md. Hasnan Sakin Sarker, and Md. Yeamin Ali

CONTENTS

12.1 INTRODUCTION

Human actions are responsible for significant changes in the character of urban ecology, as well as for decreases in the vegetated area, which point towards indicative change in the environment, particularly decreased diversity in the terrestrial ecosystem, on both local and global scales (Meyer and Turner 1992, Connors, Galletti, and Chow 2013, Moore, Gould, and Keary 2003). These environmental changes mainly occur as a consequence of the rapid expansion of urban areas, which is one of the most noteworthy human impacts (Lin et al. 2013). The growing expansion of cities and towns worldwide has now reached the point where over 54% of the global population resides in urban areas, and the UN expects that this proportion will reach up to 68% by 2050 (Alfraihat, Mulugeta, and Gala 2016, Alqurashi, Kumar, and Sinha 2016, UN 2015, Collins et al. 2013, IPCC 2014, Karim and Mimura 2008, Rimi et al. 2019). Urban areas primarily alter, transform, and replace natural vegetation and agricultural land into less- or non-evaporating and non-transpiring surfaces (Kafy, Faisal, et al. 2020, Kafy, Rahman, et al. 2020, Rahman et al. 2018, Ahmed 2018, Tomlinson et al. 2011, Xiang 2017, Zakšek and Oštir 2012). One of the significant environmental changes due to urban expansion is the increase in surface temperature (Saleh 2011). Changing the form of land use/land cover (LULC) also affects heat absorption, evaporation, and transmission (Aboelnour and Engel 2018, Swetnam et al. 2011, Wang et al. 2017). The major causes for this urban expansion are population growth, economic development, improper urban management, and inappropriate planning (Aboelnour and Engel 2018, Kafy, Mansour, Al-Belushi, and Al-Awadhi 2020, Rahman, Aldosary, and Mortoja 2017, Ullah et al. 2019). Such changes in LULC profoundly affect the urban thermal environment, which is associated with land surface temperature (LST), urban heat islands (UHI), and urban field variance index (UTFVI).

The LULC change due to rapid urbanization and its effects on LST is a matter of great concern. The increase in LST in the urban context results from the UHI effects (Ahmed 2018, Fernando 2018, Silva, da Silva, and Santos 2018, Zhang et al. 2006, Zhou et al. 2013). Climate change, induced by urban growth, is an effect of UHI, and UHI highlights the variation in the ambient temperature of the metropolitan area compared to its neighbouring rural areas (Ahmed 2018, Fernando 2018, Lai and Cheng 2010, Santamouris 2014). Many researchers nowadays use thermal remote sensing (RS) and geographic information system (GIS)-based applications for describing the relationship between LULC and LST (Bokaie et al. 2016, Fernando 2018, Ullah et al. 2019, Weng, Lu, and Schubring 2004, Kafy, Faisal, et al. 2020, Kafy, Rahman, et al. 2020). Multi-spectral and multi-temporal remote sensing-based data are most suitable for displaying the LULC impact on LST (Ahmed 2018, Kafy, Rahman, et al. 2020, Kim 2008, Mou 2019, Santamouris 2014, Sarkar, Islam, and Akter 2016, Silva, da Silva, and Santos 2018, Yang et al. 2020).

The use of RS and GIS to measure land cover changes and LST has flourished significantly in recent decades (Balogun and Ishola 2017, Lilly Rose and Devadas 2009, Ahmed 2011, Kafy et al. 2019, Kafy, Rahman, et al. 2020, Rahman et al. 2018, El-Hattab, Amany, and Lamia 2018, Rahman, Aldosary, and Mortoja 2017c, Rahman, Aldosary, and Mortoja 2017a, Silva, da Silva, and Santos 2018). GIS and RS applications today receive a great deal of attention, as they can estimate changes in the landscape, ecosystem biodiversity, and heat stress of the urban environment (Al-Hathloul and Rahman 2003, Li and Zhao 2003, Streutker 2003, Celik et al. 2019, Swetnam et al. 2011, Trolle et al. 2019). Gathering information by direct field visits to understand LULC and LST transition scenarios often takes more time, requires more labour, and involves greater chances of error (Hart and Sailor 2009, Lilly Rose and Devadas 2009, Kafy, Rahman, et al. 2020). Several studies have described the LULC influence on UHI in various parts of the word (Hart and Sailor 2009, Bokaie et al. 2016, Silva, da Silva, and Santos 2018, Gaur, Eichenbaum, and Simonovic 2018, Sejati, Buchori, and Rudiarto 2019, Huang et al. 2019, Ahmed 2018). The earliest efforts began in 1972, when Rao first tested the pattern of surface temperature distribution and UHI in American coastal towns on the Atlantic Ocean, using remote-sensing technology (Krishna 1972). Huang et al. (2019) used the land contribution index method to describe the periodic thermal contributions in UHI for China for 2005–2015 for each type of LULC (Huang et al. 2019).

Chattogram is the second largest metropolitan area in Bangladesh, as well as being its industrial and business capital. Studies have been conducted to demonstrate the LULC pattern and its impact on the LST for Chattogram (Hassan and Nazem 2016, Roy et al. 2020). Chattogram (formerly, Chittagong) is one of the fastest-expanding metropolitan cities and is the second most-populated city in Bangladesh. To date, no study has focused on the LULC change and its impact on the urban thermal environment of Chattogram, using LST, UHI, and UTFVI. Therefore, this present study was undertaken to assess the LULC change and its impact on the thermal urban environment of Chattogram City, using the Landsat data for the 1999–2019 time period.

12.2 STUDY AREA PROFILE

Chattogram is the gateway to Bangladesh, located between 21°54' and 22°59' N latitude and between 91°17' and 92°13' E longitude (Figure 12.1). It lies on the banks of the Karnaphuli river, between the Chittagong Hill Tracts and the Bay of Bengal (BBS 2013). As part of many hill branches from the Himalayas, the topography shows hilly characteristics, and the hills become lower the closer to Chattogram City they are located, breaking up into tiny hills sprawled throughout the city (ICISET 2018). India's Tripura State bounds the district on the North, with Khagrachhari, Rangamati, and Bandarban districts to the East, Cox's Bazar to the South, and the Bay of Bengal, Feni, and Noakhali to the West.

According to the 2011 Census, of the total population of the division, 7.3 million people live in the urban area, with the area of Chattogram City Corporation being 187.4 km^2 (Roy et al. 2020). During the 1991–2001 time period, the city showed a population growth rate of 4.527% annually, with the pace decelerating between the years 2001–2011, where the percentage was only 2.81% (Islam and Chowdhury 2014). As the largest seaport of Bangladesh is situated in Chattogram, this city is leading the way with respect to almost all kinds of import–export activities. Also, it is home to the nation's many large and renowned companies. The overall

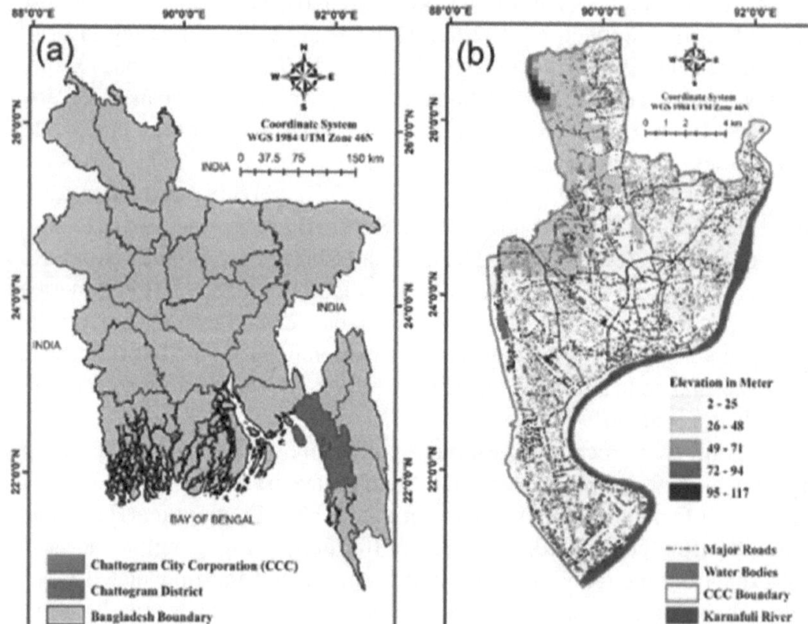

FIGURE 12.1 Location map of the study area of a) Chattogram City in Bangladesh, and b) Chattogram City Corporation.

climate of Chattogram city significantly influences the microclimate of the study area. A tropical monsoon climate characterizes the weather in this region. The mean annual rainfall is nearly 3,810 mm, with an annual average temperature of 30°C in the south and west zone of the city (Roy et al. 2020). Between November and March, the weather is dry and wintry, whereas the pre-monsoon season, April-May, is quite hot, with the rainy season, from June to October, being sunny with clouds, and moist.

12.3 MATERIALS AND METHODS

This study used the multi-temporal Landsat 4–5 Thematic Mapper (TM), and Landsat 8 Operational Land Imager (OLI) satellite images, which were obtained from the US Geological Survey (USGS) website (https://earthexplorer.usgs.gov), where the path is 136 and the row is 45 (Table 12.1). The images were collected for the years 1999, 2009, and 2019. The spatial resolution of all the images was 30 m, and the cloud coverage of each image was below 10%. Using WGS-84 datum, images are projected to UTM zone 46 North projection. To avoid seasonal variation in the assessment, selected images were taken from the same month for each year. Image processing and further evaluation, such as Land Use Land Cover (LULC) classification, Land Surface Temperature (LST) derivation, and Urban Field Variance Index (UTFVI) estimation, were completed using ERDAS IMAGINE 15, ENVI 5.3, and ArcMap 10.6 software.

12.3.1 Classification of Land-Use/Land-Cover

Using the supervised classification technique, the Support Vector Machine (SVM) classifier was used for this LULC classification procedure, using ENVI 5.3 software. In SVM choices, the Radial Basis function was used as the kernel type, and, in the kernel function, Gamma was set to 0.07, the penalty parameter was set to 120.00, pyramid level was set by default to zero, and the threshold of likelihood classification was set to 0.05. The LULC of trainee area is categorized into four classes, namely built-up area, water body, bare land, and vegetated, as

TABLE 12.1
Description of the Satellite Images

Date Acquired (M/D/Y)	Scene ID	Sensor	Cloud Cover (%)	Path/ Row
04/15/1999	LT05L1TP136045199904152016122001T1	Landsat 5 TM	< 10 %	136/45
04/10/2009	LT05L1TP136045200904102016102601T1	Landsat 5 TM		136/45
04/22/2019	LC08L1TP136045201904222019050701T1	Landsat 8 OLI		136/45

Source: US Geological Survey, 2020.

several studies found that SVM is best suited for image classification (Kafy et al., 2020). To validate the LULC classification, accuracy assessment (overall accuracy, Kappa coefficient, and accuracy validation) was employed. Using Google Earth images, 300 points were randomly selected for evaluating the accuracy of each classified map.

12.3.1.1 LULC Change Estimation

Based on the transformation of the overall territory occupied by each LULC type, four change detection maps were computed for each land-use type. Change detection was performed in ERDAS IMAGINE 15 software, using the image difference tool. From the LULC change map area under each land-use type, the converted areas from one land use to another were estimated.

12.3.2 Land Surface Temperature Estimation

Using the Digital Numbers (DN) of the thermal bands (Band 6 in Landsat 5 TM and Band 10 in Landsat 8 Thermal Infrared Scanner (TIRS)), the LST was estimated. The spectral radiances (λ) of the Landsat 5 TM and Landsat 8 TIRS bands were computed at the preliminary phase, by using Equation (12.1) and Equation (12.2), respectively. At last, L_λ was used to derive the LST in degree Celsius, using Equation (12.3).

$$L_\lambda\left(LANDSAT\ 5\ TM\right) = L_{min} + \frac{L_{max} - L_{min}}{Qcal_{max} - Qcal_{min}} \times DN \tag{12.1}$$

$$L_\lambda\left(LANDSAT\ 8\ OLI\right) = ML \times DN + AL \tag{12.2}$$

$$LST = \frac{T_B}{1 + \left(\lambda \times \frac{T_B}{\rho}\right) * \ln(\varepsilon)} - 273.15 \tag{12.3}$$

where ML (0.0003342) is a multiplicative rescaling factor (band-specific), and AL (0.1) is an additive rescaling factor (band-specific). The values for Landsat TM, L_{max}, and L_{min} were collected from the satellite metadata file. The wavelength of emitted radiance λ is 11.5 µm (Aboelnour and Engel 2018, Kafy, Rahman, et al. 2020, Kumar, Bhaskar, and Padmakumari 2012, Rahman, Aldosary, and Mortoja 2017b, Ullah et al. 2019).

$$\rho = \frac{h \times c}{\sigma} = 1.438 \times 10 - 2\ \text{mK}$$

where h represents Planck's constant which is equal to 6.626 × 10–34 J·s, c indicates the velocity of light, which is equal to 2.998×10^{-8} ms^{-2}, and σ is the Boltzmann constant (5.67×10^{-8} $W{\cdot}m^2k^{-4} = 1.38 \times 10^{-23}$ $J{\cdot}K^{-1}$); ε is the land surface emissivity which ranges between 0.97 and 0.99 (Guha et al. 2018, Mallick, Kant, and Bharath 2008, Pal and Ziaul 2017).

$$T_B = \frac{K_2}{\ln\left(\frac{K_1}{L_\lambda}+1\right)} \tag{12.4}$$

where $T_B i$ is the satellite brightness temperature, and the constants K_1 and K_2 values for (1) Landsat-5: K_1 is 607.7, and K_2 is 1260.6 and (2) Landsat 8: K_1 is 774.9 and K_2 is 1321.07 (Anbazhagan and Paramasivam 2016, Kafy, Rahman, et al. 2020, Roy et al. 2020, Ullah et al. 2019).

12.3.2.1 Standardization of LST

Chattogram is a hilly area, and there are topographic and seasonal variations associated with sloping characteristics in the thermal images of Landsat data from several years and seasons. Therefore, it is not appropriate to make comparisons of the LST different years without rearranging it into a standardized common platform. Standardization of LST is primarily conducted to bring all the factors into the proportion to each other (Roy et al. 2020, Ullah et al. 2019). Gross techniques eliminated contaminated pixels of the clouds, and the LST was standardized using Equation (12.5)

$$LST_s = LST - \frac{LST_\mu}{LST_\Omega} \tag{12.5}$$

where *LSTs* is the standardized LST, *LSTμ* is the mean of estimated LST from 1999 to 2019, LST_Ω is the standard deviation of LST from 1999 to 2019.

12.3.3 Classification of Temperature Zones

To determine the relative areas in the various LST ranges, the LST ranges were subdivided into five categories: < 20, 20– <26, 26–< 31 31– < 36, and ≥ 36°C. This division of LST into different ranges aims to evaluate the LST variation in different zones and to observe which areas fall under high-temperature zones and which areas fall into low-temperature zones.

12.3.4 Estimation of Urban Heat Island Effect

Urban heat island in the study area was estimated as those zones that fell under the standardized temperature greater than 1.5°C above the mean LST. Then, the area was categorized into two groups, namely the areas having UHI experience (value 1) and areas having no UHI experience (value 0). Equation 12.6 was used to estimate the UHI effect in the study area.

$$UHI = \frac{T_s - T_m}{SD} \tag{12.6}$$

where T_s stands for LST, T_m for the mean of the LST of the study area, and *SD* for the standard deviation.

12.3.5 Estimation of Urban Thermal Field Variance Index

There are numerous thermal satisfaction indices: Temperature Humidity Index (THI), the Physiological Equivalent Temperature (PET), the Wet-bulb Globe Temperature (WBGT), and the Urban Thermal Field Variance Index (UTFVI) (Anbazhagan and Paramasivam 2016, Ahmed 2018). Using UTFVI, this study described the effect of UHI, and it considers the socio-economic factors for describing the effects. The values of UTFVI were divided into six groups, namely Strongest, Stronger, Strong, Middle, Weak and None, each group displaying different impact of UHI effect on the environment. Equation 12.7 was used for UTFVI calculation :

$$UTFVI = \frac{LST_s - LST_\mu}{LST_s} \tag{12.7}$$

12.4 RESULTS AND DISCUSSION

12.4.1 Land Cover Change Analysis

Accuracy assessment of the categorized LULC maps was estimated using Kappa statistics, as shown in Table 12.2. For every period tested, overall accuracy was greater than 90%, and comparison among the three years showing that the highest percentage (96.15%) was exhibited in 2009. The value for the Kappa coefficient was higher than 0.85 in all three years.

If the Kappa coefficient is higher than 0.75, the degree of accuracy is sufficiently good (Pontius Jr and Millones 2011). For assessing the precision intensity of the classifications, a comparison between 300 sample points and their corresponding points in the Google Earth Image at the same period was evaluated. The validation values were more than 90% for all the years tested. Thus, it can be concluded that, overall, the accuracy level was satisfactory.

For this study, a total of four categories were selected for LULC analysis, naming "vegetated", "water body", "built-up area", and "bare land" (Figure

TABLE 12.2
Accuracy Assessment (%) of Classified LULC Images

Year	User's Accuracy	Producer's Accuracy	Overall Accuracy	Kappa Statistics
1999	96.77	94.07	92.86	89.86
2009	98.53	94.37	96.15	91.56
2019	96.67	98.31	95.45	92.36

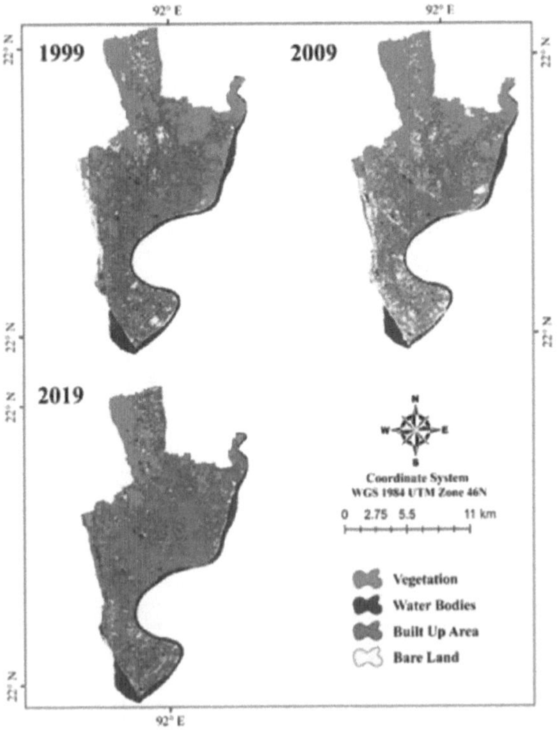

FIGURE 12.2 Multi-temporal LULC classified maps of the study area.

12.2). Two trends were apparent. Firstly, the built-up area increased at a rapid rate of 3.51%, and vegetated land, along with bare land, decreased at rates of 6.13% and 6.81%, respectively (Table 12.3). The values in the excursus in the legend of Figure 12.2 represent the percentage of change among different features in the research area. During the entire study, the most significant negative change occurred in the vegetation area during the period 1999–2019 (–24.95%), whereas the most significant positive change occurred in built-up areas during the period 1999–2019 (32.77).

Assessment of LULC showed a large increase in built-up area. Approximately 34% of the area in 1999 was the built-up area, and this increased to about 54% in 2009, and to 67% in 2019. Built-up zones were expanding from the city centre to the north-eastern side of the study area, covering up the bare land towards the north-western side, and resulting in the only positive net change for 1999–2019 (Table 12.3).

Though the decreasing rate was not as significant as for the other two features named as water body and bare land that have decreased, for a coastal region of the country, it was still substantial. A decrease of almost 2% was seen in 1999–2009 and a decrease of 1% in the next decade (Table 12.3), with the net decrease of 5% in the water body area being significant over the two-decade period.

TABLE 12.3
Area Distribution of Different LULC Classes in the Study Area

LULC	Area in km²			Change (%)			Net Change (%)
	1999	2009	2019	1999–2009	2009–2019	1999–2019	1999–2019
Vegetation land	92.09	60.54	46.00	−17.08	−7.87	−24.95	−6.81
Water bodies	11.63	8.21	6.39	−1.85	−0.99	−2.84	−5.07
Built-up area	62.70	99.24	123.22	19.78	12.99	32.77	3.51
Bare land	17.85	16.28	8.65	−0.85	−4.13	−4.98	−6.13

Gradually, the vegetated areas were converted into built-up areas. In 1999, around half of the total area was represented by the vegetated area, which decreased to nearly one-third of the total area in 2009, whereas complete conversion was 17.08% in the first decade. This trend resulted in a 6.81% net decrease in vegetation land during the time period 1999–2019 (Table 12.3).

Though the magnitude of change in bare land was not significant in the first decade (1999–2009), a decrease of almost 4% was seen in the next decade, which represented a decrease of almost 5% over the two-decade period. The net change of bare land is reduction of 6.13%, as the amount of bare land was lower than the area of the vegetated land and more built-up areas were replacing the barren lands.

12.4.1.1 LULC Transition Assessment

The spatial pattern shows significant changes among the vegetated, bare and built-up areas. The transitions of both vegetated and barren land to built-up areas were influenced mainly by the inherent characteristics of the study area (Figs. 12.3 and 12.4). Being a port city and the second capital of the country, the Chittagong area is the home of leading employment sectors and business opportunities in the country. In the 1999–2009 period, only 4.61% of the vegetated land and 2.24% of the bare land were converted to built-up areas (Figure 12.3). This trend would be considered to be a starting point for the study area to be considered for development. On the other hand, the transition of about 6% and 0.26% vegetated areas and water bodies, respectively, to bare lands, were affected more by the decrease

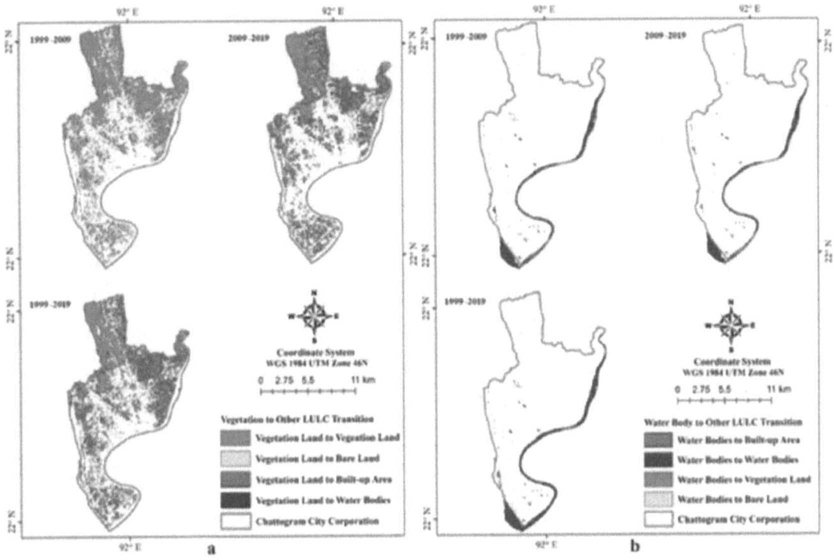

FIGURE 12.3 LULC transition maps: a) vegetated land converted to other LULC classes; b) water bodies converted to other LULC classes.

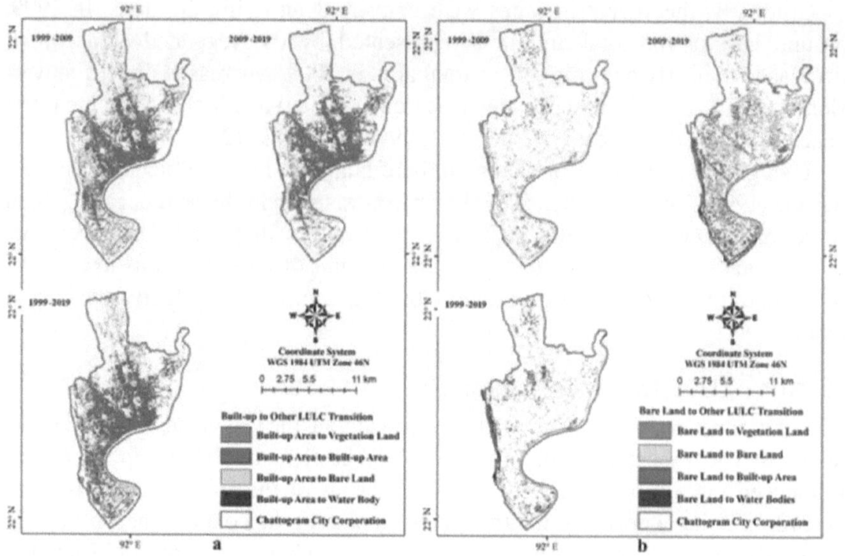

FIGURE 12.4 LULC transition maps: a) built-up areas converted to other LULC classes; b) bare land converted to other LULC classes.

in vegetated land area in the study area. The period also showed that nearly 1% of both water bodies and vegetated lands were to become bare lands, with 7.47% of vegetative land to become barren land (Figure 12.4).

The 2009–2019 period showed a different picture, as nearly 16% of vegetated land, and about 13% of bare lands were turned into built-up area. The increasing number of job availabilities in the past decade mainly encouraged this urban expansion. The pattern showed that the transition in the central study area was gradually increasing towards the northwest side of the city. This period also showed a non-significant amount of change in water body area. Contrary to this, the trend of built-up areas converting into other land-use types decreased significantly.

The overall picture of the 1999–2019 period showed that about 20% of the vegetated lands became built-up areas, where the percentage cover was only 5.76% for bare lands. About 27% of both vegetated and built-up regions remained unchanged (Table 12.4).

12.4.2 Land Surface Temperature Change Analysis

The Land surface temperature variation and aerial distribution of LST in the research area are indicated in Figure 12.5 and Table 12.4, respectively, for the years 1999, 2009, and 2019. Brownish red and yellowish hues in LST variation maps represent zones with higher temperatures, and light green to bluish zones indicate medium to lower temperature areas (Figure 12.5). These spatio-temporal trends of the LST shift illustrate the rapid transition in LULC levels.

TABLE 12.4
Percentage of Different LULC transitions in the Study Area

	Change (%)		
LULC Transition	**1999–2009**	**2009–2019**	**1999–2019**
Vegetation land to vegetated land	39.33	29.23	27.26
Vegetated land to water bodies	5.92	2.16	2.37
Vegetation Land to Built-up Area	7.61	15.95	19.99
Vegetation Land to Bare Land	0.172	0.38	0.44
Water Bodies to Water Bodies	5.71	5.7	6.35
Water Bodies to Vegetation Land	0.26	0.022	0.08
Water Bodies to Built-up Area	0.09	0.198	0.62
Water Bodies to Bare Land	1.21	0.09	0.2
Built-up Area to Built-up Area	19.82	24.5	27.46
Built-up Area to Vegetation Land	2.7	0.88	3.58
Built-up Area to Water Body	0.04	0.12	0.14
Built-up Area to Bare Land	1.47	1.26	1.85
Bare Land to Bare Land	4.9	2.37	1.46
Bare Land to Vegetation Land	2.43	2.71	1.91
Bare Land to Water Bodies	0.098	1.17	0.53
Bare Land to Built-up Area	8.24	13.26	5.76
Total	100	100	100

According to Table 12.5, in 1999, out of the total study area of 155.4 km^2, 39.83 km^2 area experienced temperatures < 20°C, whereas 91.54 km^2 area experienced temperatures in the range 26–<31°C. In 2009, 18.30 km^2 area fell under < 20°C temperature range, whereas 95.20 km^2 area had fallen within the temperature range of 26–<31°C. For the < 20°C and 26–<31°C temperature ranges, during the period 1999–2009, it shows changes of −11.65% and +1.98%, respectively, in the LST of the study area. Moreover, 2019 saw a rapid rise in the temperatures, with only 11.82 km^2 area falling under < 20°C, 40.12 km^2 within 26–<31°C, 47.89 km^2 within 31–<36°C, and a maximum of 61.87 km^2 area was experiencing the high temperatures of ≥ 36°C, in the period 2009–2019, which displays a substantial +33.30% rise in the LST. During 1999–2019, an overall −27.81% change in LST was observed for the temperature range of 26–<31°C, and a +33.42% change for temperatures ≥ 36°C. North-eastern portions of the study area exhibited higher temperatures compared to other areas, associated with rapid urbanization in these areas.

12.4.2.1 LST Distribution in Different LULC Classes

The influence of different LULC classes on LST is shown in Table 12.6, which describes the variation of mean LST in different LULC classes for the years 1999, 2009, and 2019. The mean LST of vegetated land, water bodies, built-up areas,

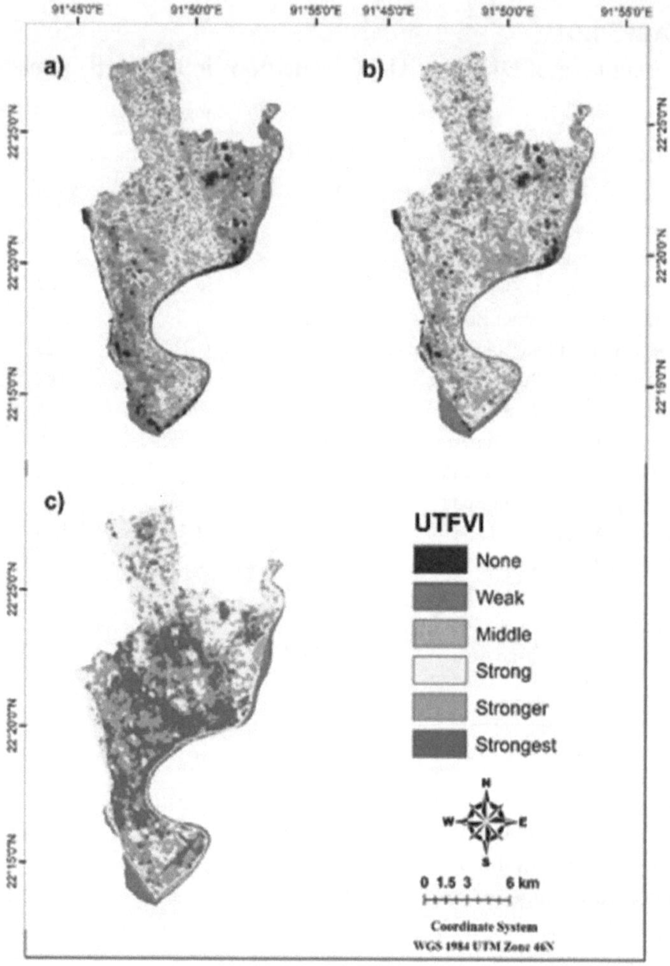

FIGURE 12.5 Land surface temperature variation in the study area.

and bare land are listed in the table. These LST values focuses on spatial patterns and temporal shift LST patterns and illustrates the transition in LULC classes. For example, the mean LST of vegetated land in 1999 was 25°C. The temperature increased to 27°C in 2009, and this increased further to 28°C in 2019.

Similarly, we can see the rise in mean LST of water bodies, built-up areas and barren land. In the case of the built-up areas, the mean LST rose drastically. The built-up areas are warmer than their peripheral parts. Due to decreases in areas associated with vegetation, agricultural lands, and water reservoirs, the mean LST is higher in the built-up areas. In the year 1999, the mean LST of the built-up areas was 29°C, which increased dramatically to 32°C in 2009. The mean LST further increased to 35°C in 2019.

TABLE 12.5
Aerial Distribution of LST in the Study Area

LST Range	LST Area in km^2 1999	2009	2019	Change in % 1999–2009	2009–2019	1999–2019
< 20	39.83	18.30	11.82	−11.65	−3.51	−15.16
20- <26	49.26	55.94	23.18	3.61	−17.72	−14.11
26- < 31	91.54	95.20	40.12	1.98	−29.80	−27.81
31- < 36	4.16	15.12	47.89	5.93	17.73	23.66
≥ 36	0.08	0.30	61.87	0.12	33.30	33.42

TABLE 12.6
Mean LST Distribution in Different LULC Classes

LULC Classes	Mean LST in Degrees Celsius 1999	2009	2019
Vegetated land	25	27	28
Water bodies	20	22	23
Built-up area	29	32	35
Bare land	27	28	29

12.4.2.2 Urban Heat Island Variation Analysis

The area was categorized into two classes, on the basis of the UHI impact in the research area (Figure 12.6), namely the areas which had UHI experience (value 1) and areas having no UHI experience (value 0). The area was calculated in km^2 (Table 12.7). The weighting of the UHI was computed by splitting the UHI affected area over the entire study area.

In 1999, only 5.67 km^2 of the study area experienced a UHI effect, with a weighting of only 3.08%. In 2009, 27.56 km^2 of the study area coming under the UHI distribution , where the weighting was increased to 14.95%, and in 2019, the UHI-distributed area a maximum of 51.42 km^2 area, with a weighting of 27.90%, which, compared with 1999 and 2009, was the highest value. From 1999–2009, the trend of UHI showed an increase of about 11.88%, and from 2009-2019, it rose to 12.95%. Through 1999 to 2019, UHI values increased by 24.82% (Table 12.7).

12.4.2.3 Analysis of Urban Thermal Field Variance Index

There are many indexes of thermal satisfaction for determining the effect of UHI on the standard of life in metropolitan environments. Among these indexes, the Urban Thermal Field Variance Index (UFTVI) has been estimated here, as of it is previously used with Landsat image data. For the years 1999, 2009, and 2019, the UFTVI variation in the research region is shown in Figure 12.7. Reddish-orange

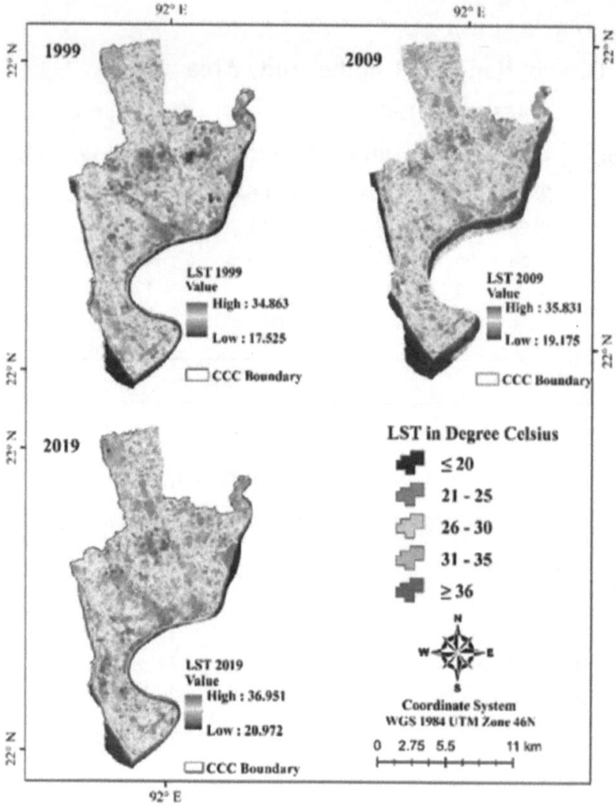

FIGURE 12.6 UHI variation in the study area for the years a) 1999 b) 2009, and c) 2019.

TABLE 12.7
UHI Distribution and Weighting in Different Years

Year	1999	2009	2019
Area in km2	5.67	27.56	51.42
Weighting (%)	3.08	14.95	27.90
Change 1999–2009 (%)		11.88	
Change 2009–2019 (%)		12.95	
Change 1999–2019 (%)		24.82	

regions display the Strongest UFTVI impact zones, followed by orange and yellow, illustrating Stronger, and Strong impact regions, respectively. The green to blue-coloured areas show Middle to No UFTVI impact zones, respectively, in the study area (Agarwal 2002, Kakon et al. 2010, Liu and Zhang 2011, Matzarakis, Mayer, and Iziomon 1999, Willett and Sherwood 2012, Zhang et al. 2006).

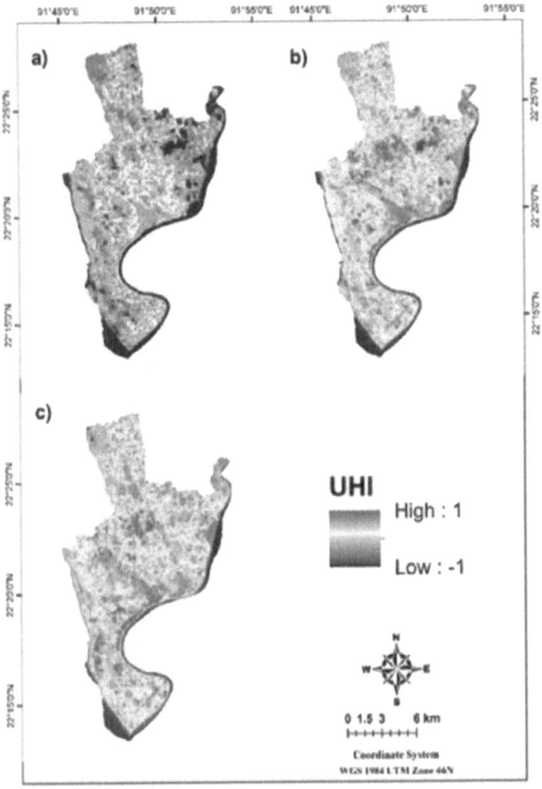

FIGURE 12.7 UTFVI variation in the study area for the years a) 1999 b) 2009, and c) 2019.

Table 12.8 exhibits the UFTVI distribution in the study area for the years 1999, 2009, and 2019. In 1999, 71.81 km^2 were under the Middle (0.005–0.010) class range, with 16.33 km^2 area under the Stronger class (0.015–0.020) region of the UFTVI. In 2009, an area of 65.54 km^2 was under the Strong (0.010–0.015) UFTVI class range, and 43.84 km^2 area had moved under the Stronger class range (0.015–0.020) along with 7.17 km^2 area under the Strongest UFTVI class range (> 0.020). In 2019, 65.41 km^2 and 50.73 km^2 of the study area now fell under the Stronger (0.015–0.020) and Strongest (> 0.020) UFTVI class range, respectively, which were higher than the corresponding figures in 1999 and 2009. Overall changes of –34.69%, +26.64%, and +25.84% were recorded in the Middle, Stronger and Strongest classes of UTFVI, respectively, from 1999 to 2019.

12.4.3 Impact of LULC Change on the Urban Environment

As the years passed, Chattogram City has become more and more crowded because of an increasing number of employment and business activities and is home to

TABLE 12.8
Aerial distribution of UTFVI in the Study Area

		Area in km²			Changes (%)		
Class Name	**Class Range**	**1999**	**2009**	**2019**	**1999–2009**	**2009–2019**	**1999–2019**
None	<0	8.03	4.68	0.03	−1.81	−2.52	−4.34
Weak	0–0.005	45.37	22.01	7.20	−12.68	−8.04	−20.72
Middle	0.005–0.010	71.81	41.02	7.88	−16.71	−17.98	−34.69
Strong	0.010–0.015	39.62	65.54	53.01	14.07	−6.80	7.27
Stronger	0.015–0.020	16.33	43.84	65.41	14.93	11.71	26.64
Strongest	>0.020	3.11	7.17	50.73	2.20	23.64	25.84

more entrepreneurs and companies, with people from all over the country, mainly the eastern side, moved to Chattogram City to have a better life. As people come here for greater opportunities, this results in a profound change in LULC pattern and this trend, following the vast rise in built-up areas with a combined reduction in vegetation and bare lands, affected the urban environment. As urbanization expands in the area, the climate has become even hotter, with the higher temperature being distributed in the urban areas. The urban areas expanded; with respect to the LST distribution of the whole city, the increasing trend of LST was no different but growing. Therefore, the built-up areas increased, along with the temperature. This statement can be validated with the UTFVI distribution, as seen in Figure 12.7. As the years passed, the built-up areas expanded gradually from the eastern side to the north-western side, following the UTFVI pattern. Areas with the Strongest class of UTFVI were rare in 1999, but showed a significant increase in 2009, resulting in a 2.20% increase, with a 14.93% increase of Stronger areas too. The same trend followed as expansion continued in the next decade, resulting in an overall rise in UTFVI values in urban areas.

12.5 CONCLUSION

The main reason behind the change in LULC patterns and its impact on UTFVI is the rapid urban expansion. As the increase in built-up (urban) areas are mainly replacing other land-use categories, principally vegetated areas and water bodies, the surface temperature of the land has increased rapidly. These increased temperatures and LULC changes are harmful to the habitat, the ecosystem, and the urban environment. The findings, that the growing urban areas were significantly replacing the water bodies, bare land, and vegetated regions were consistent from the year 1999 to 2019. The maximum positive net change in urban growth was evident for the urban areas (+3.51%), with the maximum negative change in the vegetation area (−6.81%) in the past 20 years. The maximum temperature of the city increased from 34.86°C in 1999 to 36.95°C in 2019. The assessed LST showed that lower recorded temperature zones in 1999 were converted into higher temperature zone during the period 2009-2019. From these outcomes, it is found that rapid urban expansion increased the UTFVI of the study area during the observation period. To restrict and minimize the effects of UHI as a consequence of unplanned and scattered development of urban areas, it is necessary to track and predict the changes in LULC and the LST pattern of the city. Along with the LULC and LST change tracking, measures and policies should be undertaken to limit the haphazard city development and to preserve the existing greenery of the city. Also, steps should be adopted to increase the greenery in the city by planting trees and minimizing the use of non-evaporating and non-transpiring construction materials, replacing them with eco-friendly materials. These measures will help to reduce the surface temperature.

CONFLICT OF INTEREST

The authors announce that they have no established conflicting financial interests or personal relationships which may seem to have affected the research stated in this paper.

ACKNOWLEDGEMENT

The authors gratefully acknowledge Chattogram City Corporation, the Bangladesh Meteorological Survey, and the US Geological Survey for their support in providing datasets.

REFERENCES

Aboelnour, Mohamed, and Bernard A. Engel. 2018. "Application of remote sensing techniques and geographic information systems to analyze land surface temperature in response to land use/land cover change in greater Cairo region, Egypt." *Journal of Geographic Information System* 10 (01):57.

Agarwal, Chetan. 2002. *A Review and Assessment of Land-Use Change Models: Dynamics of Space, Time, and Human Choice*. Vol. 297. US Department of Agriculture, Forest Service, Northeastern Research Station.

Ahmed, Bayes. 2011. "Urban land cover change detection analysis and modeling spatio-temporal growth dynamics using remote sensing and GIS techniques: A case study of Dhaka, Bangladesh." Dissertation, Universitat Jaume I (UJI), Castellón, Spain.

Ahmed, Bayes. 2018. "Assessment of urban heat islands and impact of climate change on socioeconomic over Suez governorate using remote sensing and GIS techniques." *The Egyptian Journal of Remote Sensing and Space Science* 21 (1):15–25.

Alfraihat, R., G. Mulugeta, and T. Gala. 2016. "Ecological evaluation of urban heat island in Chicago City, USA." *Journal of Atmospheric Pollution* 4 (1):23–29.

Al-Hathloul, S., and Mohamed Abdel Rahman. 2003. "Dynamism of metropolitan areas: The case of metropolitan Dammam, Saudi Arabia." *Journal of the Gulf & Arabian Peninsula Studies* 29:11–43.

Alqurashi, Abdullah F., Lalit Kumar, and Priyakant Sinha. 2016. "Urban land cover change modelling using time-series satellite images: A case study of urban growth in five cities of Saudi Arabia." *Remote Sensing* 8 (10):838.

Anbazhagan, S., and C.R. Paramasivam. 2016. "Statistical correlation between land surface temperature (LST) and vegetation index (NDVI) using multi-temporal landsat TM data." *International Journal of Advanced Earth Science and Engineering* 5 (1):333–46.

Balogun, I.A., and K.A. Ishola. 2017. "Projection of future changes in landuse/landcover using cellular automata/markov model over Akure city, Nigeria." *Journal of Remote Sensing Technology* 5 (1):22–31.

BBS. (2013). Chittagong District Statistics 2011.

Bokaie, Mehdi, Mirmasoud Kheirkhah Zarkesh, Peyman Daneshkar Arasteh, and Ali Hosseini. 2016. "Assessment of urban heat island based on the relationship between land surface temperature and land use/land cover in Tehran." *Sustainable Cities and Society* 23:94–104.

Celik, Bahadir, Sinasi Kaya, Ugur Alganci, and Dursun Zafer Seker. 2019. "Assessment of the relationship between land use/cover changes and land surface temperatures: A case study of thermal remote sensing." *Feb-Fresenius Environmental Bulletin* 3:541.

Collins, Matthew, Reto Knutti, Julie Arblaster, J.L. Dufresne, Thierry Fichefet, Pierre Friedlingstein, Xuejie Gao, William J. Gutowski, Tim Johns, and Gerhard Krinner. 2013. "Long-term climate change: Projections, commitments and irreversibility."

Connors, John Patrick, Christopher S. Galletti, and T.L. Winston. 2013. "Landscape configuration and urban heat island effects: Assessing the relationship between landscape characteristics and land surface temperature in phoenix, Arizona." *Landscape Ecology Chow* 28 (2):271–283.

El-Hattab, M., S.M. Amany, and G.E. Lamia. 2018. "Monitoring and assessment of urban heat islands over the Southern region of Cairo Governorate, Egypt." *The Egyptian Journal of Remote Sensing and Space Science* 21 (3):311–323.

Fernando, Gmts. 2018. "Identification of urban heat Islands & its relationship with vegetation cover: A case study of Colombo & Gampaha Districts in Sri Lanka." *Journal of Tropical Forestry and Environment* 8 (2).

Gaur, Abhishek, Markus Kalev Eichenbaum, and Slobodan P. Simonovic. 2018. "Analysis and modelling of surface Urban Heat Island in 20 Canadian cities under climate and land-cover change." *Journal of Environmental Management* 206:145–157.

Guha, Subhanil, Himanshu Govil, Anindita Dey, and Neetu Gill. 2018. "Analytical study of land surface temperature with NDVI and NDBI using Landsat 8 OLI and TIRS data in florence and naples city, Italy." *European Journal of Remote Sensing* 51 (1):667–678.

Hart, Melissa A., and David J. Sailor. 2009. "Quantifying the influence of land-use and surface characteristics on spatial variability in the urban heat island." *Theoretical and Applied Climatology* 95 (3–4):397–406.

Hassan, Mohammad Mehedy, and Mohhamad Nurul Islam Nazem. 2016. "Examination of land use/land cover changes, urban growth dynamics, and environmental sustainability in Chittagong city, Bangladesh." *Environment, Development and Sustainability* 18 (3):697–716.

Huang, Qiuping, Jiejun Huang, Xining Yang, Chuanglin Fang, and Youjia Liang. 2019. "Quantifying the seasonal contribution of coupling urban land use types on urban heat Island using land contribution index: A case study in Wuhan, China." *Sustainable Cities and Society* 44:666–675.

ICISET. (2018). ABOUT CHITTAGONG. International Conference on Innovation in Science, Engineering and Technology.

IPCC. 2014. "Mitigation of climate change." *Contribution of Working Group III to the Fifth Assessment Report of the Intergovernmental Panel on Climate Change*, 1454.

Islam, Md Kamrul, and Sudipta Chowdhury. 2014. "Analysis of changing land cover in Chittagong City Corporation Area (CCC) by remote sensing and GIS." *International Journal of Innovation and Applied Studies* 8 (3):1193.

Kafy, Abdulla Al, Md Shahinoor Rahman, and Lamia Ferdous. 2017. "Exploring the association of land cover change and landslides in the Chittagong Hill Tracts (CHT): A remote sensing perspective."

Kafy, Abdulla Al, Muhaiminul Islam, Abdur Khan, Lamia Ferdous, and Md Hossain. 2019. "Identifying most influential land use parameters contributing reduction of surface water bodies in Rajshahi City, Bangladesh: A remote sensing approach." *Remote Sensing of Land* :87–95. doi:10.21523/gcj1.18020202.

Kafy, Abdulla Al, Abdullah-Al Faisal, Soumik Sikdar, Mohammad Hasan, Mahbubur Rahman, Mohammad Hasib Khan, and Rahatul Islam. 2020. "Impact of LULC changes on LST in Rajshahi District of Bangladesh: A remote sensing approach." *Journal of Geographical Studies* 3:11–23. doi:10.21523/gcj5.19030102.

Kafy, Abdulla Al, Md Shahinoor Rahman, Abdullah Al Faisal, Mohammad Mahmudul Hasan, and Muhaiminul Islam. 2020. "Modelling future land use land cover changes and their impacts on land surface temperatures in Rajshahi, Bangladesh." *Remote Sensing Applications: Society and Environment*. https://doi.org/10.1016/j.rsase.2020.100314.

Kakon, Anisha Noori, Mishima Nobuo, S. Kojima, and T. Yoko. 2010. "Assessment of thermal comfort in respect to building height in a high-density city in the tropics." *American Journal of Engineering and Applied Sciences* 3 (3):545–551.

Karim, Mohammed Fazlul, and Nobuo Mimura. 2008. "Impacts of climate change and sea-level rise on cyclonic storm surge floods in Bangladesh." *Global Environmental Change* 18 (3):490–500.

Kim, Chang-Gil. 2008. *The Impact of Climate Change on the Agricultural Sector: Implications of the Agro-Industry for Low Carbon, Green Growth Strategy and Roadmap for the East Asian Region*. Korea Rural Economic Institute.

Krishna, Rao. 1972. "Remote Sensing of urban heat Islands from an environmental satellite."

Kumar, K. Sundara, P. Udaya Bhaskar, and K. Padmakumari. 2012. "Estimation of land surface temperature to study urban heat island effect using LANDSAT ETM+ image." *International Journal of Engineering Science and Technology* 4 (2):771–778.

Lai, Li-Wei, and Wan-Li Cheng. 2010. "0> urban heat Island and air pollution—an emerging role for hospital respiratory admissions in an urban area." *Journal of Environmental Health* 72 (6):32–36.

Li, J., and H.M. Zhao. 2003. "Detecting urban land-use and land-cover changes in Mississauga using Landsat TM images." *Journal of Environmental Informatics* 2 (1):38–47.

Lilly Rose, A., and M.D. Devadas. 2009. "Analysis of land surface temperature and land use/land cover types using remote sensing imagery - A case in chennai city, India." *The Seventh International Conference on Urban Climate*, held on 29 June – 3 July, Yokohama, Japan.

Lin, Y., Liu, A., Ma, E., Li, X., & Shi Q. (2013). Impacts of future urban expansion on regional climate in the Northeast megalopolis, USA. Advances in Meteorology, 2013. https://doi.org/10.1155/2013/362925.

Liu, Lin, and Yuanzhi Zhang. 2011. "Urban heat island analysis using the Landsat TM data and ASTER data: A case study in Hong Kong." *Remote Sensing* 3 (7):1535–1552.

Mallick, Javed, Yogesh Kant, and B.D. Bharath. 2008. "Estimation of land surface temperature over Delhi using Landsat-7 ETM+." *The Journal of Indian Geophysical Union* 12 (3):131–140.

Mansour, Shawky, Mohammed Al-Belushi, and Talal Al-Awadhi. 2020. "Monitoring land use and land cover changes in the mountainous cities of Oman using GIS and CA-Markov modelling techniques." *Land Use Policy* 91:104414.

Matzarakis, Andreas, Helmut Mayer, and Moses G. Iziomon. 1999. "Applications of a universal thermal index: Physiological equivalent temperature." *International Journal of Biometeorology* 43 (2):76–84.

Meyer, William B., and Billie L. Turner. 1992. "Human population growth and global land-use/cover change." *Annual Review of Ecology and Systematics* 23 (1):39–61.

Moore, Melinda, Philip Gould, and Barbara Keary. 2003. "Global urbanization and impact on health." *International Journal of Hygiene Environmental Health* 206 (4–5):269–278.

Mou, Ishrat Zerin Hossain. 2019. "Spatial analysis of the integration core of Comilla City corporation area and the impact of the by-pass road-A space syntax analysis." *Mist International Journal of Science And Technology* 7 (1).

Pal, Swades, and S.K. Ziaul. 2017. "Detection of land use and land cover change and land surface temperature in English Bazar urban centre." *The Egyptian Journal of Remote Sensing and Space Science* 20 (1):125–145.

Pontius Jr, Robert Gilmore, and Marco Millones. 2011. "Death to Kappa: Birth of quantity disagreement and allocation disagreement for accuracy assessment." *International Journal of Remote Sensing* 32 (15):4407–4429.

Rahman, Md Shahinoor, Hossain Mohiuddin, Abdulla-Al Kafy, Pintu Kumar Sheel, and Liping Di. 2018. "Classification of cities in Bangladesh based on remote sensing derived spatial characteristics." *Journal of Urban Management.*

Rahman, M.T., Aldosary A.S., & Mortoja, M. (2018). Modeling future land cover changes and their effects on the land surface temperatures in the Saudi Arabian eastern coastal city of Dammam. Land, 6, 36.

Rahman, Muhammad, Adel S. Aldosary, and Md. Mortoja. 2017. "Modeling future land cover changes and their effects on the land surface temperatures in the Saudi Arabian eastern coastal city of Dammam." *Land* 6 (2):36.

Rimi, Ruksana H, Karsten Haustein, Emily J. Barbour, Richard G. Jones, Sarah N. Sparrow, and Myles R. Allen. 2019. "Evaluation of a large ensemble regional climate modelling system for extreme weather events analysis over Bangladesh." *International Journal of Climatology* 39 (6):2845–2861.

Roy, Sanjoy, Santa Pandit, Eshita Akter Eva, Md Shaddam Hossain Bagmar, Mossammat Papia, Laboni Banik, Timothy Dube, Farhana Rahman, and Mohammad Arfar Razi. 2020. "Examining the nexus between land surface temperature and urban growth in Chattogram metropolitan area of Bangladesh using long term Landsat series data." *Urban Climate* 32:100593.

Saleh, S. (2011). Impact of Urban Expansion on Surface Temperature in Baghdad, IRAQ using Remote Sensing and GIS Techniques. *J. Environ. Construct. Civil Eng.,* 2. https://doi.org/10.22401/JNUS.13.1.07.

Santamouris, M. 2014. "On the energy impact of urban heat island and global warming on buildings." *Energy and Buildings* 82:100–113.

Sarkar, M., J.B. Islam, and S. Akter. 2016. "Pollution and ecological risk assessment for the environmentally impacted Turag River, Bangladesh." *Journal of Materials and Environmental Science* 7 (7):2295–2304.

Sejati, Anang Wahyu, Imam Buchori, and Iwan Rudiarto. 2019. "The spatio-temporal trends of urban growth and surface urban heat islands over two decades in the Semarang metropolitan region." *Sustainable Cities and Society* 46:101432.

Silva, Janilci Serra, Richarde Marques da Silva, and Celso Augusto Guimarães Santos. 2018. "Spatiotemporal impact of land use/land cover changes on urban heat islands: A case study of Paço do Lumiar, Brazil." *Building and Environment* 136:279–292.

Streutker, David R. 2003. "Satellite-measured growth of the urban heat island of Houston, Texas." *Remote Sensing of Environment* 85 (3):282–289.

Swetnam, Ruth D., Brendan Fisher, Boniphace P. Mbilinyi, Pantaleo K.T. Munishi, S. Willcock, Taylor Ricketts, Shadrack Mwakalila, Andrew Balmford, Neil D. Burgess, and Andrew R. Marshall. 2011. "Mapping socioeconomic scenarios of land cover change: A GIS method to enable ecosystem service modelling." *Journal of Environmental Management* 92 (3):563–574.

Tomlinson, Charlie J., Lee Chapman, John E. Thornes, and Christopher J. Baker. 2011. "Including the urban heat island in spatial heat health risk assessment strategies: A case study for Birmingham, UK." *International Journal of Health Geographics* 10 (1):42.

Trolle, Dennis, Anders Nielsen, Hans E. Andersen, Hans Thodsen, Jørgen E. Olesen, Christen D. Børgesen, Jens Chr Refsgaard, Torben O. Sonnenborg, Ida B. Karlsson, and Jesper P. Christensen. 2019. "Effects of changes in land use and climate on aquatic ecosystems: Coupling of models and decomposition of uncertainties." *Science of the Total Environment* 657:627–633.

Ullah, Siddique, Adnan Ahmad Tahir, Tahir Ali Akbar, Quazi K. Hassan, Ashraf Dewan, Asim Jahangir Khan, and Mudassir Khan. 2019. "Remote sensing-based quantification of the relationships between land use land cover changes and surface temperature over the lower Himalayan region." *Sustainability* 11 (19):5492.

UN. 2015. "Sustainable development goals." Accessed 25th November.

Wang, Haiting, Yuanzhi Zhang, Jin Yeu Tsou, and Yu Li. 2017. "Surface urban heat island analysis of Shanghai (China) based on the change of land use and land cover." *Sustainability* 9 (9):1538.

Weng, Qihao, Dengsheng Lu, and Jacquelyn Schubring. 2004. "Estimation of land surface temperature–vegetation abundance relationship for urban heat island studies." *Remote Sensing of Environment* 89 (4):467–483.

Willett, Katharine M., and Steven Sherwood. 2012. "Exceedance of heat index thresholds for 15 regions under a warming climate using the wet-bulb globe temperature." *International Journal of Climatology* 32 (2):161–177.

Xiang, Gao. 2017. "Urban development with the constraint of water resources: A case study of Gansu section of Western Longhai-Lanxin economic zone." In *Water Challenges of an Urbanizing World*. IntechOpen.

Yang, Xiaoshan, Lilliana L.H. Peng, Zhidian Jiang, Yuan Chen, Lingye Yao, Yunfei He, and Tianjing Xu. 2020. "Impact of urban heat island on energy demand in buildings: Local climate zones in Nanjing." *Applied Energy* 260:114279.

Zakšek, Klemen, and Krištof Oštir. 2012. "Downscaling land surface temperature for urban heat island diurnal cycle analysis." *Remote Sensing of Environment* 117:114–124.

Zhang, Yong, Tao Yu, Xingfa Gu, Y. Zhang, and L. Chen. 2006. "Land surface temperature retrieval from CBERS-02 IRMSS thermal infrared data and its applications in quantitative analysis of urban heat island effect." *Journal of Remote Sensing-Beijing* 10 (5):789.

Zhou, Ji, Yunhao Chen, Xu Zhang, and Wenfeng Zhan. 2013. "Modelling the diurnal variations of urban heat islands with multi-source satellite data." *International Journal of Remote Sensing* 34 (21):7568–7588.

13 Modeling Land-Use/Land-Cover Change, Using Multi-Layer Perceptron and Markov Chain Analysis

A Study on Bahir Dar City, Ethiopia

Dodge Getachew and Ravinder Kaur

CONTENTS

13.1 INTRODUCTION

Low levels of urbanization have paved the way for an accelerated rate of urbanization for the least-developed countries, mainly in Asia and Africa. The global urban population is predicted to rise by 2.5 billion persons, from 4.2 billion to 6.7 billion, between 2018 and 2050, with 90 percent of this surge estimated to be in Asia and Africa (UNDESA, 2018). Population explosion and unprecedented urban expansion have generally been the main features of most developing countries. This abrupt horizontal built-up expansion is happening at the cost of farmland and forest in the periphery of urban centers (Verma and Raghubanshi, 2019).

Ethiopia is the second most-populous country in Africa, with a population of 112,079,000 (UNDESA, 2019), of which nearly 80 percent reside in rural areas. According to the Ministry of Urban Development, Housing and Construction (MUDHCo), high urban growth rate (4.89 % per annum), coupled with high urban population growth rate (5.4 % per annum), has been posing a multi-faceted challenge to urban centers and their peripheries (MUDHCo, 2015). Unplanned rapid urban expansion, intense land transformation, loss of agricultural land and natural resource degradation has become the main features of most peri-urban areas of Ethiopian cities, which is a burning issue in urban and regional development policies.

Understanding of the earlier, present-day and future land-use/land-cover (LULC) change and land transformation process is crucial for the planning and sustainable use of natural resources (Regmi, 2014; Hamad et al., 2018; Nasiri et al., 2019). Hence, studies on the process of LULC change, its impact and prediction of future change have received growing attention from experts and various stakeholders. The past two decades were marked by the development of different models to assess and forecast land transformation, which aimed at assisting resource (land, water, forest, etc.) management and sustainable urban and regional development (Mas et al., 2014). In studying the process of land transformation and its projection, Markov Chain (MC) and Multi-Layer Perceptron Neural Network (MLPNN) models are the most widely used and well-developed models (Nasiri et al., 2019). Markov Chain works based on a transition probability matrix, by which the state of the LULC of a place at time 2 can be projected by the state of the LULC of the place at time 1. Thus, two LULC maps are needed as an input to forecast future land transformation (Eastman, 2006).

Accordingly, the main objective of this study is to predict the future (2029) LULC of Bahir Dar using MLPNN-MC models in Land Change Modeler (LCM) within a TerrSet (IDRISI) software environment.

13.2 STUDY AREA

Bahir Dar is the capital city of Amhara Region, which is positioned in the northwestern part of Ethiopia. Its location covers between 11°30' N to 11°40' N latitude and 37°17' E to 37°29' longitude. The city is situated on the lakeshore of Tana and is drained by the River Abay (Blue Nile), with an altitude of 1800 m above sea level (a.s.l.) (Figure 13.1).

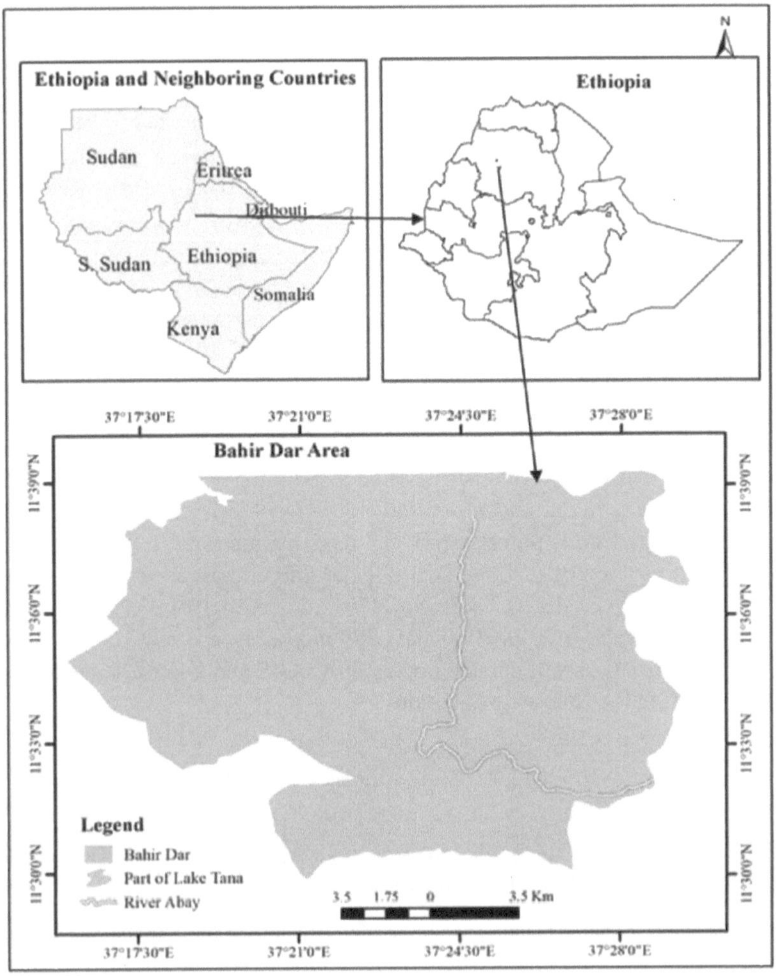

FIGURE 13.1 Map of the study area. *Source: research work (2019-2020).*

13.3 MATERIALS AND METHODS

13.3.1 Sources and Method of Data Collection

Both primary and secondary data sources were used. Landsat images of Bahir Dar city for the years 1999, 2009 and 2019 were used, accompanied by observation and collection of Ground Control Points (GCP) (Table 13.1).

13.3.2 Image Processing and Change Analysis

Image preprocessing, which included layer stacking, choice of appropriate band combinations, noise removal and histogram equalization, was performed

TABLE 13.1
Landsat Satellite Data Used in This Study

Sensor	Path and Row	Spatial Resolution	Date of Acquisition	Source
TM	170/52	30 m	08/02/1999	USGS
TM	170/52	30 m	18/11/2009	USGS
OLI	170/52	30 m	30/01/2019	USGS

to increase the quality of the images. Based on the researchers' prior knowledge and the European Union land-cover classification scheme called CORINE (Coordination of Information on the Environment), five major LULC types were recognized. These are agricultural areas, built-up areas, forest and semi-natural areas ("forest"), open spaces and water bodies. Then, the classification of each satellite image was carried out using a supervised image classification technique and the maximum likelihood algorithm in ERDAS IMAGINE 2014. Finally, using the collected GPS points (265) for the 2019 map and GCP points from Google Earth for the 2019, 2009 and 1999 classified maps, accuracy assessment was performed in ArcGIS 10.3 software. Overall accuracy of 93.9%, 93.5% and 89.6 % was achieved for 2019, 2009 and 1999 maps, respectively.

The change analysis was carried out using ArcGIS and Land Change Modeler (LCM), based on the following equations:

$$\text{Change in percent} = \frac{X - y}{y} * 100 \tag{13.1}$$

$$\text{Rate of Change (ha/year)} = \frac{X - y}{T} * 100 \tag{13.2}$$

$$\text{Net Change in percent} = \frac{X - y}{X} * 100 \tag{13.3}$$

where X is the area of LULC (ha) in the recent year, Y is the area of LULC (ha) in the earlier year, and T is the time interval (years) between X and Y (Kindu et al., 2015; Yesuph and Dagnew, 2019).

13.3.3 Preparation and Testing of Driver Variables

By taking into account the nature, coverage, availability, accessibility and effect on the modeling, four major driver variables have been selected, namely slope, distance from water bodies, distance from existing built-up areas and distance from major roads. With the exception of slope, which is a static variable, all the other variables have a dyanamic nature, which changes with time during the process of modeling. After selection of the driver variables, standardizing each

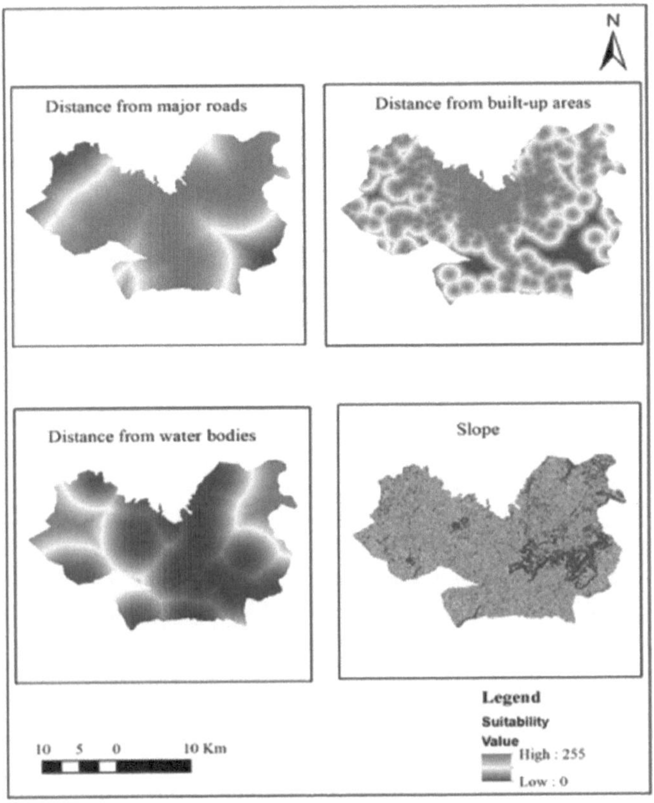

FIGURE 13.2 Map of driver variables.

variable in a continuous scale of suitability has been carried out with 0–255 bytc level of standardization, using a fuzzy module on TerrSet (IDRISI) software. Maps of the driver variables are presented in Figure 13.2, which indicates those areas with high suitability and vulnerability to change colored in red, whereas green color represents low suitability.

The explanatory power of the selected driver variables was tested using Test and Selection of the Site and Driver Variables Panel in TerrSet software. Cramer's V and P values were used to check the relevance of driver variables, by which variables with low Cramer's V and/or high P value can be discarded. Driver variables with Cramer's V of 0.15 or more were considered to be useful, while those above 0.4 were considered to be good. Hence, driver variables with high Cramer's V and zero P values have greater explanatory power, which should be incorporated into the model, whereas variables with Cramer's V of less than 0.15 can be rejected (Eastman, 2006, 2016; Jamali, 2019). Even though the Cramer's V of slope was low, it was included in the model for further testing, and was the only static variable. Moreover, an evidence likelihood map change was included in the model as a dynamic variable.

13.3.4 Training Multi-Layer Perceptron Neural Network (MLPNN)

MLPNN is the most commonly used neural network, by which signals are transmitted directly from input to output (feedforward architecture). During the learning process, MLP involves two basic steps, namely forward and backward propagations, while hidden layers play a vital role in the process (Eastman, 2006; Jamali, 2019). In this study, the default parameters were used. Thus, the minimum cells that transitioned from 1999 to 2009 were 740 per class, of which, from the requested sample size (740) per class, 50% were used for training and 50% for testing with iterations of 10,000.

13.3.5 Prediction and Validation

In this study, the default procudure was followed to forecast the future LULC for the year 2019 and 2029, using the Markov Chain. Markov Chain analysis projects the amount of change in the near future, using the earlier land cover at time t1 and the later land cover at time t2. It is based on a probability matrix, which records the probability of transition from one land-cover class to another land-cover class between the earlier land-cover map and the later land-cover maps to predict future land cover. Even if one or more of the following conditions may not be met, Levinson et al. (2012) made the following assumptions for effective functioning of the model. These are:

- in the table of the transition probability matrix, the sum of each row must be one;
- in the table of the transition probability matrix, the values must be constant for any two periods (in this study, the transition probability matrix of 1999 to 2009 must be similar to 2009 to 2019);
- the transition probabilities have no memory, which means that the LULC of the future is totally dependent on the present (the Markov condition); and
- the time interval in this study must be equal or similar to from 1999 to 2009, from 2009 to 2019 and from 2019 to 2029, which is 10 years.

Specific to the present study, the second assumption has a strong impact on the accuracy of the predicted map. Since conversion of agricultural areas to built-up areas has been intensifying over time, becoming more pronounced in the past decade, the predicted map may show lower conversion than the reality. To fill this gap, the researchers developed two scenarios: Business-as-Usual (Scenario I) and Urban Expansion (Scenario II) scenarios.

The business-as-usual scenario focuses on accepting the default Markov Chain prediction result without undertaking any change in the whole process.

The urban expansion scenario aims at modifying the probability matrix. By taking into account assumption two (similar transition matrix for any two periods), the high population growth and urban expansion of Bahir Dar in the later period, reducing the persistence of agriculture by half and increasing the conversion of agricultural land to built-up area by half, has been carried out.

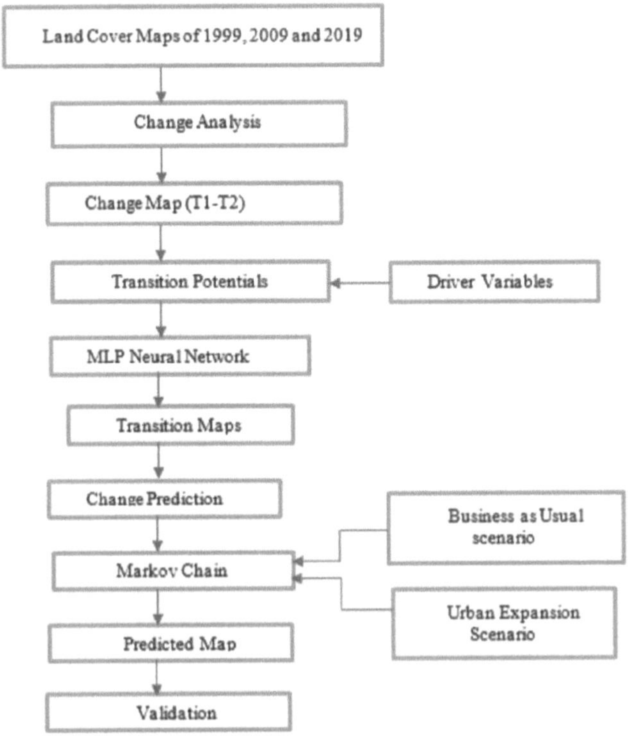

FIGURE 13.3 Flowchart of the study.

Finally, to check the accuracy and validity of the simulated map, crosstab and validate modules were used. Crosstabulation provides a comparative crosstabulation analysis of two images, namely the first image (reference map) and the second image (simulated map). It provides Kappa Index of Agreement values for each LULC category and overall Kappa (Pontius Jr. and Chen, 2006; Eastman, 2016). Thus, the similarity of the simulated/predicted map of 2019 to the reference map of 2019 was assessed. The validate module was used to check the validity of the simulated map in terms of quantity and locations. Hence, it assesses the degree of agreement between the simulated and reference maps on the basis of area (quantity) and location of cells in each land-cover class (Pontius Jr. and Chen, 2006; Eastman, 2016). Figure 13.3 demonstrates the general flowchart of the research.

13.4 RESULTS AND DISCUSSION

13.4.1 Change Analysis

As observed from the LULC map (Figure 13.4a), in 1999 the dominant LULC type was agriculture, which covered not only peri-urban Bahir Dar but also most of its inner city. Agricultural areas constituted 92% percent of the land-cover

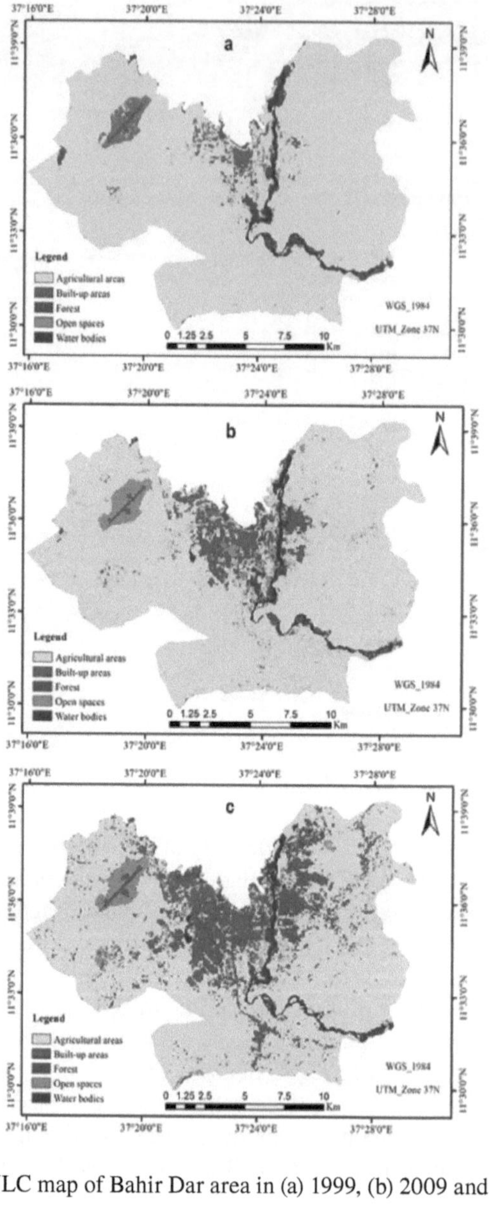

FIGURE 13.4 LULC map of Bahir Dar area in (a) 1999, (b) 2009 and (c) 2019.

classes, with other land-cover classes constituting less than 3%. The built-up area share was the lowest (1.52%), which was largely confined to the inner city, except for the airport, which was in the periphery.

In 2009, most of the agricultural area and open spaces in the inner city were encroached upon by built-up areas (Figure 13.4b), which caused the share of agricultural land to decrease (to 86.39%) and that of the built-up areas to rise to

TABLE 13.2
LULC Change Statistics from 1999 to 2009

	1999		2009		Change	
LULC Type	**Area (ha)**	**%**	**Area (ha)**	**%**	**Area (ha)**	**%**
Agricultural areas	23820.14	92.04	22358.87	86.39	−1461.27	−6.13
Built-up areas	394.17	1.52	1415.94	5.47	1021.77	259.22
Forest	614.46	2.37	618.42	2.39	3.96	0.64
Open spaces	579.56	2.24	1025.99	3.96	446.43	77.03
Water bodies	472.07	1.82	461.18	1.78	−10.88	−2.31
Total	25880.41	100	25880.41	100		

5.47%. Forest, open spaces and water bodies did not show any significant change (Table 13.2).

In 2019, whereas the agricultural areas were shrinking (to 77.28%), built-up areas engulfed peri-urban Bahir Dar (amounting to 15.54%), which caused intense land transformation. The share of other land cover classes remained almost constant: forest (3.01%), open spaces (3.96%) and water bodies (1.48%), with slight variations (Figure 13.4c, Table 13.2).

13.4.1.1 Land-Use/Land-Cover Changes from 1999 to 2009 in Bahir Dar Area

As shown from Table 13.1, the relative area of agricultural land declined from 92 percent (23,820.14 ha) in 1999 to 86 percent (22,358.87 ha) in 2009. However, built-up areas increased from 1.52 percent (394.17 ha) to 5.47 percent (1415.94 ha) in the same period. The change recorded by other land-cover classes between 1999 and 2009 varied from 2.37 percent (614.46 ha) to 2.39 percent (618.42 ha) for forest, from 2.24 to 3.96 percent for open spaces and from 1.82 to 1.78 percent for water bodies.

13.4.1.2 Land-Use/Land-Cover Changes from 2009 to 2019 in the Bahir Dar Area

In the past decade, the process of land transformation has been more intense and rapid, especially along the major routes and near to existing built-up areas (Figure 13.4c). As in the previous decade, agricultural areas have decreased further, from 86.39 percent in 2009 to 77.28 percent in 2019. In contrast, following the same trend, built-up areas have skyrocketed from 5.47 percent in 2009 to 15.54 percent in 2019. Transitions of the other land-cover classes have been insignificant (less than 1% each) (Table 13.3).

13.4.2 Determining the Driver Variables and Training the Model

The input images used for the modeling were LULC maps of 1999 and 2009. After the multiple trials, four transitions were selected, which yielded greater accuracy

TABLE 13.3
LULC Change Statistics from 2009 to 2019

	2009		2019		Change (from 2009 to 2019)	
LULC Type	Area (ha)	%	Area (ha)	%	Area (ha)	%
Agricultural areas	22358.87	86.39	20000.95	77.28	–2357.92	–10.55
Built-up areas	1415.94	5.47	4022.75	15.54	2606.81	184.11
Forest	618.42	2.39	779.17	3.01	160.74	25.99
Open spaces	1025.99	3.96	693.89	2.68	–332.10	–32.37
Water bodies	461.18	1.78	383.65	1.48	–77.54	–16.81
Total	25880.41	100	25880.41	100		

TABLE 13.4
Cramer's V of Driver Variables

No.	Driver Variables	Cramer's V	P Value
1	Slope	0.0297	0
2	Distance from water bodies	0.1822	0
3	Distance from major roads	0.1296	0
4	Evidence of likelihood of map-change	0.4395	0
5	Distance from built-up areas	0.3715	0

during the training of the MLP Neural Network, namely agricultural areas to built-up areas, agricultural areas to forest, forest to agricultural areas and open spaces to agricultural areas; these transitions are shown in maps (Figure 13.6). Driver variables with Cramer's V of 0.15 and above, which were considered useful, were incorporated into the model. As a result, the static variable "slope" was discarded, while all the other driver variables were included as dynamic variables (Table 13.4).

With all the parameters used as default as given by the model, the MLPNN result showed an accuracy rate of 81.78% and a training root-mean-square (RMS) of 0.2163%, which is acceptable (Eastman, 2016). The whole Run Transition Sub-Model panel, which shows the MLP Neural Network result, is presented in Figures 13.5 and 13.6.

13.4.3 Change Prediction

The projection of LULC change for 2019 was based on the 1999 and 2009 LULC maps, as an input, using the default Markov Chain model within LCM.

Scenario I (business-as-usual scenario) assumes the projection of future LULC change built on the historical trend of land-cover transitions (Hamad et al., 2018; Larbi et al., 2019). The transition probability matrix of LULC maps of 1999 (t1)

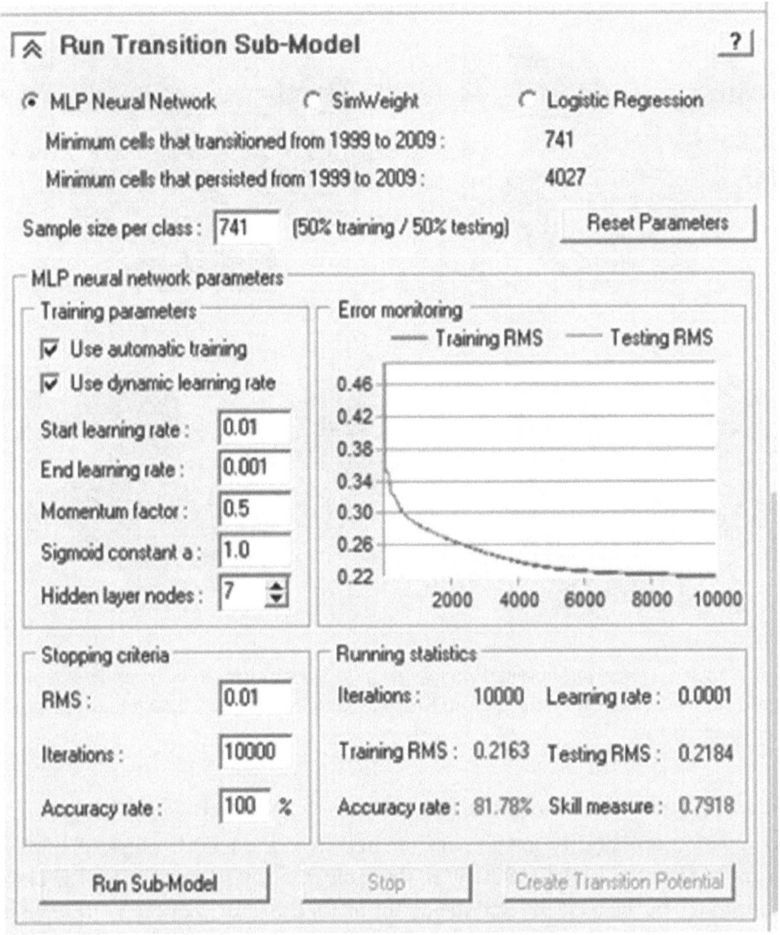

FIGURE 13.5 Training result of MLP Neural Network.

and 2009 (t2) were used to predict the LULC of 2019 (t3). Similarly, the transition probability matrix of LULC maps of 2009 (t1) and 2019 (t2) were used to predict the LULC of 2029 (t3). Hence, the rate and probability of transitions from one land-cover class to another remained similar throughout the period under study.

Scenario II (urban expansion scenario) is adapted from the afforestation scenario of Larbi et al. (2019). It assumes that, with the increasing population and urban expansion of the study area, the transition probability changes, especially where there has been a higher transition probability rate from agricultural land to built-up area. As a result, altering the probability matrix values for agricultural land and built-up area was carried out. Thus, increasing the probability of agricultural land being transformed to built-up area by 150 % and reducing the value for agriculture persistence by the corresponding proportion was performed.

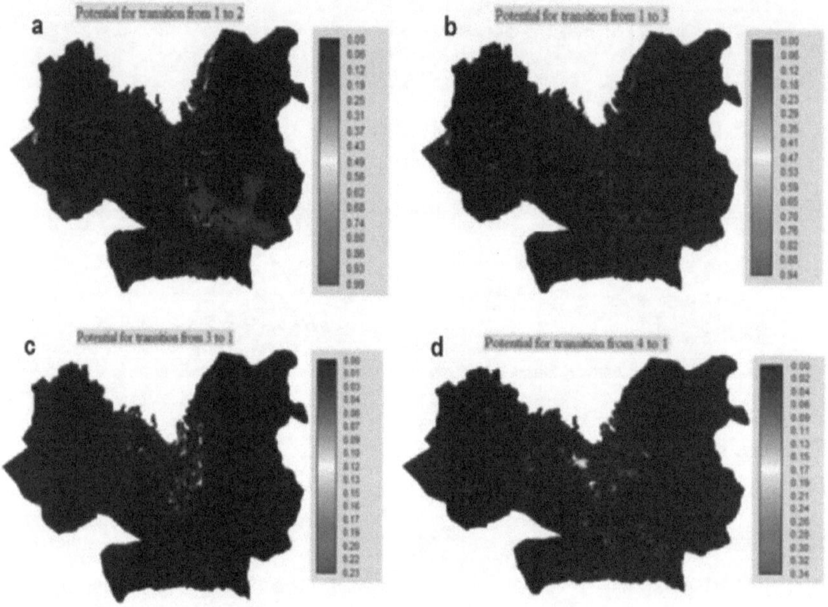

FIGURE 13.6 Transition potential maps from (a) agricultural to built-up land, (b) from agricultural land to forest area, (c) from forest area to agricultural land and (d) from open spaces to agricultural land.

13.4.3.1 Prediction Using Markov Chain – Scenario I

The transition probability matrix was created from the LULC maps of 1999 and 2009 to forecast the LULC of 2019. In the table, the rows stand for LULC classes of the initial year map (1999) and the columns for the later year map (2009). Thus, the agricultural areas recorded the highest persistence (0.9261) next to built-up areas (1.000) and the agricultural areas had the higher probability of converting to built-up areas (0.0423) than to other land-cover classes (Table 13.5).

The prediction result showed an increase of 1,483.22 ha (7.42 %) in agricultural area and a decrease in built-up areas by 1,666.91 ha (41.44 %), compared with the LULC map of 2019 (reference map). This variation is attributed to the high persistence probability value of agricultural areas from 1999 to 2009 LULC. Open spaces and water bodies increased by 121.9 ha (15.64 %) and 95.26 ha (24.83 %), respectively, whereas forest cover decreased by 121.9 ha (15.64 %) (Table 13.6, Figure 13.7b). Based on the assumptions and functions of the Markov Chain model, the result is satisfactory.

13.4.3.2 Prediction Result – Scenario II

By altering the probability matrix only for agriculture persistence and transition from agricultural land to built-up area, the LULC for 2019 was modeled, using the Markov Chain analysis.

TABLE 13.5
Transition Probability Matrix from 1999 to 2009 LULC

LULC Type	Agriculture	Built-up	Forest	Open Spaces	Water Bodies
Agriculture	0.9261 (0.8625)	0.0423 (0.1058)	0.0093	0.0198	0.0025
Built-up	0.0000	1.0000	0.0000	0.0000	0.0000
Forest	0.2721	0.0019	0.5891	0.0897	0.0472
Open spaces	0.1139	0.0137	0.0297	0.8427	0.0000
Water bodies	0.1351	0.0002	0.0334	0.0010	0.8303

TABLE 13.6
Area Statistics of 2019 (Reference) and 2019 (Scenario I and II) Maps

	2019 Reference Map		2019 Scenario I (II)		Change Scenario I (II)	
LULC Type	Area (ha)	%	Area (ha)	%	Area (ha)	%
Agricultural areas	20000.95	77.28	21484.17 (20064.78)	83.01 (77.53)	1483.22 (63.83)	7.42 (0.32)
Built-up areas	4022.75	15.54	2355.84 (3775.23)	9.10 (14.59)	−1666.91 (−247.52)	−41.44 (−6.15)
Forest	779.17	3.01	657.27	2.54	−121.90	−15.64
Open spaces	693.89	2.68	905.22	3.50	211.33	30.46
Water bodies	383.65	1.48	478.91	1.85	95.26	24.83
Total	25880.41	100	25881.41	100		

The model result shows that agricultural areas increased by only 63.83 ha (0.32%) and built-up areas decreased by 247.52 ha (6.15 %), which shows that the modification made in the urban expansion scenario (scenario II) gave a more accurate result. Since the modification of the transition probability matrix was carried out on persistent agricultural areas and the transition from agricultural areas to built-up areas, the remaining LULC categories showed no change (Table 13.6). Moreover, the LULC map of 2019 (reference) depicts a pattern of LULC distribution almost identical to that of the projected map of 2019, using scenario II (Figure 13.7c).

13.4.4 Model Validation

To evaluate the accuracy of the model, both the validate and crosstab modules were used. The predicted map of 2019 (scenario I) was crosschecked against the LULC map of 2019 (reference), using the validate module and the Kappa (K)

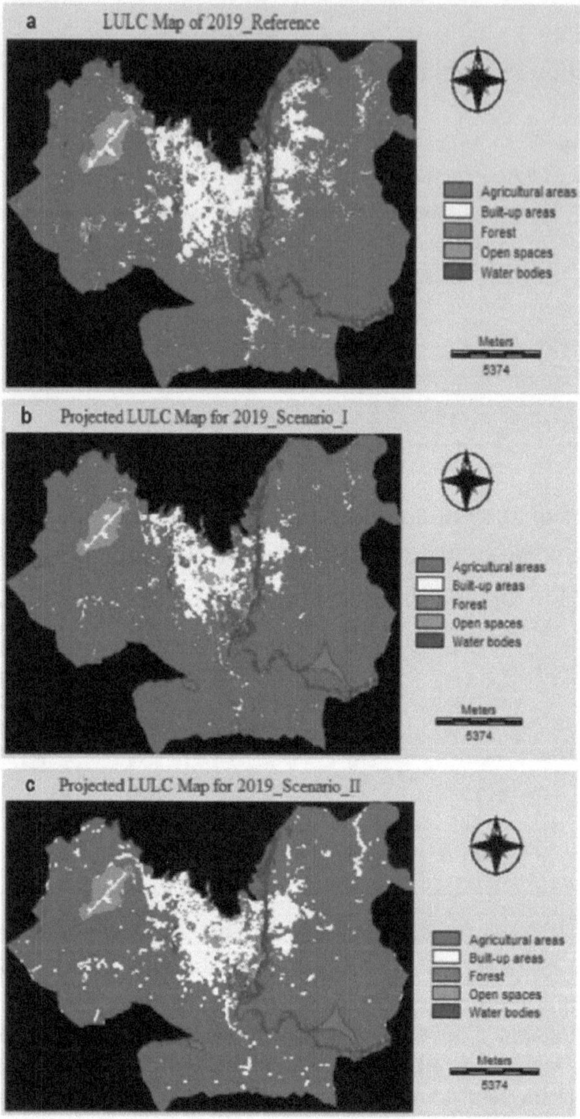

FIGURE 13.7 LULC map of (a) 2019, (b) Scenario I, and (c) Scenario II.

coefficient of quantity and location was obtained. The statistical report shows that Kno is 0.8836, Klocation is 0.9020 and Kstandard is 0.8391 (Table 13.7). Similarly, the crosstab report shows an overall Kappa index of 0.8391, with the lowest Kappa index of agreement recorded for forest (0.2631) followed by built-up area (0.4596). The higher quantity disagreement in built-up classes leads to a lower Kappa index of agreement (Table 13.7). In scenario II, the crosstab result

TABLE 13.7
Kappa Index of Agreement (KIA) for 2019 LULC Map (Scenario I and II)

Validate			Crosstab		
K indicators	2019 (I)	2019 (II)	LULC Type	2019 Scenario I (KIA)	2019 Scenario II (KIA)
Kno	0.8836	0.8739	Agricultural areas	0.8806	0.8082
Klocation	0.9020	0.8412	Built-up areas	0.4596	0.5769
KlocationStrata	0.9020	0.8412	Forest	0.2631	0.2631
Kstandard	0.8391	0.8293	Open spaces	0.6069	0.6069
			Water bodies	0.8525	0.8525
			Overall Kappa	0.8391	0.8293

TABLE 13.8
Transition Probability Matrix of Scenario I (II) from 2009 to 2019 LULC

LULC Type	Agriculture	Built-up	Forest	Open Spaces	Water Bodies
Agriculture	0.8626 (0.7007)	0.1079 (0.2698)	0.0159	0.0114	0.0022
Built-up	0.0000	1.0000	0.0000	0.0000	0.0000
Forest	0.4447	0.0503	0.4819	0.0057	0.0175
Open spaces	0.3468	0.1555	0.0583	0.4245	0.0148
Water bodies	0.1828	0.0064	0.1404	0.0002	0.6702

shows an improved Kappa index of agreement for built-up areas (0.5769) and an overall Kappa index of 0.8293 (Table 13.7). The overall accuracy results of both the validate and crosstab modules (TerrSet) for scenarios I and II predicted maps which show small quantification and location errors, fulfilling the minimum standard (Pontius Jr. and Chen, 2006; Eastman, 2016).

13.4.5 LULC Prediction for 2029

In the first (business-as-usual) scenario (I), the default projection was executed, using the Markov Chain model. Therefore, the LULC of 2029 is predominantly dependent on the trend of land-cover transitions from 2009 to 2019. Compared with the previous decade, the transition probability of agricultural to built-up area has been higher (0.1079) and the persistence of agriculture has decreased to 0.8626 between 2009 and 2019 (Table 13.8). In the second (urban expansion) scenario (II) the transition probability matrix was modified to forecast the LULC for 2029. Thus, in the transition probability from agricultural land to built-up

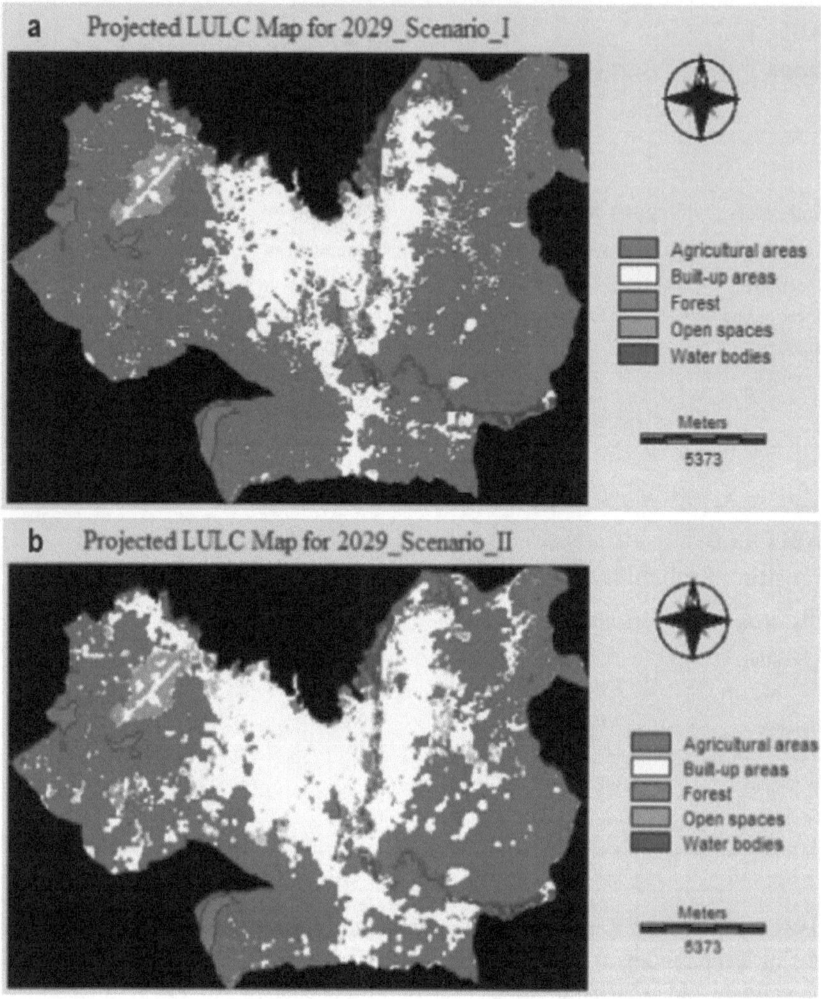

FIGURE 13.8 Predicted LULC map of 2029 by (a) Scenario I and (b) Scenario II.

area, 150% was added and the same value was deducted from agriculture persistence (Table 13.8). Difficult to validate, visually both the business-as-usual and the urban expansion scenario LULC maps of 2029 showed acceptable patterns of land transformation (Figure 13.8a and b).

13.4.6 Land-Use/Land-Cover Changes from 2019 to 2029 in the Bahir Dar Area

Change detection analysis shows that there will be a reduction in agricultural areas from 20,000.95 ha (77.28 %) in 2019 to 17,366 ha (67.1%) in 2029. In the

TABLE 13.9
LULC Change Statistics from 2019 and 2029 (Scenario I and II)

	2019		2029 Scenario I (II)		Change Scenario I (II)	
LULC Type	**Area (ha)**	**%**	**Area (ha)**	**%**	**Area (ha)**	**%**
Agricultural areas	20000.95	77.28	17366.97 (14128.88)	67.10 (54.59)	−2633.98 (−5872.08)	−13.17 (−29.36)
Built-up areas	4022.75	15.54	6288.83 (9526.92)	24.30 (36.81)	2266.07 (5504.17)	56.33 (136.83)
Forest	779.17	3.01	1097.15	4.24	317.98	40.81
Open spaces	693.89	2.68	813.98	3.15	120.09	17.31
Water bodies	383.65	1.48	313.48	1.21	−70.16	−18.29
Total	25880.41	100	25880.41	100		

same period, built-up areas will increase from 4,022.75 ha (15.54%) to 6,288.83 ha (24.3%). Likewise, other land-cover classes, including forest and open spaces, are expected to witness incremental increases of 317.98 ha and 120 ha, respectively, in the upcoming decade. In contrast, the water body area will show a very small reduction (70 ha), with a significant change compared to its total area (Table 13.9). Based on the 2029 urban expansion (scenario II) LULC result, agricultural areas will record a dramatic decline from 20,000.95 ha (77.20%) in 2019 to 14,128.88 ha (54.59%) in 2029, leading to a loss of 5,872.08 ha of agricultural land within a decade. Conversely, built-up areas will witness an increase in area of 5,504.17 ha of land, which will upgrade its total share to 36.81%. Other land-cover classes will remain almost the same as with the scenario I result (Table 13.9).

13.5 CONCLUSION

For sustainable urban and regional development, the accuracy of past, present and future spatio-temporal LULC information of urban centers has become vital. With the advances of geospatial technologies, a range of models (such as the LCM, MPLPNN and MC) have been developed and integrated with different software programs (like ArcGIS, IDRISI and TerrSet) to monitor and predict LULC change. The current study predicted and assessed the LULC of the Bahir Dar area from 1999 to 2029, using the MLP Neural Network and MC integrated in LCM of the TerrSet (IDRISI) software environment. The model validation (crosstab and validate modules), using the reference map of 2019, delivered a satisfactory result. The change analysis result revealed that the study area has been experiencing rapid land transformation since 1999, especially from agricultural land to built-up area. In 1999, the agricultural area covered 92% of the study area, which decreased to 86.39% in 2009, whereas the built-up areas had the lowest coverage (1.52%) in 1999, which increased to 5.47% in 2009. The change and conversion among other LULC classes were very small. With the major conversion happening between agricultural land and built-up area, between 2009 and 2019, agricultural areas decreased from 86.39% to 77.28%, while built-up areas increased from 5.47% to 15.54%.

Likewise, the prediction results of 2029 (scenarios I and II) showed extensive land transformations in the upcoming decade. Based on the predicted result of scenario I (business-as-usual), agricultural areas are expected to reduce from 77.28% (20,000.95 ha) in 2019 to 67.1% (17,366 ha) in 2029, whereas built-up areas will increase from 15.54% (4,022.75 ha) to 24.30% (6,288.83 ha). The change in other LULC classes will be very small (less than 1%).

Generally, the process of LULC change in Bahir Dar city and its periphery has been escalating since 1999. Following the increase in population and concomitant rapid urban expansion, huge amounts of farmland have been transformed to built-up area, with more conversion expected in the following decade (to 2029). Moreover, the study showed the LULC prediction power of MLPNN and MC integrated with Land Change Modeler (LCM), which could be replicated on other cities under similar situations.

REFERENCES

Eastman, J.R. (2006). *Guide to GIS and Image Processing.* Clark Labs, Clark University. April, 0–327.

Eastman, J.R. (2016). *TerrSet Geospatial Monitoring and Modeling System Manual.* Clark University.

Hamad, R., Balzter, H., & Kolo, K. (2018). Predicting land use/land cover changes using a CA-Markov model under two different scenarios. *Sustainability (Switzerland), 10*(10), 1–23. https://doi.org/10.3390/su10103421

Jamali, A.A. (2019). Jamali-GhorbaniKalkhajeh2019_Article_UrbanEnvironmentalAndLandCover.pdf.

Kindu, M., Schneider, T., Teketay, D., & Knoke, T. (2015). Drivers of land use/land cover changes in Munessa-Shashemene landscape of the south-central highlands of Ethiopia. *Environmental Monitoring and Assessment, 187*(7). https://doi.org/10.1007/s10661-015-4671-7

Larbi, I., Forkuor, G., Hountondji, F.C.C., Agyare, W.A., & Mama, D. (2019). Predictive land use change under business-as-usual and afforestation scenarios in the Vea catchment, West Africa. *International Journal of Advanced Remote Sensing and GIS*, 8(1), 3011–3029. https://doi.org/10.23953/cloud.ijarsg.416

Levinson, D., El-geneidy, A., Building, M., & Wasfi, R. (2012). A Markov chain model of land use change in the twin cities, 1958-2005. *Journal of Land Use, Mobility and Environment* 8(3), 1–24.

Mas, J.F., Kolb, M., Paegelow, M., Camacho Olmedo, M.T., & Houet, T. (2014). Inductive pattern-based land use/cover change models: A comparison of four software packages. *Environmental Modelling and Software, 51*, 94–111. https://doi.org/10.1016/j.envsoft.2013.09.010

MUDHCo. (2015). *State of Ethiopian Cities Report.* http://www.mudc.gov.et/c/document_library/get_file?uuid=cf0ef7a3-66c9-429c-b921-be76b9291851&groupId=10136

Nasiri, V., Darvishsefat, A.A., Rafiee, R., Shirvany, A., & Hemat, M.A. (2019). Land use change modeling through an integrated multi-layer perceptron neural network and Markov chain analysis (case study: Arasbaran region, Iran). *Journal of Forestry Research, 30*(3), 943–957. https://doi.org/10.1007/s11676-018-0659-9

Pontius, Jr. R.G., & Chen, H. (2006). *GEOMOD Modeling*. Clark Labs, 1–44.

Regmi, R.R., S.K.S. and M.K.B. (2014). Geospatial analysis of land use land cover change predictive modeling at Phewa Lake Watershed of Nepal. *4*(4), 2617–2627.

Regmi, R.R., Saha, S. K. and Balla, M. K. (2014). Geospatial analysis of land use land cover change modeling at Phewa Lake Watershed of Nepal by using Cellular Automata Markov Model. *Int. J. Curr. Eng. Tech. 4*(4), 2617–2627.

UNDESA. (2018). World urbanization prospects. *Demographic Research*, 12. https://doi.org/10.4054/demres.2005.12.9

UNDESA. (2019). World population prospects. In *World Population Prospects 2019: Vol. I*. http://www.ncbi.nlm.nih.gov/pubmed/12283219

Verma, P., & Raghubanshi, A.S. (2019). Rural development and land use land cover change in a rapidly developing agrarian South Asian landscape. *Remote Sensing Applications: Society and Environment, 14*(March), 138–147. https://doi.org/10.1016/j.rsase.2019.03.002

Yesuph, A.Y., & Dagnew, A.B. (2019). Land use / cover spatiotemporal dynamics, driving forces and implications at the Beshillo catchment of the Blue Nile Basin, North Eastern Highlands of Ethiopia. *Environmental Systems Research*. https://doi.org/10.1186/s40068-019-0148-y

14 Geoinformatics Approach to Desertification Evaluation

Using Vegetation Cover Changes in the Sudano-Sahelian Region of Nigeria from 2000 to 2010

O.Y. Ekundayo, E.C. Okogbue, F.O. Akinluyi, A.M. Kalumba, and I.R. Orimoloye

CONTENTS

14.1 INTRODUCTION

The United Nations has defined desertification processes as the "degradation of the land into arid, semi-arid and sub-humid areas, resulting from various factors, including climatic variations and human activity" (UNEP 1992). The importance of vegetation as a type of land cover over any area cannot be overestimated, since, in some cases, it is the initial land cover of the area (Kabisch et al., 2019; Nguyen et al., 2020). The extent to which vegetation is being depleted in an area or region, be it periodically, seasonally, or yearly, is a pointer to the occurrence of desertification in the area. Human activities, such as over-farming, over-grazing, overpopulation of marginal lands, or deforestation, are some reasons for the loss or removal of vegetation from a region (Atem, 2011). The effects of climate change, such as increased air temperature or decreased precipitation, will cause drought conditions and prevent natural vegetation establishment and spread (Orimoloye et al., 2019; Pugnaire et al., 2019; Snyder et al., 2019). The vegetation loss caused by desertification lowers carbon sinks, resulting in the release of larger amounts of atmospheric greenhouse gases, which, in turn, leads to climate change and then again to desertification (forming a cycle) (Zeng and Neelin, 1999).

Globally, desertification is one of the major problems, especially in dryland areas (including significant areas of Africa, such as the regions of Sudan and the Sahel Savanna). Approximately 40 per cent of the Earth's land area is covered by drylands, and over 2 billion people live in these areas. Drylands, due to the low soil water content, are highly vulnerable to natural and human destruction (UNCCD, 2014). There is a thin line between drylands and deserts, and it is hard to return once that line has been crossed. Preventing drylands from degradation is much more cost-effective than reversing it, since land degradation often creates desert-like conditions. Globally, desertification affects some 35 million km^2 of land cover (UN, 1980). Between the early 1970s and the mid-1990s, the African Sahel experienced one of the world's most dramatic long-term climate changes in the 20th century, with rainfall declining by more than twenty per cent on average (Hulme et al., 2001; Nicholson, 2001; Kruger and Shongwe, 2004). The United Nations Convention on Combating Desertification (UNCCD) recently indicated that desertification's climatic effects on land occur at the ecosystem and landscape level. Individual and community efforts to rehabilitate the land are therefore most effective when they are part of an effort at country or regional level to preserve and rehabilitate the landscape (UNCCD, 2014).

14.2 MATERIALS AND METHODS

The study area is the Sudan Savanna and the Sahel Savanna region of the Agro-Ecological zone of Nigeria, also known as the Sudano-Sahelian region. The area is geographically located between latitudes 10°–14°N and longitudes 4°E–14°E. The annual total rainfall of the area is in the range 250–500 mm and the temperature is in the range 15°C–40°C. The Moderate-resolution Imaging Spectroradiometer (MODIS) Vegetation Indices data, MOD13A3 monthly Terra imageries and the map of the area was used in this study. The study area map of the region was produced by geo-referencing, digitizing the Nigerian Ecological map and clipping out the region needed. The MODIS Terra imageries were converted to float using the 'Float Function' of ArcGIS software. The scale factor for MOD13A3 (Terra) imagery was used by converting MODIS vegetation index to Normalized Difference Vegetation Index (NDVI) (with values from −1 to +1). The monthly NDVI maps were produced by overlaying the MODIS NDVI imageries onto the study area map. The month NDVI maps were used to generate the annual NDVI

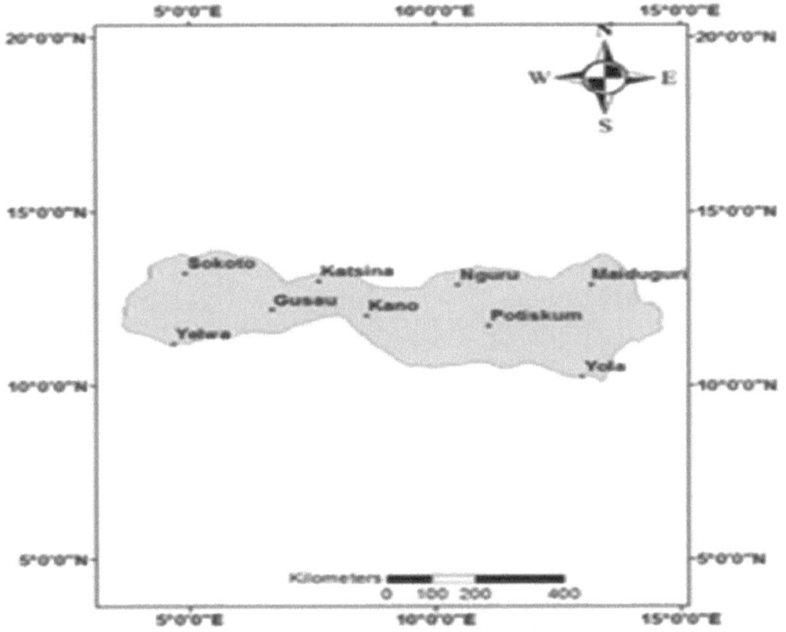

FIGURE 14.1 Study area map showing some locations in the Sudano-Sahelian region of Nigeria.

maps through the use of the Raster Calculator Tool (Figure 14.1). The yearly NDVI was classified into six classes, A, B, C, D, E, and F, representing water bodies, desert and bare-land, semi-desert, steppe, shrub and grassland, dense vegetation and forest, respectively, from the year 2000 to 2010, adapted from the classification of Weier and Herring (1999) and Tahira (2010). The corresponding NDVI values and their classes, used for this work, are seen in Table 14.1.

14.3 RESULTS AND DISCUSSION

The NDVI, an indicator of vegetation health and density, showed the annual changes in the vegetation cover of the Sudano-Sahelian region of Nigeria over the period of study. Figures 14.2(a–b)–14.3(a–i) show the annual coverage distribution of NDVI classes in the Sudano-Sahelian region from 2000 to 2010. Figure 14.2(a-b) showed annual NDVI in this region from 2000 to 2003. The NDVI for the year 2000 varied from −0.175 to 0.778 (Figure 14.2a) and, for the year 2001, from −0.17 to 0.778 (Figure 14.2b). NDVI values for the year 2002 varied from −0.172 to 0.797 (Figure 14.3a) and for the year 2003 from −0.167 to 0.79 (Figure 14.3b). The years 2004 and 2005 showed varying NDVI from −0.178 to 0.78 (Figure 14.3c) and from −0.17 to 0.763 (Figure 14.3d), respectively. The NDVI values varied from −0.181 to 0.745 for 2006 (Figure 14.3e), from −0.175 to 0.733 for 2007 (Figure 14.3f). For 2008, the NDVI varied from

TABLE 14.1
NDVI Value and Its Interpretation

NDVI Value	Classification
−1.0–0	Water bodies
0–0.1	Desert
0.1–0.2	Semi-desert
0.2–0.3	Semi-desert
0.3–0.4	Steppe
0.4–0.6	Shrub and grassland
0.6–1.0	Dense vegetation and forest

Source: Adapted from Weier and Herring (1999) and Tahira (2010).

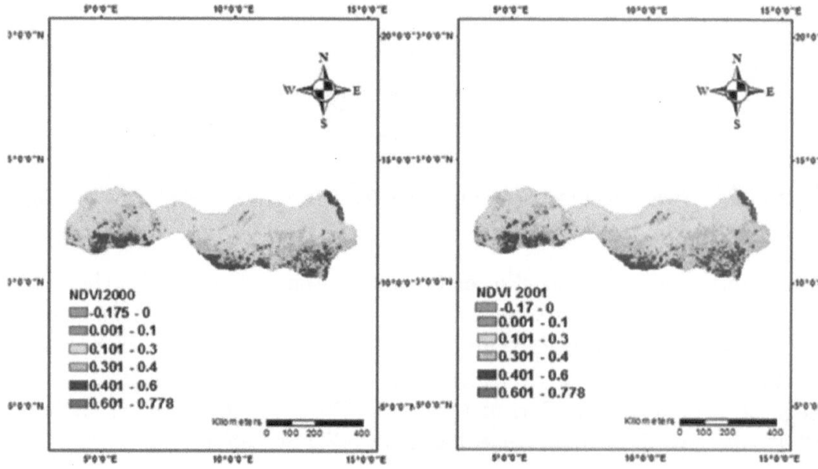

FIGURE 14.2(A–B) Yearly spatial images of MODIS NDVI from (a) 2000 and (b) 2001 for the Sudano-Sahelian region of Nigeria.

−0.171 to 0.725 (Figure 14.3g), whereas, for 2009, the NDVI varied from −0.166 to 0.717 (Figure 14.3h) and for 2010, from −0.172 to 0.710 (Figure 14.3i). Throughout these years, the NDVI value ranges of −1.0 to 0.0, 0.0 to 0.1, 0.1 to 0.3, 0.3 to 0.4, 0.4 to 0.6, and 0.6 to 1.0 were denoted as classes A, B, C, D, E, and F, respectively, these representing water bodies, desert, and bare-land, semi-desert, steppe, shrub and grassland, dense vegetation and forest, respectively, as shown in Table 14.1. The variation in NDVI might have been influenced by multiple factors, such as nitrogen deposition, CO_2, fertilization, forest regrowth, and climatic changes, which affect vegetation productivity in the study area (Milesi et al., 2010; He et al., 2012). Increased temperature has the ability to alter the composition of the landscape and the characteristics of

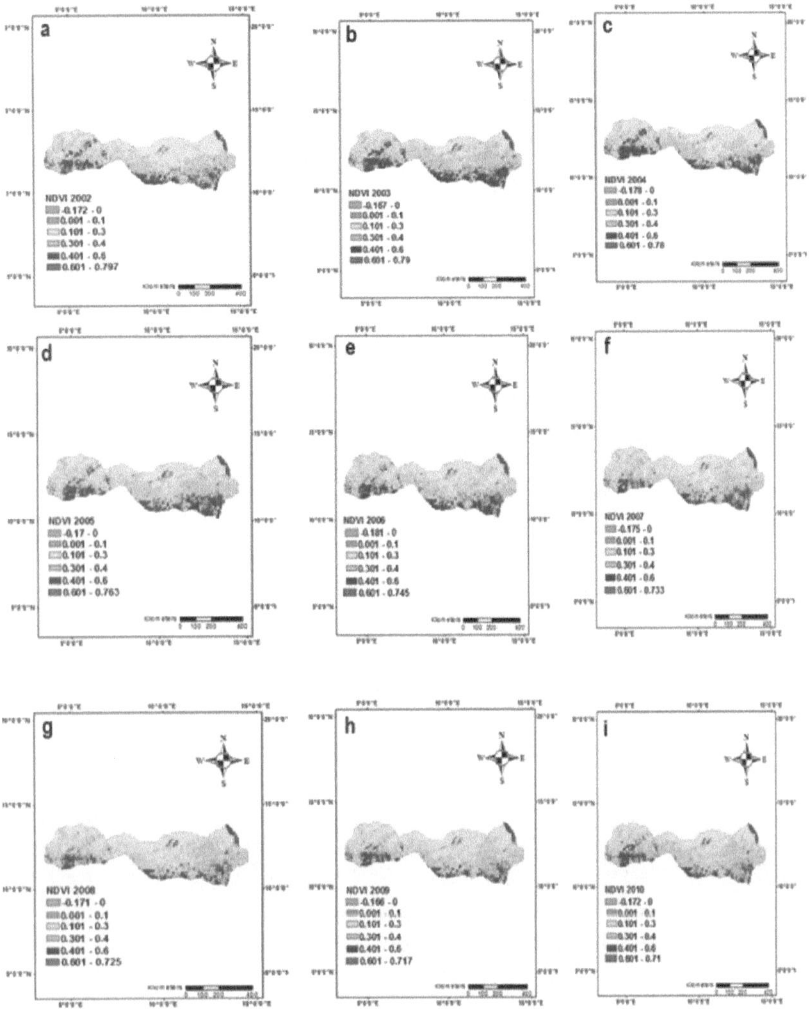

FIGURE 14.3(A–I) Yearly spatial images of MODIS NDVI from (a) 2002, (b) 2003, (c) 2004, (d) 2005, (e) 2006, (f) 2007, (g) 2008, (h) 2009, and (i) 2010 for the Sudano-Sahelian region of Nigeria.

the vegetation cover (e.g., infiltration of tall shrubs and forests on tundra) (He et al., 2012). Other potential consequences may include the loss or disappearance of some ice-associated animals (e.g., polar bears) (Molnar et al., 2010). Published research on a range of topics strongly suggests that, at northern middle to high latitudes, many effects are already being felt. Studies have shown, for example, that a warming climate causes deeper thawing of the active layer or a greater effect on vegetation cover (Zhang et al., 2018; Rasouli et al., 2019; Huang et al., 2020), and increased biological productivity and changes in vegetative communities.

The semi-desert, NDVI class C, covered the study area to a wide extent through all the years, especially 2000, 2001, and 2006, but retreated northwards in 2003, 2004, and 2010. Similarly, in 2008, the semi-desert class retreated northwards at the east side of Potiskum and in 2010 at the west of Gusau. The NDVI class D, which is steppe, also covered a wide extent of the Sudano-Sahelian region, occupying mainly the South of the semi-desert area throughout the years. The area around Sokoto, Nguru, Katsina, and Gusau was dominated by either semi-desert or steppe throughout the study period. Table 14.1 shows that, in some years, there was more semi-desert than steppe, and *vice versa*; these two classes were the main NDVI classes in the Sudano-Sahelian region, displacing other vegetative NDVI classes. The NDVI class E, denoting the shrub and grassland, dominated the area around Yola (close to the south-east boundary of the study area), Yelwa, the south-east of Kano close to the boundary of the study area and the south of Potiskum in 2003, 2005, and 2006, but was displaced in some areas by NDVI class D, steppe, in 2008, 2009, and 2010. Generally, NDVI class E was found very close to the south boundary of the region and immediately next to the NDVI class F (dense vegetation and forest). Only areas around Lake Chad, close to Maiduguri, were in class F, which was more of a gallery forest, usually found around water bodies (Beard, 1955; Onamuti et al., 2017). The areas covered by NDVI class A, 'waterbodies', (–1.0 to 0.0), and NDVI class B, 'desert and bare-land', (0.0 to 0.1), were not so obvious on all the spatial maps because their coverage was so low.

Table 14.2 shows the area covered by the different individual NDVI classes while Table 14.3 shows the increase and decrease in area coverage of the NDVI classes in the Sudano-Sahelian region over the years of the study period, where a negative sign denotes a decrease, whereas a positive sign denotes an increase in area covered by the particular NDVI class. The NDVI class A, representing water bodies, which range from –1 to 0, varied from 317.69 km^2 to 261.025 km^2 in the region during the study period (Table 14.2). A net decrease of 56.67 km^2 in area covered by the NDVI class A was observed during the period under study, as seen in Table 14.3.

The NDVI class B, representing desert and bare-land (range 0–0.1), in the Sudano-Sahelian Savanna occupied an area ranging from 94.449 km^2 to 146.827 km^2 of the region in the period from years 2000 to 2010 (Table 14.2). NDVI class B exhibited a net increase of 34.346 km^2 in its coverage within the region (Table 14.3). NDVI class C, representing semi-desert (range 0.1–0.3), occupied a total area ranging from 111,055.811 km^2 to 170,123.009 km^2 (Table 14.2.), a total net decrease of 44,643.402 km^2 in its area was observed as seen in Table 14.3 throughout the study period. This represented a decrease in the area covered by the semi-desert.

The NDVI Class D is the steppe class. A steppe can be defined as a vast flat area of land with xerophilous vegetation (treeless tracts and very few grasses which require little moisture to thrive), usually found in regions of extreme temperature and loess soil. The area of the Sudano-Sahelian Savanna region occupied by steppe ranged from 86,650.837 km^2 to 13,208.351 km^2 within the study period (2000 to 2010). There was an increase in the steppe coverage over the study years, with a net increase of 45,435.514 km^2 (Table 14.3). Short trees, low woody plants

TABLE 14.2
NDVI Classes and the Area (km^2) Covered by Each Class for the Sudano-Sahelian Savanna Region

	NDVI Classes					
Years	**A (−1 to 0.0)**	**B (0.0 to 0.1)**	**C (0.1 to 0.3)**	**D (0.3 to 0.4)**	**E (0.4 to 0.6)**	**F (0.6 to 1.0)**
2000	317.695	112.481	170123.009	86650.837	29932.864	1191.785
2001	297.088	118.492	140462.332	113184.366	33110.671	1156.581
2002	285.067	104.753	136229.2632	117468.095	33260.932	981.419
2003	291.077	106.471	111055.811	125866.401	49705.504	1281.083
2004	291.077	94.449	136505.744	110017.722	40381.589	1037.231
2005	274.763	102.178	121205.732	125438.801	40368.71	932.478
2006	270.469	96.167	134978.232	116818.109	35309.634	854.342
2007	273.904	109.095	144729.745	112864.955	29350.701	999.451
2008	290.219	97.884	148438.189	113609.391	25161.431	732.454
2009	282.491	125.361	157745.789	103446.591	26050.977	674.887
2010	261.025	146.827	125488.602	132086.351	29618.604	728.122

NB: Class A = waterbodies; class B = desert and bare-land; class C = semi-desert; class D = steppe; class E = shrub and grassland; class F = dense vegetation and forest.

TABLE 14.3
Changes in Area (km²) Covered by Individual NDVI Classes Over the Period 2000 to 2010 in the Sudano–Sahelian Region of Nigeria

	NDVI Classes					
Period	A (–1 to 0.0)	B (0.0 to 0.1)	C (0.1 to 0.3)	D (0.3 to 0.4)	E (0.4 to 0.6)	F (0.6 to 1.0)
2000–2001	–20.607	+6.011	–2966 0.677	+26533.529	+3177.807	–35.204
2001–2002	–12.021	–13.739	–4233. 069	+4283.729	+150.261	–175.162
2002–20003	+6.010	+1.717	–25173.452	+8398.306	+164–44.5 72	+299.664
2003–2004	0.0	–12.022	+25449.93 3	–15848.679	–9323.915	–243.852
2004–2005	–16.314	+7.729	–1 5300.01 2	+15421.079	–12.879	–104.753
2005–2006	–4.294	–6.011	+13772.500	–8620.692	–5 05 9.076	–78.136
2006–2007	+3.435	+12.928	+9751.513	–3953.154	–5958.933	+145.109
2007–2008	+16.315	–11.211	+3708.444	+744.436	–4189.270	–266.997
2008–2009	–7.728	+27.477	+9307.600	+10162.800	+889.546	–57.567
2009–2010	–21.466	+21.466	–32257.187	+28639.760	+3567.637	+53.235
Total increase	+25.760	+77.328	+61989.990	+84020.839	+24229.823	+498.008
Total decrease	–82.430	–4.2983	–106624.397	–38585.325	–24544.073	–961.671
Net increase	0.0	+24.345	0.0	+45435.514	0.0	0.0
Net decrease	–52.670	0.0	–44634.407	0.0	–314.250	–463.663

with several stems and grasses, representing 'shrub and grassland', represent the NDVI class E (0.4 to 0.6). As shown in Table 14.2, the area covered by shrub and grassland varied from 40,368.71 km^2 to 49,705.504 km^2 during the study period. A net decrease of about 314.25 km^2 was noted in the NDVI class E over the study period (Table 14.3). This decrease in the area covered by the shrub and grassland could be attributed to anthropogenic activities, such as deforestation, bush burning, and so on. The NDVI class F (0.6 to 1.0) represents the vegetation combination, 'dense vegetation and forest'. NDVI class F occupied a range from 674.887 km^2 to 1281.083 km^2 (Table 14.2) of the region during the study period, exhibiting a net decrease of 463.663 km^2 in its area coverage from 2000 to 2010 (Table 14.3) There was a large reduction in the area covered by dense vegetation and forest between 2000 and 2010, although the extent of coverage of this class was wide, compared with some other classes, as observed in Table 14.2.

14.4 CONCLUSION

The areas covered by NDVI class A, 'waterbodies', and NDVI class B, 'desert and bare-land', were not obvious on the spatial maps, because their coverage was low; however, they are clearly represented in the tables. The dominant NDVI classes were class C (NDVI range 0.1–0.3) and class D (0.3–0.4) from 2000 to 2010, representing semi-desert and steppe, respectively. It is observed that the decrease in the semi-desert coverage during the study period was replaced by an increase in steppe coverage. It was found that there was a trend of decreasing area occupied by 'shrub and grassland' and by 'dense vegetation' in the region for the period of study, denoting desertification in the region. Many of the issues related to the debate on desertification in this part of the world, as in others, are surrounded by confusion and contradictory results. Many of those controversies underlie a fundamental tension between two alternate views of vegetation dynamics. Globally, including the study region in question, vegetation recovery continues to be endorsed by government policies, as it is regarded as an ongoing land-use activity that is an increasingly sustainable solution to the highly volatile natural environment, whereas livestock dynamics follow the conventional 'boom and bust' process, in line with the theory of non-equilibrium.

REFERENCES

Atem, P. A. (2011). *The Forms and Effects of Soil Degradation on Agricultural Land: Case Study Bor Town, Jonglei State*. South Sudan.

Beard, J. S. (1955). A note on gallery forests. *Ecology, 36*(2), 339–340. doi:10.2307/1933242

He, Y., Guo, X., Dixon, P., & Wilmshurst, J. F. (2012). NDVI variation and its relation to climate in Canadian ecozones. *The Canadian Geographer/Le Géographe Canadien, 56*(4), 492–507.

Huang, S., Zheng, X., Ma, L., Wang, H., Huang, Q., Leng, G., ... Guo, Y. (2020). Quantitative contribution of climate change and human activities to vegetation cover variations based on GA-SVM model. *Journal of Hydrology, 584*, 124687.

Hulme, M., Doherty, R., Ngara, T., New, M., & Lister, D. (2001). African climate change: 1900–2100. *Climate Research, 17*, 145–168.

Kabisch, N., Selsam, P., Kirsten, T., Lausch, A., & Bumberger, J. (2019). A multi-sensor and multi-temporal remote sensing approach to detect land cover change dynamics in heterogeneous urban landscapes. *Ecological Indicators*, *99*, 273–282.

Kruger, A. C., & Shongwe, S. (2004). Temperature trends in South Africa: 1960–2003. *International Journal of Climatology: A Journal of the Royal Meteorological Society*, *24*(15), 1929–1945.

Milesi, C., Samanta, A., Hashimoto, H., Kumar, K. K., Ganguly, S., Thenkabail, P. S., ... Myneni, R. B. (2010). Decadal variations in NDVI and food production in India. *Remote Sensing*, *2*(3), 758–776.

Molnár, P. K., Derocher, A. E., Thiemann, G. W., & Lewis, M. A. (2010). Predicting survival, reproduction and abundance of polar bears under climate change. *Biological Conservation*, *143*(7), 1612–1622.

Nguyen, L. H., Joshi, D. R., Clay, D. E., & Henebry, G. M. (2020). Characterizing land cover/land use from multiple years of Landsat and MODIS time series: A novel approach using land surface phenology modeling and random forest classifier. *Remote Sensing of Environment*, *238*, 111017.

Nicholson, S. E. (2001). Climatic and environmental change in Africa during the last two centuries. *Climate Research*, *17*(2), 123–144.

Onamuti, O. Y., Okogbue, E. C., & Orimoloye, I. R. (2017). Remote sensing appraisal of Lake Chad shrinkage connotes severe impacts on green economics and socio-economics of the catchment area. *Royal Society Open Science*, *4*(11), 171120.

Orimoloye, I. R., Ololade, O. O., Mazinyo, S. P., Kalumba, A. M., Ekundayo, O. Y., Busayo, E. T., ... Nel, W. (2019). Spatial assessment of drought severity in Cape Town area, South Africa. *Heliyon*, *5*(7), e02148.

Pugnaire, F. I., Morillo, J. A., Peñuelas, J., Reich, P. B., Bardgett, R. D., Gaxiola, A., ... Van Der Putten, W. H. (2019). Climate change effects on plant-soil feedbacks and consequences for biodiversity and functioning of terrestrial ecosystems. *Science Advances*, *5*(11), eaaz1834.

Rasouli, K., Pomeroy, J. W., & Whitfield, P. H. (2019). Are the effects of vegetation and soil changes as important as climate change impacts on hydrological processes?. *Hydrology and Earth System Sciences*, *23*(12), 4933–4954.

Snyder, K. A., Evers, L., Chambers, J. C., Dunham, J., Bradford, J. B., & Loik, M. E. (2019). Effects of changing climate on the hydrological cycle in cold desert ecosystems of the Great Basin and Columbia Plateau. *Rangeland Ecology & Management*, *72*(1), 1–12.

Tahira, A. (2010). *Detection and Analysis of Changes in Desertification in the Caspian Sea Region.* Masters Thesis submitted to the Department of Physical Geography and Quaternary Geology, Stockholm University.

UNEP (1992). *The Status of Desertification and Implementation of the United Nations Plan of Action to Combat Desertification.* Nairobi: UNEP.

United Nations (1980). *Desertification*, M. K. Biswas and A. K. Biswas, Eds., Oxford: Pergamon Press.

Weier, J., & Herring, D. (1999), Measuring vegetation (NDVI & RVI). http://earthobservatory.nasa.gov/Features/MeasuringVegetation/, Accessed on 10/02/2010.

Zeng, N and Neelin, J.D (1999). A land-atmosphere interaction theory for the tropical deforestation problem. *Journal of Climate*, *12*, 857–872.

Zhang, Z., Chang, J., Xu, C. Y., Zhou, Y., Wu, Y., Chen, X., ... Duan, Z. (2018). The response of lake area and vegetation cover variations to climate change over the Qinghai-Tibetan Plateau during the past 30 years. *Science of the Total Environment*, *635*, 443–451.

Section IV

Resource Analysis and Bibliometric Studies

15 Role of Geospatial Technology in Crop Growth Monitoring and Yield Estimation

P. K. Kingra, Raj Setia, Jatinder Kaur, R. K. Pal and Som Pal Singh

CONTENTS

15.1 INTRODUCTION

Climate change refers to a statistically significant variation in the weather patterns over many decades, resulting in a regional change in temperature and weather patterns. Climate change is a global issue which occurs due to increase in the concentration of greenhouse gases in the atmosphere, mainly because of human activities, like the burning of fossil fuels, intensive agricultural practices, deforestation and decomposition of food, plant waste and sewerage (IPCC 2014). Climate change-triggered increases in the intensity and frequency of extreme weather events are resulting in more severe heat waves, longer droughts, more storms, stronger cyclones, rises in sea levels and decreased availability of usable water throughout the world. It is also affecting wildlife and biodiversity, as well as unbalancing natural ecosystems. Average surface air temperature over the globe was observed to increase by 0.85°C during the past century, which had direct effects on water vapour content and atmospheric circulation

(Jhajharia et al. 2012). All these changes are likely to have severe implications for crop productivity over the Indian sub-continent. Sinha and Swaminathan (1991) predicted decreases in rice yield of about 0.75 t ha^{-1} in the high-yielding areas with an increase in temperature of 2°C and a decrease in wheat yield by 0.45 t ha^{-1} with an increase in winter temperature of 0.5°C in India. Similarly, Aggarwal and Sinha (1993), using the WTGROWS model, simulated a decrease in wheat yield over most places in India with an increase in temperature of 2°C. Saseendran (2000) reported a 6% reduction in rice yield with each 1°C rise in temperature. There are contrasting reports of the effects of CO_2 fertilization on crop yield. For example, Rao and Sinha (1994) reported wheat yield changes, due to CO_2 fertilization, from −28 to −68% without CO_2 fertilization effects and from +4 to −34% with the CO_2 fertilization effect, but Lal et al. (1996) concluded that the adverse effects of higher-temperature scenarios on rice yield might not be fully counter-balanced by CO_2 fertilization.

Because of fluctuating crop yields due to recent climatic changes, its prediction at different spatial scales has attracted much attention. The forecasting of crop productivity in advance of harvest, using crop modeling and remote-sensing technology, is operational in many regions globally to ensure food security and to provide early warnings to avert food shortages. Various operations and strategies, which can be adopted by governments, such as import–export policies, allocation of subsidies to farmers, strategic planning and decision making, etc., depend on crop yield forecasts (Baez-Gonzalez et al. 2002; Doraiswamy et al. 2003, Baez-Gonzalez et al. 2005). Providing real-time, quantitative and instantaneous information on crops non-destructively over a large area is the unique capability of remote-sensing technology. Although conventional methods are also in use for estimating crop yields, their dependence on labour-intensive, costly, time-consuming field reports/surveys, which are usually available too late to make strategic decisions and are highly prone to errors, leads to inaccurate yield estimations.

In view of the variable nature of agriculture and highly variable weather conditions, forecasting crop production in India is quite a challenging task, but the unique potential of satellite remote sensing for assessing crop acreage and productivity has made such forecasting possible at various spatial scales (Sharma and Sood 2003). Remote-sensing data has the capability to not only improve the accuracy of estimates based on ground surveys, but also to reduce the volume of field data collection needed (Dadhwal et al. 2002).

Remote sensing is a powerful tool for assessing spatio-temporal changes in crop conditions for site-specific management of the crop during its growing period. It can provide developmental information that is time-critical for site-specific crop management schedules. Spectral reflectance characteristics of plants are able to estimate the level of biochemicals, such as chlorophylls, which can readily be detected due to increases in reflectance of visible radiation at a particular wavelength as a result, say, of nitrogen deficiency (Moran et al. 1997; Hatfield and Pinter 1993; Blackmer et al. 1994). As a result, monitoring of plant health and estimation of yield and total biomass can be undertaken by remote-sensing technology. Most previous reviews have discussed remote-sensing data and types of sensors used for prediction of crop yield through satellite remote sensing, but there is a need to review the various indices coupled with crop growth simulation models to predict crop yield.

15.2 CONVENTIONAL METHODS OF CROP YIELD ESTIMATION

The crop yield is generally estimated by collecting data conventionally based on field reports, which generally use empirical statistical and crop growth models (Reynolds and Yittayew 2000) (Figure 15.1).

Empirical statistical models take into account the historical yield data and climate parameters. By using statistical procedures, significantly correlated parameters are identified, and an empirical equation is developed. Crop growth models directly assimilate the effect of weather, soil and other environmental parameters on plant development (Wiegand and Richardson 1990). They are based on physiological concepts and complex designs with too-detailed requirements for weather, soil and crop management data, and, as a result, give low predictability at large spatial scales when spatial variability in soil and management practices is high. Hence, regression models are used quite extensively for crop yield prediction (Thompson 1962; Irwin and Good 2010). Sehgal et al. (2002) reported crop

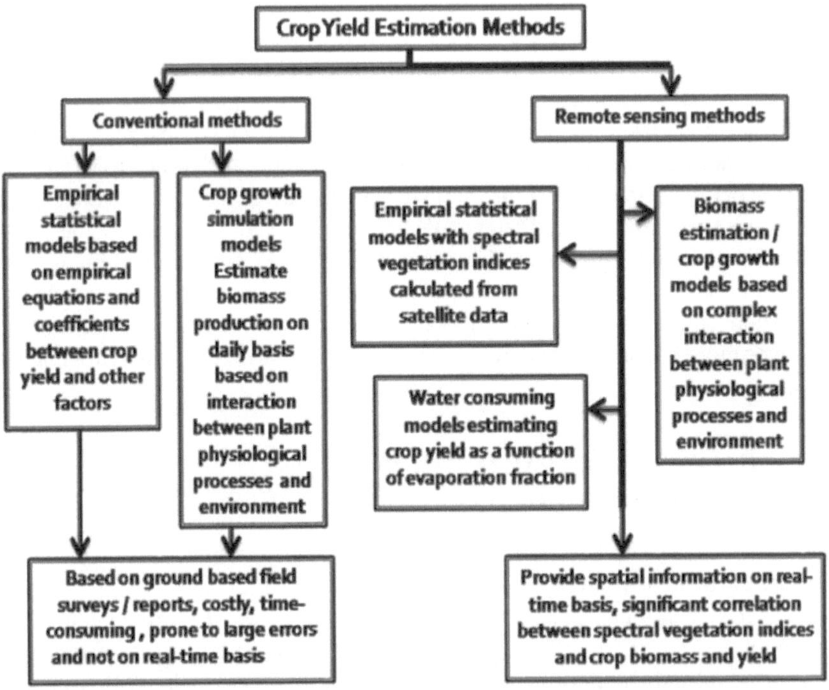

FIGURE 15.1 Conventional and remote-sensing crop yield estimation methods

simulation models to be useful tools for predicting crop growth and yield under variable environmental and cultural factors, but noted the lack of availability of detailed inputs for application at a regional scale.

Weather and crop growth models also have many limitations. For example, if yield prediction is based on monthly temperature or rainfall averages, it may not reflect the actual scenario of weather conditions, as these parameters might not be evenly distributed throughout the month. Secondly, the use of weather and crop growth models fails to accurately characterize the local weather conditions when weather data is aggregated over large spatial areas. Thus, lack of spatial information on the actual conditions in each field or region pose a major limitation to crop growth models. However, filling gaps caused by missing spatial information required by crop models for improved yield prediction is possible through remote-sensing technology (Kasampalis et al. 2018).

15.3 SATELLITE REMOTE SENSING-BASED METHODS OF CROP YIELD ESTIMATION

Identification of and discrimination between different surface features through remote sensing depends on their spectral reflectance characteristics. The advent of multispectral and hyperspectral remote-sensing technology has made it possible to monitor crop health, discriminate between different types of crop stresses and forecast crop yield with greater accuracy. These techniques are highly applicable as they are related to plant physiological processes, especially photosynthesis and evapotranspiration. Absorption in blue and red and reflection in green wavelengths of light by chlorophyll pigments play a major role in crop growth monitoring and yield estimation (Chappelle et al. 1992). Thus, a decrease in chlorophyll concentration due to any type of biotic or abiotic stress leads to variations in the spectral characteristics of the leaf, which may be used to estimate crop yield. The percentage of radiation reflected from a leaf is higher in the near-infrared (NIR) regions than in the green (GRN) region of the electromagnetic spectrum (Gausman et al. 1971). Thus, spectral reflectance characteristics depict crop health status, and the accuracy of information is improved when reflectance in two or more bands is considered collectively in the form of spectral vegetation indices.

15.4 CROP GROWTH MONITORING USING SPECTRAL VEGETATION INDICES

The indices based on spectral features of plants derived from remotely sensed data are correlated with crop health status and are used for assessing growth and yield. It has been found that spectral vegetation indices, which are computed by employing arithmetic operations in various spectral bands or their combinations, are important indicators of crop health and yield (Tucker et al. 1980; Weigand and Richardson 1990) (Table 15.1), as they are affected by even the fraction

TABLE 15.1
Use of Remote-Sensing-Derived Spectral Vegetation Indices for Crop Growth Modeling and Yield Prediction

Sr. no.	Vegetation Index	Formula	Satellite Data Used	Crop	Parameter Studied	References
1.	Normalized Difference Vegetation Index (NDVI)	(NIR−R) (NIR+R)	NOAA AVHRR MODIS LANDSAT IKONOS QUICKBIRD IRS LISS III	Wheat	Vegetation changes, yield, water productivity	Tucker et al. 1983; Labus et al. 2002; Wright et al. 2003; Ren et al. 2008; Gontia and Tiwari 2011; Mkhabela et al. 2011; Dempewolf et al. 2014, Zhang et al. 2014
2.	Green Normalized Difference Vegetation Index (GNDVI)	(NIR−GRN) (NIR+GRN)	VIRAF Spectroradiometer	Beans	Estimation of chlorophyll content in leaves	Buschmann and Nagel 1993
3.	Ratio Vegetation Index (RVI)/Simple Ratio Index (SRI)	NIR/R	Spectroradiometer	Forest	LAI, above- ground dry phytomass	Jordan 1969
4.	Green Reflectance- based Simple Ratio Index (GSRI)	NIR/GRN	Spectroradiometer	Rice	LAI, yield	Shibayama and Akiyama, 1989; Huete 1988
5.	Soil-Adjusted Vegetation Index (SAVI)	$\frac{(NIR-R)}{(NIR+R+L)} * (1+L)$	IKONOS QUICKBIRD IRS LISS III	Wheat	Soil background effects, yield	Huete 1988; Wright et al. 2003
6.	Wide Dynamic Range Vegetation Index (WDRVI)	$(\alpha \times R_{800nm} - R_{670nm})/(\alpha \times R_{800nm} + R_{670nm})$ $\alpha = 0.1$	MODIS Landsat Field hyperspectral data	Maize, wheat	Green LAI, yield, water productivity	Gontia and Tiwari 2011; Guindin Garcia et al. 2012; Sakamoto et al. 2013; Sibley et al. 2014; Dempewolf et al. 2014; Jin et al. 2016

(Continued)

TABLE 15.1 (CONTINUNED)
Use of Remote-Sensing-Derived Spectral Vegetation Indices for Crop Growth Modeling and Yield Prediction

Sr. no.	Vegetation Index	Formula	Satellite Data Used	Crop	Parameter Studied	References
7.	Enhanced Vegetation Index (EVI2)	$2.5 \times ((NIR-R)/(NIR + (2.4 \times R+1))$	Landsat MODIS	Wheat	Yield	Dempewolf et al. 2014
8.	Saturation-Adjusted Normalized Difference Vegetation Index (SANDVI)	$(NIR-R) \times (1+L)/ (NIR+R+L)$	Landsat MODIS	Wheat	Yield	Dempewolf et al. 2014
9.	Normalized Difference Matter Index (NDMI)	$(R_{1649nm}-R_{1722nm})/ (R_{1649nm}+R_{1722nm})$	Field hyperspectral data	Wheat	Biomass, yield	Jin et al. 2016
10.	Vegetation Condition Index (VCI)	$100(NDVI-NDVI_{min})/ (NDVI_{max}-NDVI_{min})$	AVHRR	Wheat	Yield	Salazar et al. 2007
11.	Temperature Condition Index (TCI)	$100(BT_{max}-B)/(BT_{max}-BT_{min})$	AVHRR	Wheat	Yield	Salazar et al. 2007
12.	Water Index (970, 900) (WI I)	R_{970}/R_{900}	Field hyperspectral data	Wheat	Biomass, yield	Jin et al. 2016
13.	Water Index (1300, 1450) (WI II)	R_{1300}/R_{1450}	Field hyperspectral data	Wheat	Biomass, yield	Jin et al. 2016
14.	Normalized Difference Infrared Index (NDII)	$(R_{850}-R_{1650})/ (R_{850}+R_{1650})$	Field hyperspectral data	Wheat	Biomass, yield	Jin et al. 2016
15.	Three Band Water Index (TBWI)	$(R_{973}-R_{1720})/ R_{1447}$	Field hyperspectral data	Wheat	Biomass, yield	Jin et al. 2016

(Continued)

TABLE 15.1 (CONTINUNED)
Use of Remote-Sensing-Derived Spectral Vegetation Indices for Crop Growth Modeling and Yield Prediction

Sr. no.	Vegetation Index	Formula	Satellite Data Used	Crop	Parameter Studied	References
16.	Enhanced Vegetation Index (EVI)	$2.5\times(R_{800}-R_{660})/(1+R_{800}+2.4\times R_{660})$	Field hyperspectral data	Wheat	Biomass, yield	Jin et al. 2016
17.	Transformed Chlorophyll Absorption in Reflectance Index (TCARI)	$3\times((R_{700}-R_{670})-0.2\times(R_{700}-R_{550})\times(R_{700}/R_{670}))$	Field hyperspectral data	Wheat	Biomass, yield	Jin et al. 2016
18.	Optimized Soil-Adjusted Vegetation Index (OSAVI)	$1.16\times(R_{800}-R_{670})/(R_{800}+R_{670}+0.16)$	Field hyperspectral data	Wheat	Biomass, yield	Jin et al. 2016
19.	Combined Index II (TCARI/OSAVI)	TCARI/OSAVI	Field hyperspectral data	Wheat	Biomass, yield	Jin et al. 2016
20.	MERIS Terrestrial Chlorophyll Index MTCI	$(R_{750}-R_{710})/(R_{710}-R_{680})$	Field hyperspectral data	Wheat	Biomass, yield	Jin et al. 2016
21.	Red Edge Model ($CI_{red\ edge}$)	$(R_{750}/R_{720})-1$	Field hyperspectral data	Wheat	Biomass, yield	Jin et al. 2016
22.	Double-peak Canopy Nitrogen Index I (DCNI I)	$(R_{750}-R_{700})/(R_{700}-R_{670})/(R_{750}-R_{670}+0.09)$	Field hyperspectral data	Wheat	Biomass, yield	Jin et al. 2016
23.	Combined Index I OSAVI $-CI_{red\ edge}$	OSAVI $-CI_{red\ edge}$	Field hyperspectral data	Wheat	Biomass, yield	Jin et al. 2016

of photosynthetically active radiation absorbed by the plant. Previous studies have related yield with spectral vegetation indices (like Normalized Difference Vegetation Index, NDVI) at a specific phenophase (e.g., vegetative and reproductive stages) or with cumulative values of spectral vegetation indices over the entire growing season (Labus et al. 2002; Wall et al. 2008; Lobell and Burke 2010; Mkhabela et al. 2011; Lobell 2013). These techniques require sufficient long-term remotely acquired data and its correlation with the historical regional values under study. The historical values of spectral vegetation indices (like NDVI) for a specific region are compared with current values to detect any anomalies or deviations from historical data, and then the data are used to estimate yields.

The Ratio Vegetative Index (RVI), computed by dividing NIR reflectance by red light, is one of the earliest remote-sensing techniques used to assess the condition of vegetation (Jordan 1969), due to the strong absorption (low reflectance) of light by chlorophyll in the red region and low absorption (high reflectance) of light in the NIR (Avery and Berlin 1992). Other spectral vegetation indices (SVIs) include Simple Ratio Index (SRI), the ratio of NIR to red (Jordan 1969), and the NDVI, the ratio of (NIR–R)/(NIR+R). Shibayama and Akiyama (1989) suggested the Green Simple Ratio Index (GSRI), the ratio of NIR to green, for estimating the leaf area index (LAI).

Many researchers have related NDVI to yield. NDVI derived from visible and NIR bands of the NOAA AVHRR satellite is used to monitor temporal changes in vegetation at a spatial scale (Tucker et al. 1983). Labus et al. (2002) derived NDVI from an AVHRR time-series for Montana in the United States and found it to be significantly positively correlated with wheat yield. Ren et al. (2008) predicted winter wheat yield at a regional level from 10-day MODIS NDVI composites in China and they concluded that an accurate prediction (within 5% of the yields from official statistics) could be made 40 days before harvest date of winter wheat. Mkhabela et al. (2011) predicted the yield of major crops in the Canadian Prairies, including spring wheat, using 10-day MODIS NDVI composites. Temporal changes in NDVI were related to net primary production (Malingreu 1986; Goward et al. 1987; Prince 1991). NDVI became less sensitive when chlorophyll content, vegetation fraction or leaf area index reached moderate to high values. Under such conditions, the green band was observed to be more sensitive to the chlorophyll content than was the red band. Buschmann and Nagel (1993) used the Green Normalized Difference Vegetation Index (GNDVI), where GNDVI = (NIR–GRN)/(NIR+GRN) for evaluating variation in biomass. Research has shown that the green band (in combination with the NIR band) is closely related to leaf chlorophyll, N content and grain yield (Blackmer et al. 1994). Spectral vegetation indices reduce other effects, e.g., Soil-Adjusted Vegetation Index (SAVI) reduces the effect of the soil background (Huete 1988). Green reflectance, based on the simple ratio (GSRI), showed the highest r^2 with grain yield but prediction of leaf area index (LAI) was best achieved by the use of NDVI. Guindin-Garcia et al. (2012), Sakamoto et al. (2013) and Sibley et al. (2014) used WDRVI (Wide Dynamic Range Vegetation Index) derived from time-series MODIS 16- and 32-day composites and found a positive significant linear relationship between

WDRVI and field measurements of maize green leaf area index, with the forecast of yield for most of the cases being within 10% of the final reported values. Dempewolf et al. (2014) reported that wheat yield could be estimated successfully from satellite remote-sensing data. They used freely available Landsat and MODIS satellite data about six weeks prior to harvest. WDRVI provided more consistent and accurate yield predictions than did NDVI, EVI2 or saturation-adjusted normalized difference vegetation index (SANDVI).

15.5 COUPLING OF CROP MODELS WITH REMOTE SENSING AND GIS

As the green biomass, water and protein content in the leaf affect its spectral reflectance, spectral indices are used for real-time monitoring of crop health and productivity (Anup et al. 2006). The spectral reflectance of crops can be used to detect environmental stresses, like assessing plant nutrient status, estimating plant growth and physiology, monitoring plant condition at various scales and predicting crop yields. A number of studies suggested that spectral crop yield models give more accurate predictions than do agro-meteorological models. Yield forecasting on a regional basis can be improved by using a combination of crop simulation models, remote sensing, geographic information system (GIS) technology and climatic data (Figure 15.1). Analysis of satellite data coupled with a crop simulation model can give real-time trends by which to arrive at a conclusion which may affect policy making at the district and state levels and therefore can be replicated at the national level.

Crop yield prediction is one of the foremost applications of the crop simulation models. Several models have been developed to predict the behaviour of a crop sown on any soil at any time of the year (Jones and Kiniry 1986). Recently, the use of dynamic simulation models has increased markedly to predict crop growth on a daily basis. The main aim of modeling crop growth is to simulate day-to-day assimilation of photosynthetic material, based on the exchange of energy and mass among all the growth processes taking place in a plant. The purpose of dynamic simulation models is to forecast crop productivity before the actual harvesting of the crop. In the recent past, several dynamic simulation models have been presented by different investigators, with the Crop Environment Resources Synthesis (CERES) models being used successfully all over the world.

The CERES-Maize model has been found to be useful to assess the effects of abiotic stresses on maize at an early stage of growth so that policy makers can declare areas to be drought stricken, in order to implement subsidy schemes on a fair basis (Matthews et al. 2002). The CERES-Maize model simulated phenology and yield with close agreement with the field- observed data (Kumar et al. 2010). Singh et al. (2010) evaluated the SOYGRO model for simulating the yield of a soybean crop. The model overestimated grain and straw yield at all the sowing dates, except for straw yield under late-sown conditions, whereas grain and straw yield were underestimated. Reddy et al. (2008) evaluated the performance of the CERES-Rice and WOFOST models for prediction of phenology and yield of rice

TABLE 15.2
Crop Yield Prediction Using Crop Simulation Models and Geospatial Technology

Sr. no.	Crop	Simulation Model Used	Satellite Data Used	References
1.	Barley, durum wheat	AquaCrop	MODIS	Panday 2014
2.	Wheat	EPIC	MODIS	Yang et al. 2005
3.	Maize	CSWB model	SPOT	Rojas et al. 2005
4.	Wheat	WTGROWS	WiFS	Sehgal et al. 2002
5.	Maize	CSWB model	METEOSAT, NOAA-AVHHR	Reynolds and Yittayew 2000
6.	Cereal crops	SPATIAL-EPIC	ARC/INFO	Satya et al. 1998
7.	Peanut	DSSAT	ArcView-2	Engel et al. 1997
8.	Wheat, sugar beet	WOFOST	ARC/INFO	Bouman et al. 1997
9.	Potato	SIMPOTATO	ARC/INFO GIS	Han et al. 1995
10.	-	WOFOST	ARC/INFO	Meyer-Roux and Vossen 1994
11.	Wheat	WTGROWS	ARC/INFO	Aggarwal 1993
12.	-	DSSAT	ARC/INFO	Calixte et al. 1992
13.	Wheat	Aquacrop	Field Hyperspectral data	Jin et al. 2016

and found that they performed better under delayed planting up to 30 August. Kumar et al. (2008) observed that the APSIM model predicted the yield of sorghum correctly under conditions of normal sowing dates, whereas the CERES model predicted phenology correctly but overestimated grain yield under late-sowing conditions. Jadhav et al. (2008) reported that the CERES-Sorghum model predicted yield attributes very accurately and grain yield very well but underestimated predicted LAI, biomass and straw yield under all sowing dates.

Although they have wide applicability, the major limitation of using crop simulation models is that they require huge datasets, making it difficult to run such models for a number of stations. In addition, they are able to generate only point data, which are difficult to be used at a regional level. Thus, coupling of crop simulation models with geospatial techniques could be helpful in enhancing the applicability of the models for regional yield estimations (Table 15.2).

The methods for predicting crop yield, using remote sensing, are mainly divided into two categories, namely remote-sensing methods based on either empirical statistics models or on biomass estimation models

15.5.1 Remote-Sensing Methods Based on Empirical Statistical Models

A number of studies have reported various yield prediction models in which remote-sensing techniques have been used. It has been found that agro-meteorological-based crop yield prediction models mainly take into account the weather conditions prevailing during the life cycle of the crops. However, the predictability of agro-meteorological models can be enhanced by incorporating various crop indices, such as spectral indices and biological indices. Even the trend-predicted yields can be introduced, to account for the technological trend. The important predictive models are given below (Mahi and Kingra 2013):

1) Trend-Predicted Yield Models
2) Spectral Yield Models
3) Agro-Meteorological Spectral Yield Models
4) Agro-Meteorological Spectral Trend Yield Models

Although these predictive models are generally based on conventional techniques, spectral vegetation indices, such as NDVI, computed through satellite remote sensing, are incorporated as additional parameters, so that these models are known as spectral yield models. Previous research investigations have reported the increased precision of agro-meteorological models when spectral indices, derived from remote-sensing data, are included as auxiliary variables in these models. The fusion of trend-predicted yields (estimated from yield time-series provided by statistical agencies at national and regional scales) and spectral indices helps to improve the accuracy of yield predictions at a district level.

Kalubarme et al. (1995) developed agro-meteorological spectral trend-yield models for two zones in Punjab (India) for predicting wheat yield. Jand (1999) developed spectral agro-meteorological yield models, using NDVI and meteorological indices (maximum and minimum temperature, temperature difference (TD), growing degree-days (GDD) and photothermal units (PTU)) for south-western districts of Punjab. That study suggested that trend-predicted yield, NDVI and PTU were important parameters in the prediction of wheat yield. Singh et al. (1999) reported that incorporation of spectral indices (NIR/red ratio and NDVI), along with agro-meteorological (stress degree-days and radiation use efficiency) and biological indices (LAI and accumulated dry matter at anthesis), improved wheat yield prediction. The results suggested that spectral yield, agrometeorological yield and biological models had coefficients of determination (r^2) of 80%, 75% and 81%, respectively. On the other hand, a spectral agro-meteorological biological model explained 84% of the variation in yield (i.e., r^2), showing better yield prediction than did the other three models. Bazgeer (2005) reported that the prediction of wheat yield in two districts of Punjab was improved by using spectral indices in addition to agro-metorological parameters.

15.5.2 Remote-Sensing Methods Based on Biomass Estimation Models

Remote sensing involves measuring radiation reflected from or emitted by vegetation at specific wavelengths (Maas 1988) and is used to predict various crop parameters, namely rate of photosynthesis, interception of photosynthetically active radiation (IPAR), biomass, LAI, etc. (Aparicio et al. 2002). Previous studies have incorporated remote-sensing data into crop models to improve the yield predictions. Maas (1988) described ways to combine crop growth models with satellite data, which were revisited by Delecolle et al. (1992). Three methods of data integration have been identified: direct use of a driving variable estimated from remote-sensing information in the model; updating of a state variable of the model (for example, LAI) derived from remote-sensing data; and the calibration of model variables by using satellite images (the assimilation method).

Baez-Gonzalez et al. (2005) predicted maize yield by integrating LAI estimated from spectral data with a simple LAI-based yield model under irrigated conditions in Sinaloa, Mexico, with a mean error of −9.2% with maximum ground LAI (gLAI) and −11.2% with satellite-derived LAI (sLAI). Doraiswamy et al. (2003) incorporated parameters (LAI, soil moisture, etc.), derived from satellite data, in a crop growth model to estimate the yield of spring wheat. Crop simulation models (CERES models) are used successfully for estimation of crop yields at the field level (Engel et al. 1997; Sinclair and Muchow 2001). However, large input data requirements have restricted their use on a regional scale. Thus, incorporating input parameters derived from remotely sensed data in these models may provide real-time calibration of simulations of model parameters (Guerif et al. 1993; Maas 1993; Moulin et al. 1998; Doraiswamy et al. 2000).

15.6 CONCLUSIONS

Conventional methods, relying on ground surveys and reports, are highly prone to errors, leading to poor yield estimates. Remote-sensing methods are able to remove the disadvantages of such conventional methods. Crop models are complete, taking into account many parameters, can be run on a large scale, but require lots of parameters and are able to generate only point data. Thus, the coupling of crop simulation models with remote sensing and GIS can be highly beneficial for regional yield estimations with greater accuracy. Remotely sensed data, with a synoptic view and repeat coverage, can assess crop conditions on a real-time basis on a spatio-temporal scale, so are able to enhance the consistency of the predictions and provide better yield estimates prior to harvest.

REFERENCES

Aggarwal, P. K. 1993. Agro-ecological zoning using crop growth simulation models: characterization of wheat environments of India. In: *Systems Approaches for Agricultural Development*, Vol. 2, ed. F. W. T. Penning de Vries, P. Teng and K. Metselaar, 97–109. Kluwer Academic Publishers, Dordrecht.

Aggarwal, P. K. and S. K. Sinha. 1993. Effect of probable increase in carbon dioxide and temperature on productivity of wheat in India. *Journal of Agrometeorology* 48:811–14.

Anup, K., Prasad Lim, Chai Ramesh, P. Singh and Kafatos Menas. 2006. Crop yield estimation model for Iowa using remote sensing and surface parameters. *International Journal of Applied Earth Observation and Geoinformation* 8:26–33.

Aparicio, N., D. Villegas, J. L. Araus, J.Casadesus and C. Royo. 2002. Relationship between growth traits and spectral vegetation indices in durum wheat. *Crop Science* 42:1547–55.

Avery, T. E. and G. L. Berlin. 1992. *Fundamentals of Remote Sensing and Air Photo-Interpretation*. 5th ed. Macmillon, New York.

Baez-Gonzalez, A. D., P. Cheri, M. Tiscareno-Lopez and R. Sriskosan. 2002. Using satellite and field data with crop growth modelling to monitor and estimate corn yield in Mexico. *Crop Science* 42:1943–49.

Baez-Gonzalez, A. D., J. R. Kiniry, S. J. Maas et al. 2005. Large-area maize yield forecasting using leaf area index based yield model. *Agronomy Journal* 97:418–25.

Bazgeer, S. 2005. *Land-Use Change Analysis in the Sub-Mountainous Region of Punjab Using Remote Sensing, GIS and Agrometeorological Parameters*. Ph. D. Dissertation, PAU, Ludhiana, India.

Blackmer, T. M., J. S. Schepers and G. E. Varvel. 1994. Light reflectance compared with other nitrogen stress measurement in corn leaves. *Agronomy Journal* 86:934–38.

Bouman, B. A. M., C. A. van Dipen, P. Vossen and T. van Der Wal. 1997. Simulation and systems analysis tools for crop yield forecasting,. In: *Applications of Systems Approaches at the Farm and Regional Levels*, Vol. 1, ed. P. S. Teng, M. J. Kropff, H. F. M. ten Berge, J. B. Dent, F. P. Lansigan and H. H. van Laar, 325–40. Kluwer Academic Publishers, Dordrecht.

Buschmann, C. and E. Nagel. 1993. In vivo spectroscopy and internal optics of leaves as basis for remote sensing of vegetation. *International Journal of Remote Sensing* 14:711–22.

Calixte, J. P., F. J. Beinroth, J. W. Jones and H. Lai. 1992. Linking DSSAT to a geographic information system. *Agrotechnology Transfer* 15:1–7.

Chappelle, E. W., M. S. Kim and J. E. McMurtrey. 1992. Ratio analysis of reflectance spectra (RARS) – An algorithm for the remote estimation of the concentrations of chlorophyll-a, chlorophyll-b, and carotenoids in soybean leave. *Remote Sensing of Environmental* 39:239–47.

Dadhwal, V. K., R. P. Singh, S. Dutta and J. S. Parihar. 2002. Remote sensing based crop inventory: A review of Indian experience. *Tropical Ecology* 43(1):107–22.

Delecolle, R., S. J. Maas, M. Guerif and F. Barat. 1992. Remote sensing and crop production models: present trends. *ISPRS Journal of Photogrammetry and Remote Sensing* 47:145–61.

Dempewolf, J., B. Adusei, I. Becker-Reshef *et al.* 2014. Wheat yield forecasting for Punjab province from vegetation index time series and historic crop statistics. *Remote Sensing* 6:9653–75.

Doraiswamy, P. C., P. A. Pasteris, K. C. Jones, R. P. Motha, and P. Nejedlik, 2000. Techniques for methods of collection, database management and distribution of agrometeorological data. *Agricultural and Forest Meteorology* 103(1–2):83–97.

Doraiswamy, P. C., S. Moulin, P. W. Cook and A. Stern. 2003. Crop yield assessment from remote sensing. *Photogramm Eng Remote Sensing* 69:665–74.

Engel, T., G. Hoogenboom, J. W. Jones and P. W. Wilkens. 1997. AEGIS/WIN – a program for the application of crop simulation models across geographic areas. *Agronomy Journal* 89:919–28.

Gausman, H. W., W. A. Allen, R. Cardenas and A. J. Richardson. 1971. Effects of leaf nodal position on absorption and scattering coefficients and infinite reflectance of cotton leaves (*Gossypium hirsutum* L.) *Agron J* 63(1):87.

Gontia, Narendra Kumar and K. N. Tiwari. 2011. Yield estimation model and water productivity of wheat crop (*Triticum aestivum* L.) in an irrigation command area using remote sensing and GIS. *Journal of the Indian Society of Remote Sensing* 39(1):27–37.

Goward, A. Kerber, D. G. Dye and V. Kalb. 1987. Comparison of North and South American biomes from AVHRR observations. *Geocarto* 2:27–40.

Guérif, M., S. D. Brisis, and B. Seguin. 1993. Combined NOAA-AVHRR and SPOT-HRV data for assessing crop yields of semi-arid environments. EARSeL *Advances in Remote Sensing* 2:110–23.

Guindin-Garcia, N., A. A. Gitelson, T. J. Arkebauer, J. Shanahon and A. Weiss. 2012. An evaluation of MODIS 8 and 16 day composite products for monitoring maize green leaf area index. *Agricultural and Forest Meteorology* 161:15–25.

Han, S., R. G. Evans, T. Hodges and S. L. Rawlins. 1995. Linking geographic information system with a potato simulation model for site-specific crop management. *Journal of Environmental Quality* 24:772–77.

Hatfield, J. L. and P. J. Pinter Jr. 1993. Remote sensing for crop production. *Crop Production* 12:403–11.

Huete, A. R. 1988. A soil-adjusted vegetation index (SAVI). *Remote Sensing of Environmental* 25:295–309.

IPCC. 2014. Summary for policymakers: Climate change impacts, adaptation, and vulnerability. Part A: Global and sectoral aspects. In: *Contribution of Working Group II to the Fifth Assessment Report of the Intergovernmental Panel on Climate Change*, ed. C. B. Barros, V. R. Dokken, D. J. Mach *et al.*, 1–32. Cambridge University Press, Cambridge.

Irwin, S. and Good, D. 2010. *Alternative 2010 Corn Production Scenarios and Policy Implications, Marketing and Outlook Brief 10-01.* Department of Agricultural and Consumer Economics, University of Illinois at Urbana-Champaign, March.

Jadhav, M. G., M. C. Varshneya, S. S. Salunke and P. V. Thanedar. 2008. Validation of growth subroutine of CERES-Sorghum model for *kharif* sorghum. *Journal of Agrometeorology (Special Issue – Part II)*:436–38.

Jand, S. 1999. *Spectral Agromet Yield Modelling of Cotton and Wheat for the South-Western Districts of Punjab.* M. Sc. Thesis, PAU, Ludhiana.

Jhajharia, D., Y. Dinpashoh, E. Kahya, V. P. Singh and A. Fakheri-Fard. 2012. Trends in evapotranspiration in the humid region of north-east India. *Hydrol Process* 26 (3):421–35.

Jin, Xiuliang, Lalit Kumar, Zhenhai Li, Xingang Xu, Guijun Yang and Jihua Wang (2016). Estimation of winter wheat biomass and yield by combining the AquaCrop model and field hyperspectral data. *Remote Sensing* 8:972, doi:10.3390/rs8120972 1-15

Jones, C. A. and J. R. Kiniry (ed.). 1986. *CERES-Maize: A Simulation Model of Maize Growth and Development.* Texas A & M University Press, College Station.

Jordan, C. F. 1969. Derivation of leaf area index from quality of light on the forest floor. *Ecology* 50:663–66.

Kalubarme, M. H., R. K. Mahey, S. S. Dhaliwal *et al.* 1995. Agromet spectral wheat yield modelling in Punjab. In: *Proc. National Symp. on Remote Sensing of Environment with Special Emphasis on Green Revolution*, 11–17. PAU, Ludhiana.

Kasampalis, Dimitrios A., Alexandridis, Thomas K., Deva Chetan, Challinor Andrew, Moshou Dimitrios and Zalidis Georgios. 2018. Contribution of remote sensing on crop models: A review. *Journal of Imaging* 4:52, doi:10.3390/jimaging4040052

Kumar, A., K. K. Singh, R. Balasubramaniyan, A. K. Baxla, P. Tripathi and B. N. Mishra. 2010. Validation of CERES-Maize model for growth, yield attributes and yield of kharif maize for NEPZ of eastern U.P. *Journal of Agrometeorology* 12 (1):118–20.

Kumar, V. R., S. R. Kumar, M. S. Raut, G. Sreeniwas G. and D. Raji Reddy. 2008. Validation of APSIM and CERES-Sorghum model for prediction of rabi sorghum yields in Solapur region of Maharashtra. *Journal of Agrometeorology (Special Issue – Part I)*:146–49.

Labus, M. P., G. A. Nielsen, R. L. Lawrence, R. Engel and D. S. Long. 2002. Wheat yield estimates using multi-temporal NDVI satellite imagery. *International Journal of Remote Sensing* 23:4169–80.

Lal, M., G. Srinivasan and U. Cubasch. 1996. Implications of increasing green house gases and aerosols on the diurnal temperature cycle of the Indian sub-continent. *Current Science* 71 (10):746–52.

Lobell, D. 2013. The use of satellite data for crop yield gap analysis. *Field Crop Research* 143:56–64.

Lobell, D. B. and M. B. Burke. 2010. On the use of statistical models to predict crop yield responses to climate change. *Agricultural and Forest Meteorology* 150 (11):1443–52.

Maas, S. J. 1993. Within season calibration of modelled wheat growth using remote sensing and field sampling. *Agronomy Journal* 85(3):669–72.

Maas, S. J., 1988. Use of remotely-sensed information in agricultural crop growth models. *Ecological Modeling* 41:247–68

Mahi, G. S. and P. K. Kingra. 2013. *Comprehensive Agrometeorology,* 355. Kalyani Publishers, New Delhi.

Malingreau, J. P. 1986. Global vegetation dynamics: satellite observations over Asia. *International Journal of Remote Sensing* 7:1121–46.

Matthews, R. B., W. Stephans, T. Hess, T. Middleton and A. Graves. 2002. Application of crop/soil simulation models in tropical agricultural systems. *Advances in Agronomy* 76:31–124.

Meyer-Roux, J. and P. Vossen. 1994. The first phase of the MARS project, 1988–1993: overview, methods and results. In: *Proceedings of the Conference on the MARS Project: Overview and Perspectives,* 33–85. Commission of the European Communities, Luxembourg.

Mkhabela, M., P. Bullock, S. Raj, S. Wang and Y. Yang. 2011. Crop yield forecasting on the Canadian prairies using MODIS NDVI data. *Agricultural and Forest Meteorology* 151:385–93.

Moran, M. S., Y. Inove and E. M. Barnes. 1997. Opputtunities and limitations for image based remote sensing in precision crop management. *Remote Sensing of Environmental* 61:319–46.

Moulin, S., A. Bondeau and R. Delecolle. 1998. Combining agricultural crop models and satellite observations: from field to regional scales. *International Journal of Remote Sensing* 19:1021–36.

Panday, S. G. 2014. *Use of Remote Sensing Data in a Crop Growth Model to Estimate Actual Crop Yields: Testing Aquacrop with NDVI Time Series for Two Crops in Spain.* M.Sc. Thesis, University of Twente.

Prince, S. D. 1991. A model of regional primary production for use with coarse resolution satellite data. *International Journal of Remote Sensing* 12:1313–30.

Rao, G. D. and S. K. Sinha. 1994. Impact of climate change on simulated wheat production in India. In: *Implications of Climate Change for International Agriculture: Crop Modelling Study*, ed. C. Rosenzweig, I. Iglesias, 1–10. EPA, USA.

Reddy, D. R., G. Sreenivas, S. G. Mahadevappa, S. B. S. N. Rao and N. R. G. Verma. 2008. Performance of CERES and WOFOST models in prediction of phenology and yield of rice in Telangaga region of Andhra Pradesh. *Journal of Agrometeorology (Special Issue – Part I)*:109–10.

Ren, J., Z. Chen, Q. Zhou and H. Tang. 2008. Regional yield estimation for winter wheat with MODIS NDVI data in Shandong, China. *International Journal of Applied Earth Observation and Geoinformation* 10:403–13.

Reynolds, M. and D. Yittayew. 2000. Estimation of crop yields and production by integrating the FAO crop specific water balance model with real-time satellite data and ground-based ancillary data. *International Journal of Remote Sensing* 21(18):3487–08.

Rojas, O., F. Rembold, A. Royer and T. Negre. 2005. Real-time agrometeorological crop yield monitoring in Eastern Africa. *Agronomy for Sustainable Development* 25:63–77 doi:10.1051/agro:2004056.

Sakamoto, T., A. A. Gitelson and T. J. Arkebauer. 2013. MODIS based corn grain yield estimation model incorporating crop phenology information. *Remote Sensing of Environmental* 131:215–31.

Salazar, L., F. Kogan and L. Roytman. 2007. Use of remote sensing data for estimation of winter wheat yield in the United States. *International Journal of Remote Sensing* 28(17):3795–11.

Saseendran, Smith Matson. 2000. Ecological and evolutionary responses to climate change. *Science* 284:1943–47.

Satya P, R. Shibasaki and S. Ochi. 1998. Modelling spatial crop production: A GIS approach. In: *Proceedings of the 19th Asian Conference on Remote Sensing*, 16–20. Manila, November, pp. A-9–1 – A-9–6.

Sehgal, V. K., D. R. Rajak, K. N. Chaudhary and V. K. Dadhwal. 2002. Improved regional yield prediction by crop growth monitoring system using remote sensing derived crop phenology. IAPRS & SIS, Vol.34, Part 7, "Resource and Environment Monitoring", Hyderabad, India, 2002. *The International Archives of the Photogrammetry, Remote Sensing and Spatial Information Sciences* 34 (7):329–34.

Sharma, P. K. and Anil Sood. 2003. Remote sensing and GIS techniques in agricultural development – A case study of Punjab. *Journal of Agricultural Physics* 3 (1&2):171–81.

Shibayama, M. and T. Akiyama. 1989. Seasonal visible, near-infrared and mid-infrared spectra of rice canopies in relation to LAI and above-ground dry phytomass. *Remote Sensing of Environmental* 27:119–27.

Sibley, A. M., P. Grassini, N. E. Thomas, K. G. Cassman and D. B. Lobell. 2014. Testing remote sensing approaches for assessing yield variability among maize fields. *Agronomy Journal* 106:24–32.

Sinclair, T. R. and R. C. Muchow. 2001. System analysis of plant traits to increase grain yield on limited water supplies. *Agronomy Journal* 93:263–70.

Singh, R. P., V. K. Dadhwal and R. R. Navalgund. 1999. Wheat crop inventory using high spectral resolution IRS-P3 MOS-B spectrometer data. *Journal of Indian Society of Remote Sensing* 27:167–73.

Singh, R., Diwan Singh, Chander Shekhar and Jugal K. Mani 2010. Evaluation of SOYGRO model for soybean crop under Hisar conditions. *Journal of Agrometeorology* 12 (1):121–22.

Sinha, S. K. and M. S. Swaminathan. 1991. Deforestation, climate change and sustainable nutrition security. *Climate Change* 16:33–45.

Thompson, L. M. 1962. *An Evaluation of Weather Factors in the Production of Corn (Report No. 12T).* Iowa State University, Center for Agricultural and Rural Development, Ames, IA.

Tucker, C. J., B. N. Holben, J. H. Elgin and J. E. Mc Murtrey. 1980. Relationship of spectral data to grain yield variation. *Photogrammetric Engineering and Remote Sensing* 46 (5):657–66.

Tucker, C. J., C. Vanpraet, E. Boerwinkel and A. Gatson. 1983. Satellite remote sensing of total dry matter production in the Senegalese Sahel. *Remote Sensing of Environmental* 13:461–74.

Wall, L., D. Larocque and Léger Pierre-Majorique. 2008. The early explanatory power of NDVI in crop yield modeling. *International Journal of Remote Sensing* 29:2211–25.

Wiegand, C. L. and A. J. Richardson. 1990. Use of spectral vegetation indices to infer leaf area, evapotranspiration and yield – I: Rationale. *Agronomy Journal* 82:623–29.

Wright Jr Dennis L., Douglas Ramsey R., Doran J. Baker and V. Philip Rasmussen Jr (2003). A comparison of two geospatial technologies in non-uniform wheat fields: Yield monitors and remote sensing. In: *ASPRS 2003. Annual Conference Proceedings May 2003*, Anchorage, Alaska.

Yang, P., G. X. Tan, Y. Zha and R. Shibasaki. 2005. *Integrating Remotely Sensed Data with An Ecosystem Model to Estimate Crop Yield in North China*. Commission VII, WG VII/2.

Zhang, Miao, Bingfang Wu, Mingzhao Yu, Wentao Zou and Yang Zheng. 2014. Crop condition assessment with adjusted NDVI using the uncropped arable land ratio. *Remote Sensing* 6: 5774–94; doi:10.3390/rs6065774.

16 Remote Sensing Applications for Crop Residue Burning Assessments

A Review

Parmod Kumar, Sattyam and Ripudaman Singh

CONTENTS

16.1 INTRODUCTION

India is one of major agrarian countries in the world and large amounts of agricultural wastes/crop residues are produced every year. To cater to the food grain demands of an increasing population, it is necessary to increase food productivity, which, in parallel, would increase the crop residue/wastes generated. Rice residue burning is a recurring problem in north-west India, especially in Punjab and Haryana states; farmers burn their crop residue after harvesting the rice crop to clear their field for the immediate sowing of the next crop. Major problems associated with crop residue burning include an increase in air pollution due to harmful gases and particulate matter released during this process, loss of soil fertility, due in part to the killing of beneficial microbes, and loss of soil moisture, due to the heating effect. Burning biomass affects the life of millions of people and results in the non-availability/loss of considerable amounts of essential plant nutrients, particularly nitrogen.

Crop residue is the biomass which is left after harvesting the crop, especially rice and wheat. To clear the residue, farmers set fire to their fields, emitting large volumes of gases, like methane, carbon dioxide, etc. and large quantities of particulate matter, which is particularly harmful for people with respiratory illness. Crop residue burning is a source of greenhouse gases and aerosols (Crutzen and Andreae 1990, Streets et al. 2003). At least 34% of the global emissions from the burning of biomass is from crop residue burning (Streets et al. 2004). The present paper reviews the applications of remote-sensing data for the mapping and monitoring of crop residue burning, such as near-real-time identification of active fire incidences and aerosols in India, as well as in a world context. Remote sensing is a valuable tool, which is able to estimate cropping area, crop yield and production, as well as biomass estimation before harvesting of a particular crop, which will be helpful for policy makers and for the preparation of residue management (Singh and Kumar 2019). Availability of satellite-based temporal data with various radiometric and spatial resolutions, along with many indices, helps to estimate the Earth's resources, especially in agriculture. Such an approach may also help to manage the crops as well as facilitating disposal of the residue, which is usually burned in the fields. The estimation of harvesting dates of particular crops is also possible, using remote-sensing techniques.

16.2 MONITORING OF RESIDUE BURNING USING REMOTE SENSING

Monitoring of crop residue burning at regular intervals, particularly prior to each harvest season, would ensure effective implementation of proactive measures to curb crop residue burning practices. Remote-sensing technology has been used effectively for agronomic purposes for quite a while now. Visible Infrared Imaging Radiometer Suite (VIIRS) is a whiskbroom scanning radiometer, and Moderate Resolution Imaging Spectroradiometer (MODIS) collects imaging and radiometric measurements of the Earth, atmosphere and cryosphere, as well as detecting fires, using visible and mid-infrared electromagnetic spectrum regions, and scoring the number of fire events. Spatial, spectral and temporal domains of remote-sensing imagery allow delineation and monitoring over large burned areas quickly. Singh et al. (2009) found that monitoring of the spatial and temporal distribution of rice residue burning is possible, using remote-sensing satellites like MODIS, which provides coarse resolution. Temporal datasets of LISS-III, LISS-IV, MODIS and AVHRR can also be used in the quantitative estimation of burned areas. The small-burn patches can be identified using high spatial resolution optical data from the Sentinel-2 satellite. Koutsias et al. (2013) reported that aerial photography can also be used in those areas where satellite data is not available due to cloudy skies. Different indices are used for crop conditions and burned area mapping in different studies all over the world. Several indices, like NDTI (Normalized Difference Tillage Index), which is used for tillage index to estimate residue cover, and NDVI (Normalized Difference Vegetation Index), which shows crop condition, with values between −1 and 1. For mapping burned

TABLE 16.1
Summary of Satellite Data and Techniques Used for Monitoring Residue Burning

Category	Methods/Index	Data Used	Resolution	Reference
Crop Health	NDVI	MODIS-Terra	250 m	Kumar et al. (2019)
Burned area	ISODATA	AWiFS	56 m	Kumar et al. (2019)
Burned location	Active fire detection algorithm	Landsat-8 OLI	30 m	Schroeder et al. (2016)
Spatial and temporal counts of burning	MODIS burned product	MODIS-Terra/Aqua	1 km	Li et al. (2016)
Burned location	Active fire detection algorithm	VIIRS	375 m	Kumar et al. (2019)
Crop residue cover	DFI, NDTI, STI, PVI, NDVI, GNDVI	Landsat-8 OLI	30 m	Zahra et al. (2020)

areas, the Normalized Burn Ratio ($NBR = \rho$ near-infrared$-\rho$ mid-infrared /ρ near-infrared + ρ mid-infrared) has been used in various studies (Table 16.1).

Kontoes et al. (2009), and Koutsias and Karteris (2000) stated that the reflectance is highest in the near-infrared region for healthy vegetation, the value decreasing for burned vegetation, and this helps in mapping burned areas. García and Caselles (1991) and Kontoes et al. (2009) reported that fire shows high reflectance in short-wave infrared and mid-wave infrared regions of the electromagnetic spectrum. Low reflectance in MODIS band 5 (1.230–1.250 μm) was reported after a burning event, especially in an African context. Various sensors of the National Oceanic and Atmospheric Administration (NOAA) satellite provide fire products, which capture the thermal energy of the fire, using mid-wave and thermal infrared bands. Fraser and Li (2002) stated that several possible algorithms have been developed, which use threshold values over the spectral bands to detect fires. Transplanting dates and the harvesting period of *kharif* (monsoon season) rice can be estimated using multi-temporal microwave satellite data, which can help in prior planning to take steps to reduce residue burning. Kumar et al. (2019) derived expected harvesting dates based on transplanting dates, which were validated by the actual spread of fire events and were found to be accurate for Punjab and Haryana.

16.3 CROP BIOMASS ESTIMATION

Estimation of rice straw, residue and post-harvest biomass by conventional methods is very costly and labour intensive, whereas remote sensing has been used, due to its large synoptic view and its consistent coverage, to derive these factors efficiently in recent years. There are various strategies used in the field, for example, photographic methods, visual estimation or line transect; however, they are long and tedious to carry out (Bannari et al. 2006). Zheng et al. (2014) reported that a high spatial and temporal resolution image is very important to generate a highly accurate crop residue cover map. It can be performed by using multispectral data obtained by sensors. Hyperspectral remote-sensing data are also useful in assessing crop residue cover on fields. It allows the calculation of some dry vegetation indices, such as the Cellulose Absorption Index (Daughtry et al. 2006). Wiseman et al. (2014) used RADARSAT-2 C-band microwave data for estimation of crop production and compared twenty-one polarimetric parameters with dry biomass of soybean, canola and winter wheat (Table 16.2).

The TerraSAR-X data have been used to analyse the morphological parameters of rice (e.g., stem height, leaf area) by Yuzugullu et al. (2017) in the north-west part of Turkey. Ndikumana et al. (2018) reported that Sentinel-1 synthetic aperture radar (SAR) data could be used to obtain measures of rice biomass and height. Xiao et al. (2005, 2006) reported that Normalized Difference Vegetation Index (NDVI), Enhanced Vegetation Index (EVI) and Normalized Difference Water Index (NDWI) could be used to estimate rice parameters. Gebhardt et al. (2012) and Bouvet et al. (2014) used X-band of SAR for estimation of rice yield or biomass. Thus, based on a review of the above studies, it is clear that remote sensing plays a significant role in estimating crop biomass, which, in turn, plays an important role in residue management and prior planning to minimize crop residue burning.

16.4 ESTIMATION OF POLLUTANTS AND GASES USING REMOTE SENSING

Crop residue burning produces greenhouse gases and aerosols in large quantities (Crutzen and Andreae 1990, Streets et al. 2003). Crop residue burning makes a 34 percent contribution to the world's emission from biomass burning (Streets et al. 2004). Burning of crop residues not only degrades the quality of the atmosphere but also has an effect on the climate and on human health. Burning of crop residue is a major contributor of carbon monoxide (CO), nitrogen oxides, sulphur dioxide, methane (CH_4), carbon dioxide (CO2) and volatile organic compounds (VOCs). Aerosol recovery from MODIS data was applied *via* offshore and ocean surface visualizations on two separate algorithms described within the literature (Kaufman and Tanre 1998). Aerosol Index (AI) values, provided by different satellites, have been used for different studies. The Cloud-aerosol Lidar and Infrared Pathfinder Satellite Observation (CALIPSO) system provides new insights into the role of clouds and atmospheric aerosols in affecting the atmosphere and the air quality of the Earth. CALIPSO incorporates a 98-degree inclination orbit and flies at an altitude of 705 km; it provides daily global information on the vertical distribution of aerosols and clouds. Badarinath et al. (2009) retrieved information regarding aerosol loadings and pollution from carbon dioxide due to anthropogenic activities over the Indo-Gangetic Plains (IGP) in India. Therefore, remote-sensing satellites are capable of monitoring changes in atmospheric composition, and play an important role in the estimation of gas emissions (Table 16.3).

TABLE 16.2
Summary of Satellite Data and Techniques Used for Biomass Estimation

Category	Study Area	Methods	Data Used	Resolution	References
Accumulation of dry biomass	Southern Manitoba (Canada)	Back-scatter response	Radarsat-2	8 m (nominal)	Wiseman et al. (2014)
Rice height and biomass	Camargue (southern France)	Multiple Linear Regression (MLR), Support Vector	Sentinel-1	20 m	Ndikumana et al. (2018)
Above-ground dry biomass	Ottawa, (Canada)	Monteith's radiation use efficiency model	Compact Airborne Spectrographic Imager and the Landsat-5,7	2 m, 30 m	Liu et al. (2018)
Crop biomass estimation	Winnipeg, Manitoba (Canada)	Multiple regressionanalysis and neural networks	UAVSAR PolSAR time series data	0.6 m × 1.6 m pixel spacing	Reisi-Gahrouei et al. (2019)
Above-ground dry phytomass accumulation	Greenbelt Farm of Agriculture and Agri-Food Canada	LAI, MTVI2 indices	CASI, hyper spectral data		Liu et al. (2004)

TABLE 16.3
Summary of Satellite Data and Techniques Used for Estimation/Monitoring of Air Pollutants and Gases

Category	Methods/ Techniques	Data Used	Resolution	Reference
Spatial and temporal distribution of emissions	Algorithm based	MODIS-Terra/Aqua	500 m and 1 km	Junpen et al. (2018)
Aerosol index and depth, carbon monoxide estimation	Back-trajectory model and algorithm	MODIS/OMI/MOPITT/IRS-P4OCM	$1^o \times 1^o/13 \times 24Km^2/22km/360 \times 250m^2$	Badarinath et al. (2009)
Impacts of burning on air quality	HYSPLIT	HYSPLIT/MODIS	MODIS 250–1 km	Li et al. (2010)
Aerosol optical depth (AOD) variations, fire radiative power (FRP) and fire radiative energy (FRE)	NASA Terra/ Aqua satellite algorithms	MODIS-Terra/Aqua AOD product	10 km	Vadrevu et al. (2011)
Emissions of gases from residue burning	Bottom-up approach	AWiFS, MODIS, ASTER, Landsat-TM	56 m,500 m–1 km,15 m, 30 m.	McCarty (2011)

16.5 CONCLUSION

Analysis of the available literature indicates that remote-sensing data are commonly used for the estimation and tracking of crop residue burning, involving the location of residue-burning sites and near-real-time residue fire sites, and providing planners and compliance agencies with the appropriate steps to reduce/stop the practice of open field residue burning. Remote-sensing plays a significant part in the assessment and monitoring of burning crop residues. Several satellites have been developed and placed into orbit, which have different spectral, spatial and temporal resolutions and provide different information about residue burning and related aspects. MODIS and Suomi NPP-VIIRS satellites identify active fire points with spatial resolutions of 375 m × 375 m and 750 m × 750 m, respectively, with datasets from these satellites being freely available to provide the near-real-time monitoring of the active crop residue fire locations. MODIS and some other satellites also provide information on environmental parameters such as atmospheric activity, e.g., aerosol optical depth, etc. Researchers find the new elements of remote sensing and satellite data important to estimate and monitor gas emission due to residue burning. Temporal resolution, that helps in identification of the date of crop transplantation and expected harvest dates, may help policymakers in residue management and monitoring. Microwave data is known as "all-weather" data, because it has the ability to capture data in any weather (at day or night), even under the cloudy conditions of the rainy season, and can provide data from cloud-free skies needed by researchers to extract more information about crop assessment, crop residue and gas emissions, etc.

REFERENCES

Abdou, B., Pacheco, A., Staenz, K., McNairn, H., & Omari, K. (2006). Estimating and mapping crop residues cover on agricultural lands using hyperspectral and IKONOS data. *Remote Sensing of Environment*, 104, 447–459. doi:10.1016/j.rse.2006.05.018.

Badarinath, K. V. S., Kharol, S. K., Sharma, A. R., & Prasad, V. K. (2009). Analysis of aerosol and carbon monoxide characteristics over Arabian Sea during crop residue burning period in the Indo-Gangetic Plains using multi-satellite remote sensing datasets. *Journal of Atmospheric and Solar-Terrestrial Physics*, 71(12), 1267–1276.

Bannari, A., Pacheco, A., Staenz, K., McNairn, H., & Omari, K. (2006). Estimating and mapping crop residues cover on agricultural lands using hyperspectral and IKONOS data. *Remote Sensing of Environment*, 104(4), 447–459.

Bouvet, A. et al. (2014). Estimation of agricultural and biophysical parameters of rice fields in Vietnam using X-band dual polarization. *SAR 2014 IEEE Geoscience and Remote Sensing Symp. 1504–7.*

Crutzen, P.J., & Andreae, M. O. (1990). Biomass burning in the tropics: Impact on atmospheric chemistry and biogeochemical cycles. *Science*, 250, 1669–1678.

Daughtry, C. S. T., Doraiswamy, P. C., Hunt Jr., E. R., Stern, A. J., & McMurtrey III, J. E., Prueger, J. H. (2006). Remote sensing of crop residue cover and soil tillage intensity. *Soil Tillage Research*, 91, 101–108.

Fraser, R. H., & Li, Z. (2002). Estimating fire related parameters in boreal forests using SPOT VEGETATION. *Remote Sensing of Environment*, 82, 95–110.

Gebhardt, S., Huth, J., Lam-dao, N., Roth, A., & Kuenzer, C. (2012). A comparison of TerraSAR-X Quadpol backscattering with RapidEye multispectral vegetation indices over rice fields in the Mekong Delta, Vietnam. *International Journal of Remote Sensing*, 33, 7644–7661.

Junpen, A., Pansuk, J., Kamnoet, O., Cheewaphongphan, P., & Garivait, S. (2018). Emission of air pollutants from rice residue open burning in Thailand, 2018. *Atmosphere*, 9(11), 449.

Kaufman, Y. J., & Tanre, D. (1998). *Algorithm For Remote Sensing of Tropospheric Aerosol from MODIS, Algorithm Theoretical Basis Document, ATBD-MOD-02*. NASA Goddard Space Flight Center, 85.

Kontoes, C. C., Poilvé, H., Florsch, G., Keramitsoglou, I., & Paralikidis, S. (2009). A comparative analysis of a fixed thresholding vs. a classification tree approach for operational burn scar detection and mapping. *International Journal of Applied Earth Observation and Geoinformation*, 11(5), 299–316.

Koutsias, N., & Karteris, M. (2000). Burned area mapping using logistic regression modeling of a single post-fire Landsat-5 thematic mapper image. *International Journal of Remote Sensing*, 21(4), 673–687.

Koutsias, N., Pleniou, M., Mallinis, G., Nioti, F., & Sifakis, N. I. (2013). A rule-based semi-automatic method to map burned areas: Exploring the USGS historical Landsat archives to freconstruct recent fire history. *International Journal of Remote Sensing*, 34(20), 7049–7068.

Kumar, P., Rajpoot, S. K., Jain, V., Saxena, S., & Ray, S. S. (2019). Monitoring of rice crop in Punjab and Haryana with respect to residue burning. *International Archives of the Photogrammetry, Remote Sensing & Spatial Information Sciences*. The International Archives of the Photogrammetry, Remote Sensing and Spatial Information Sciences, Volume XLII-3/W6, 2019.pp. 31-36.

Li, H., Han, Z., Cheng, T., Du, H., Kong, L., Chen, J., & Wang, W. (2010). Agricultural fire impacts on the air quality of Shanghai during summer harvesttime. *Aerosol and Air Quality Resarch*, 10(2), 95–101.

Li, J., Bo, Y., & Xie, S. (2016). Estimating emissions from crop residue open burning in China based on statistics and MODIS fire products. *Journal of Environmental Sciences*, 44, 158–170.

Liu, J., Miller, J. R., Pattey, E., Haboudane, D., Strachan, I. B., & Hinther, M. (2004, September). Monitoring crop biomass accumulation using multi-temporal hyperspectral remote sensing data. In IGARSS 2004. 2004 IEEE international geoscience and remote sensing symposium (Vol. 3, pp. 1637–1640). IEEE.

Liu, T., Marlier, M. E., DeFries, R. S., Westervelt, D. M., Xia, K. R., Fiore, A. M., Mickley, L. J., Cusworth, D. H., & Milly, G. (2018). Seasonal impact of regional outdoor biomass burning on air pollution in three Indian cities: Delhi, Bengaluru, and Pune. *Atmospheric Environment*, 172, 83–92, ISSN 1352-2310, https://doi.org/10.1016/j.atmosenv.2017.10.024

López-Garcia, M. J., & Caselles, V. (1991). Mapping burns and natural reforestation using Thematic Mapper data. *Geocarto International*, 1, 31–37.

McCarty, J. L. (2011). Remote sensing-based estimates of annual and seasonal emissions from crop residue burning in the contiguous United States. *Journal of the Air & Waste Management Association*, 61(1), 22–34, doi:10.3155/1047-3289.61.1.22

Ndikumana, E., Minh, H. T. D., Dang Nguyen, H., Baghdadi, N., Courault, D., Hossard, L., & El Moussawi, I. (2018). Estimation of rice height and biomass using multitemporal SAR Sentinel-1 for Camargue, Southern France. *Remote Sensing*, 10(9), 1394.

Reisi Gahrouei, Omid, Homayouni, Saeid, McNairn, Heather, Hosseini, M., & Safari, A. (2019). Crop biomass estimation using multi regression analysis and neural networks from multitemporal L-band polarimetric synthetic aperture radar data. *International Journal of Remote Sensing*, 1–19. doi:10.1080/01431161.2019.1594436.

Schroeder, W., Oliva, P., Giglio, L., Quayle, B., Lorenz, E., & Morelli, F. (2016). Active fire detection using Landsat-8/OLI data. *Remote Sensing of Environment*, 185, 210–220.

Singh, G., Kant, Y., & Dadhwal, V. K. (2009). Remote sensing of crop residue burning in Punjab (India): A study on burned area estimation using multi-sensor approach. *Geocarto International*, 24(4), 273–292.

Singh, R., & Kumar, P. (2019). Potential of remote sensing in the assessment of crop residue burning and its related aspects: A review. *Our Heritage*, 67(4), 62–69.

Streets, D. G. et al. (2004). On the future of carbonaceous aerosol emissions. *Journal of Geophysical Research: Atmospheres*, 109, D24.

Streets, D. G., Yarber, K. F., Woo, J. H., & Carmichael, G. R. (2003). An inventory of gaseous and primary aerosol emissions in Asia in the Year 2000. *Journal of Geophysical Research*, 108, 8809–8823, doi:10.1029/2002JD003093

Vadrevu, K. P., Ellicott, E., Badarinath, K. V. S., & Vermote, E. (2011). MODIS derived fire characteristics and aerosol optical depth variations during the agricultural residue burning season, north India. *Environmental Pollution*, 159(6), 1560–1569.

Wiseman, G., McNairn, H., Homayouni, S., & Shang, J. (2014). RADARSAT-2 polarimetric SAR response to crop biomass for agricultural production monitoring. *IEEE Journal of Selected Topics in Applied Earth Observations and Remote Sensing*, 7(11), 4461–4471.

Xiao, X., Boles, S., Frolking, S., Li, C., Babu, J. Y., Salas, W., & Moore, B. (2006). Mapping paddy rice agriculture in South and Southeast Asia using multi-temporal MODIS images. *Remote Sensing of Environment*, 100, 95–113.

Xiao, X., Boles, S., Liu, J., Zhuang, D., Frolking, S., Li, C., Salas,W., & Moore, B. (2005). Mapping paddy rice agriculture in southern China using multi-temporal MODIS images. *Remote Sensing of Environment,* 95, 480–492.

Yuzugullu, O., Erten, E., & Hajnsek, I. (2017). A multi-year study on rice morphological parameter estimation with X-Band Polsar data. *Applied Sciences*, 7, 602. doi:10.3390/app7060602.

Zahra, K., Raoufat, M. H., Dehghani, M., Kazemeini, S. A., & Nazemossadat, M. J. (2020). Feasibility of satellite and drone images for monitoring soil residue cover. *Journal of the Saudi Society of Agricultural Sciences*, 19(2020), 56–64.

Zheng, B., Campbell, J. B., Serbin, G., & Galbraith, J. M. (2014). Remote sensing of crop residue and tillage practices: Present capabilities and future prospects. *Soil Tillage Research,* 138, 26–34.

17 Indian Water Resource Research Using Remote Sensing, as Reflected by the Web of Science during 2009–2018

A Bibliometric Study

Shiv Singh, Pawan Agrawal, and Neha Munjal

CONTENTS

17.1 INTRODUCTION

Water is one of the most important resources on Earth and the dependency of all living beings on it makes it all the more important. It has already been stated by many that, if a third world war happens, water will be the main reason behind it, and Fergusson (2015) has claimed that soon world would be at war over water. The World Economic Forum's Global Risk Reports have identified water availability as one of the three greatest challenges worldwide (World Economic Forum, 2020). It is impossible for humans to substitute water with any other resource for most of its uses, and it is expensive to transport and difficult to de-pollute; hence, it is a precious gift from nature to us. Many countries are investing huge amounts of research and money to find water on other planets of the universe, a goal which underlines the significance of water to human beings. There are two important type of usable water, i.e. underground water and surface water. Both surface and underground water play vital roles in all types of development, such as agriculture, fisheries, forestry, hydropower, recreational activities, livestock production, etc. Understanding of the Earth's surface is essential to understand the hydrological cycle, especially in a climate change environment. Researchers throughout the world are working on freshwater resources, developing models to achieve sustainable water management for a continuously growing world population. (Tiwari, Wahr, and Sweson, 2009).

17.2 WATER RESOURCES OF INDIA

India receives annual precipitation of about 4000 billion m^3 of water, of which about 60% is usable. Of the usable water, about 40% is ground water, while 60% is surface water (Sorsa, Nag, and Kettunen, 2018; Kumar, Singh, and Sharma, 2005). Of the total water used, about 20% is treated and reused, whereas the rest is lost. The available water in India is less than that required, and water scarcity will not decrease, even if all the precipitation was used. This leads to the various hydrological organisations to work around models of water conservation and water treatment to increase the availability of water resources. A country is "water stressed" if it has less than 1700 m^3 water available per person per annum and "water scarce" if that availability is less than 1000 m^3 per person per annum. Per capita surface water availability in India in the years 1991 and 2001 were 2309 and 1816 m^3, respectively, further decreasing to 1544 m^3 in 2011. It is estimated that it will further dip down to 1486 m^3 by 2021. When the value fell below 1700 m^3 sometime between 2001 and 2011, then India was officially a water-scarce country. Although, presently, India doesn't fall into the category of a "water-scarce country", there are several states which face drought, especially in the summer. Thus, there is a need to have proper water management planning, water-harvesting techniques and effective models of water recycling.

As discussed above, water is a vital natural resource for all living beings. It is a very well-known fact that the availability of freshwater is constantly reducing. Researchers all over the world are working towards the development of effective models for water management and conservation. The current review will be a useful study for both researchers and academics, as it will provide important indicators in the form of bibliometric data, such as chronological trends of publication in particular journals, most-cited authors, most-cited papers, most effective countries at research in this field, most popular water resource journals, etc. The various indexes provided here will also be useful to show the influence of a specific article, author or journal in a given field. The bibliometric approach has already been applied to countless scientific fields (Gorraiz and Gumpenberger 2015).

17.3 METHODOLOGY

The Web of Science is a web technology-based platform, developed in 1960 by Clarivate Analytics. It includes a bibliographic database of the journals indexed, as well as information analysis tools, which are helpful in the evaluation and analysis of the research. Since 1800, it has built a vast collection currently standing at at least 34,502 reputed journals from all over the world (Clarivate Analytics, 2020).

The authors selected the Web of Science due to its vast coverage and universal acceptance. The query used was 6811–SU= ("Water Resources") and CU= (India) 2009–2018) which was used to search Titles, Abstract, and Keywords of documents published during 2009–2018 in journals, conference proceedings, reviews, and letters.

This review paper is an attempt to understand the research being conducted on the water resources of India (WRI) through bibliometric analysis. Noyon et al. (1999) discussed two major procedures of evaluative bibliometrics, i.e., performance analysis and scientific mapping. These can be achieved through analysis of citations given and received, performance of the nation-wise research, institutional performance, etc. The present paper used both procedures to draw the characteristics of the research output in WRI. To analyse the data, Microsoft Excel and R software were used.

17.4 RESULTS AND DISCUSSION

As a result of the search, 6311 records were obtained for the study period 2009–2018. The bibliographic records were downloaded with all necessary details, such as title, author, country, citation, journal, keyword, references, etc. These records were categorised into 6070 articles, 89 proceedings papers, one retracted publication, three letters, and 148 reviews. In addition, these data were compared with the total literature on water resources published worldwide, without taking any geographic area into consideration. Table 17.1 lists the yearly record of all published articles. In addition, it shows the annual number of papers on water resources published worldwide, regardless of their geographical coverage. Research into water resources in India increased from 417 in 2009 to 782 in 2018 in a zig-zag, inconsistent manner. With reference to the global context, the proportion of global water resource research which dealt with India increased (though not linearly with time) from 2009 to 2018, relative to the previous part of the study period (Table 17.1).

Table 17.2 showcases the growth of the number of research papers on water resources in India, along with mean total number of citations per article and mean total number of citations per year. The mean total citation per article was highest in the year 2010, i.e., 23.1805 followed by 23.09223 in 2011. With respect to the mean total citation number per year, the value ranges between 2 and 2.8.

Table 17.3 presents the top sources, or journals, publishing water resource research, along with the number of records worldwide, with respect to Indian and universal contributions, as well as source metrics. At the top of the listing is the journal *Desalination and Water Treatment*, with 1027 Indian contributions (9%) of universal contributions which were 11,248 in total. In second place, the journal is Environmental Earth Science, with a total of 588 Indian records and 7457 universal contributions. In third position, the source is Natural Hazards, with 516 Indian papers and 6920 papers worldwide. In the context of h-index and g-index, the top two journals are "Desalination" and "Journal of Hydrology" with h indices of 55 and 37, respectively, and g indices 85 and 55, respectively. The journal

TABLE 17.1
Year-wise Growth of Research on Indian Water Resources by Indian Authors *vs* Global Research on Water Resources

	Indian Research		World Research	
Year	Articles	(%)	Articles	(%)
2009	417	6.61	9522	7.10
2010	482	7.64	9819	7.32
2011	412	6.53	10867	8.10
2012	269	4.26	11659	8.69
2013	615	9.74	13072	9.75
2014	713	11.29	13478	10.05
2015	804	12.74	14945	11.15
2016	1034	16.38	17382	12.96
2017	783	12.40	16041	11.96
2018	782	12.39	17242	12.86
Total	6311		134027	

TABLE 17.2
Yearly Trends of Number of Publications, Citations per Article, Citations per Year

Year	Number of Papers	Mean Total Citations per Article	Mean Total Citations per Year
2009	417	20.58	2.05
2010	482	23.18	2.57
2011	412	23.09	2.88
2012	269	17.47	2.49
2013	615	14.26	2.37
2014	713	11.50	2.30
2015	804	9.917	2.47
2016	1034	6.88	2.29
2017	783	4.42	2.21
2018	782	2.26	2.26

Desalination is at the top in all metrics, except for the number of papers, which may be because of its high rejection rate. Among these top sources, the majority of them have impact factors in the range of 1–2. The top three journals, which have received the highest number of citations are Desalination, Environmental Earth Sciences and the Journal of Hydrology, with 11954, 5202 and 5039 citations, respectively.

TABLE 17.3
Top Journal Impact Along with Different Metrics

Name of Journal	Globally	Total Citations	Number of Indian Papers	*PY_start	**Impact Factor
Desalination and Water Treatment	11249	6306	1027	2009	1.383
Environmental Earth Sciences	7457	5202	588	2009	1.435
Natural Hazards	6920	4215	516	2009	1.901
Desalination	6711	11954	349	2009	6.603
Water Resources Management	6550	3698	240	2009	2.644
Journal of Hydrology	5266	5039	232	2009	3.727
Water Science and Technology	5223	1204	192	2009	1.247
CLEAN–Soil, Air, Water	4255	2524	180	2009	1.338
Journal of Water Process Engineering	4181	1929	172	2014	3.371
Journal of Hydrologic Engineering	4042	1332	155	2009	1.576

*PY_start=publication year start. Impact Factor=as per JCR 2018

Figure 17.1 represents the growth of the top five journals, with respect to water resource research in India. The top journal is 'Desalination and Water Treatment', although no clear growth trend over time can be seen from 2009 to 2018. In 2009, there were 45 articles, then 36 in 2010, then only 19 papers in each of 2011 and 2012. But, in 2013, the number of papers reached 54. This trend of increasing papers continued in 2014 (104 papers), in 2015 (173) and in 2016 (332 articles), before decreases in 2017 and 2018. In total, 1027 papers were published in this journal from 2009 to 2018.

The second-ranked journal is Environmental Earth Sciences. In this journal, 588 papers were published during the study period 2009 to 2108, with no consistent trend of the number of papers published each year over time. In 2009, there were only eight articles, then in 2010 and 2011 there were 31 and 30 articles each, respectively, followed by a decrease in 2012 to 20 articles, followed by increases in the number of papers published from 2013 to 2016, with subsequent decreases in 2017 and 2018.

Natural Hazards is the third-ranked journal with 516 papers during the study period, in the initial years of the study, very few articles were published, but, from 2011 to 2013, an increase was obvious. In 2014, the number decreased but it was followed by an increase in 2015 again jump has been seen further from 2016 to 2017 article reduces to 81 to 38 and finally, in the year 2018, there were 62 articles.

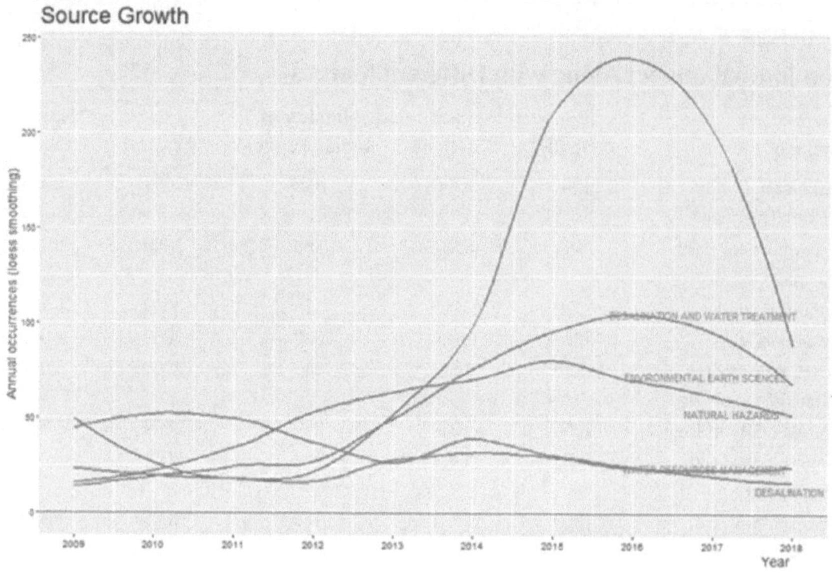

FIGURE 17.1 Growth of top five sources (journals).

The fourth-rated journal is "Desalination", with the number of papers published each year showing an inconsistent upward trend from 2009 to 2018. In 2009, there were 43 articles, which increased in 2010 and 2011 (78 articles). In 2012, there was a tremendous fall in the number of articles published, with only 14 papers. In 2013, the number of articles increased to 39 and remained in the 30s until 2015. In 2016, the number of papers fell by almost half, to 17, then rose to 23 in 2017 and 16 in 2018, with a final total of 349 articles.

The journal Water Resource Management published the fifth-highest number of articles on water resources in India between 2009 and 2018, with 240 articles, with the pattern being inconsistent over the ten years of the study period.

Table 17.4 represents the most-cited Indian papers in the area of remote sensing and water resources, the top three papers who received the most citations being by VK Gupta (2011) in Water Research-A, S Chowdhury (2011) in Desalination, and PS Kumar (2010) in Desalination, with the total number of citations per paper being 461, 416, and 391, respectively. In terms of total citations per year, the top three papers contributed by Indian authors are VK Gupta (2014) in Water Research, VK Gupta (2011) in Water Research-A, and PV Nidheesh (2012) in Desalination, with 62.2, 57.6, and 53.6 citations per year, respectively.

Table 17.5 presents the papers with the most citations in the field of remote sensing and water resources. The references which received more than 150 citations are H Freundlich (1906) Z Phys Chem-Stoch Ve, Vol. 57, pp 385–98. Lagergren S. K., 1898, Kungliga Svenska Vet, V24, P1, Doi Doi 10.1007/Bf01501332, Langmuir I, 1918, J Am Chem Soc, V40, P1361, Doi 10.1021/Ja02242A004, Ho Ys, 1999, Process Biochem, V34, P451, Doi 10.1016/S0032-9592(98)00112-5, Nash J. E.,

TABLE 17.4
Most-Cited Papers on water resources and remote sensing

Paper	Total Citations	Total Citations per Year
VK Gupta (2011), Water Research-A	461	57.62
Chowdhury S, 2011, Desalination	416	52.00
Kumar Ps, 2010, Desalination	391	43.44
Nidheesh Pv, 2012, Desalination	375	53.57
Gupta Vk, 2014, Water Res	311	62.20
Padaki M, 2015, Desalination-A	299	74.75
Montanari A, 2013, Hydrolog Sci J	294	49.00
Ganesh Bm, 2013, Desalination	259	43.16
Rockstrom J, 2010, Agr Water Manage	237	26.33
Qadir M, 2010, Agr Water Manage	237	26.33
Gupta Vk, 2011, Water Res	216	27.00
Dinpashoh Y, 2011, J Hydrol	177	22.12
Gupta A, 2009, Water Res	155	15.50
Sonali P, 2013, J Hydrol	151	25.16
Shenvi Ss, 2015, Desalination	150	37.50
Kumar V, 2010, Hydrolog Sci J	143	15.88
Tiwari Mk., 2010, J Hydrol	139	15.44
Kumar R, 2011, Desalination	138	17.25
Jhaveri Jh, 2016, Desalination	135	45.00
Mane Vs, 2011, Desalination	130	16.25

1970, J Hydrol, V10, P282, Doi 10.1016/0022-1694(70)90255-6, with 168, 167, 161, 158, and 154 citations per paper, respectively (Figure 17.2).

Figure 17.3 represents the top 30 countries having collaborations with respect to water resource research with India. The nine countries with the highest collaborations with India are the United States, Australia, the United Kingdom, Germany, South Korea, Canada, Malaysia, the People's Republic of China, and France, with 380, 151, 134, 133, 116, 107, 103, 93, and 92 papers together, respectively.

Figure 17.4 shows the country collaboration map, working together with India on water resources. The United States is at the top with 28,647 records, followed by the People's Republic of China, Germany, Canada, Australia, India, Italy, the United Kingdom, Spain, France, Iran, Netherlands, South Korea, and Japan, with 24978, 8119, 7486, 7354, 6311, 6246, 6208, 5979, 5650, 5161, 4156, 4039, and 3418 records, respectively.

17.5 CONCLUSION

Water resources are the most important natural resources, particularly in India. With a growing population, the consumption of water has also increased, in the form of either drinking water or water needed for other household or industrial

TABLE 17.5
Most Locally Cited References

Cited References	Citations
H Freundlich H (1906) Z Phys Chem-Stoch Ve, 57: 385–98.	168
Lagergren S K (1898) Kungliga Svenska Vet, V24, P1, Doi Doi 10.1007/Bf01501332	167
Langmuir I (1918) J Am Chem Soc, V40, P1361, Doi 10.1021/Ja02242A004	161
Ho Ys (1999) Process Biochem, V34, P451, Doi 10.1016/S0032-9592(98)00112-5	158
Nash J E (1970) J Hydrol, V10, P282, Doi Doi 10.1016/0022-1694(70)90255-6	154
Bagstad Kj (2013) Ecosyst Serv, V5, Pe27, Doi 10.1016/J.Ecoser.2013.07.004	139
Weber W J (1963) J Sanitary Engineeri, V89, P31, Doi Doi 10.1016/J.Desal.2010.08.036	131
Eaton A D (2005) Standard Methods Exa	129
American Public Health Association Apha, (1998) Standard Methods Exa	113
Langmuir I. (1916) J Am Chem Soc, V38, P2221, Doi 10.1021/Ja02268A002	104
Lowry Oh, (1951) J Biol Chem, V193, P265	97
Gibbs Rj, (1970) Science, V170, P1088, Doi 10.1126/Science.170.3962.1088	74
Arnold Jg, (1998) J Am Water Resour As, V34, P73, Doi 10.1111/J.1752-1688.1998.Tb05961.X	69
[Anonymous] (2011) Guid Drink Wat Qual	65
Freeze R. A., (1979) Groundwater	64
Piper Am (1944) Eos T Am Geophys Un, V25, P914	63
Robinson T (2001) Bioresource Technol, V77, P247, Doi 10.1016/S0960-8524(00)00080-8	61
Goswami Bn, (2006) Science, V314, P1442, Doi 10.1126/Science.1132027	60
Govindaraju Rs, (2000) J Hydrol Eng, V5, P115	60
Rodell M, (2009) Nature, V460, P999, Doi 10.1038/Nature08238	60

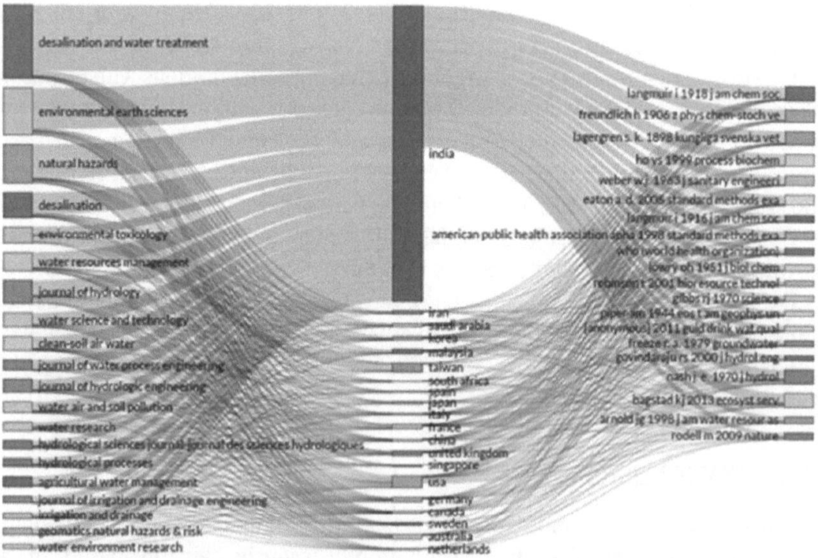

FIGURE 17.2 The top 20 items in each of three fields.

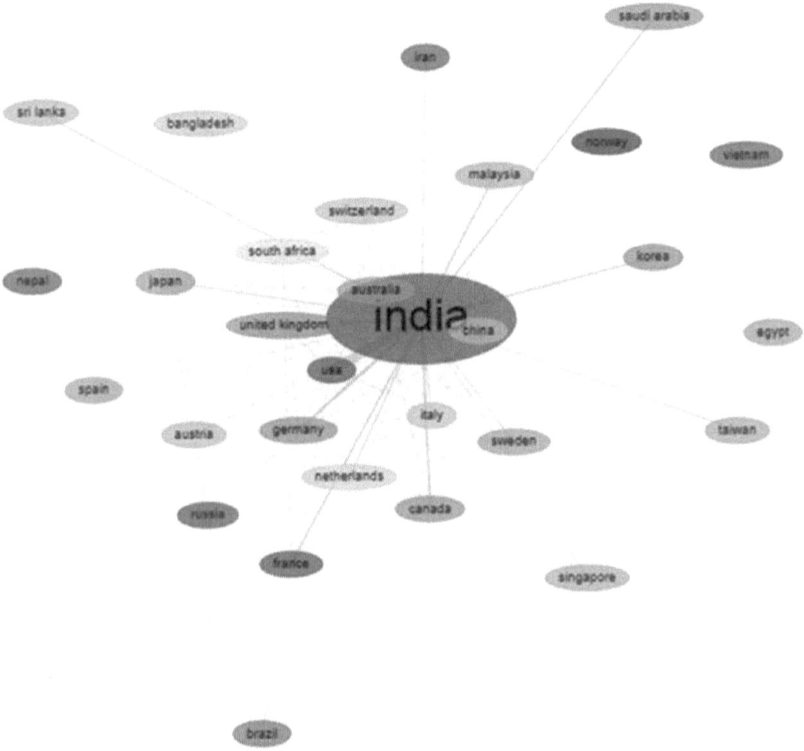

FIGURE 17.3 The top 30 country collaboration tree.

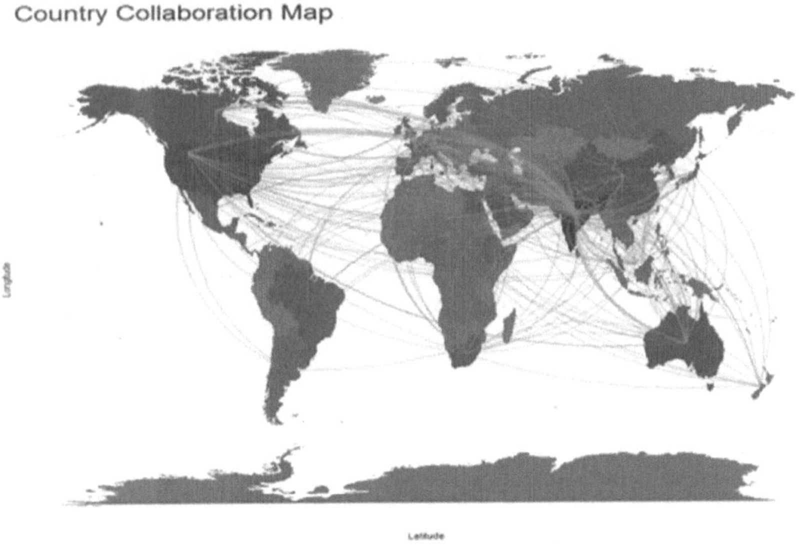

FIGURE 17.4 Country collaboration map.

uses. In the present work, the research output with respect to water resources has been explored by using 'bibliometric indicators'. For the period of 2009 to 2018, a total of 6311 documents were retrieved from 95 sources, mainly journals. The average number of citations per article was 11.29. The total number of authors was 11343, and the majority of papers was produced in collaboration with authors from other countries, with a mean of 3.66 co-authors per document. Extensive research has been carried out on 'water resources' but there is still a need to explore further for the coming generation, otherwise the next world war may be over water.

REFERENCES

Clarivate Analytics. 2020. *Web of Science platform: Web of Science: Summary of Coverage*. Retrieved February 20, 2020, from https://clarivate.libguides.com/webofscienceplatform/coverage

Fergusson, J. 2015. *The World Will Soon be at War Over Water, Newsweek*. Retrieved from https://www.newsweek.com/2015/05/01/world-will-soon-be-war-over-water-324328.html

Gorraiz, J., and Gumpenberger, C. 2015. A flexible bibliometric approach for the assessment of professorial appointments. *Scientometrics*, 105, 1699–1719. https://doi.org/10.1007/s11192-015-1703-6

Kumar, R., Singh, R., and Sharma, R.D. 2005. Water resources of India. *Current Science*, 89(5), 794–811.

Ministry of Jal Shakti, Government of India. 2019. *Per Capita availability of Water*. New Delhi: Press Information Bureau. Retrieved February 20, 2020, from https://pib.gov.in/newsite/PrintRelease.aspx?relid=195633

Noyons, E., Moed, H. F., and Luwal, M. 1999. Combining mapping and citation analysis for evaluative bibliometric purposes: A bibliometric study. *Journal of American Society for Information Science*, 50(2), 115–131. https://doi.org/10.1002/(SICI)1097-4571(1999)50:2%3C115::AID-ASI3%3E3.0.CO;2-J

Sorsa, K., Nag, T., and Kettunen, J. 2018. Exploring private participation in Indian water sector: Issues and options. In K. Sorsa, and M. Chaudhuri Eds., *Sustainable Engagement in the Indian and Finnish Business* (pp. 6–18). Turku: Turku University of Applied Sciences. Retrieved February 20, 2020, from http://julkaisut.turkuamk.fi/isbn9789522167040.pdf

Tiwari, V. M., Wahr, J., and Sweson, S. 2009. Dwindling groundwater resources in northern India, from satellite gravity observations. *Geophysical Research Letters*, 36(L18401). Doi:10.1029/2009GL039401

World Economic Forum. 2020. *The Global Risks Report 2020*. Geneva: World Economic Forum. Retrieved February 15, 2020, from https://www.weforum.org/reports/the-global-risks-report-2020

18 Research Practice on Remote Sensing
A Bibliometric Study of Indian Scholars

Shiv Singh, Sanjay Kataria, Neha Munjal

CONTENTS

18.1 INTRODUCTION

Remote sensing is a widely used technology to gather knowledge about our 'Mother Earth'. The remote-sensing technique involves the capturing and observing of the physical features of a specified zone of the Earth's surface, or any other specified object, by computing its reflected and discharged emission at a distance (usually carried out with the help of a satellite or a specially equipped aircraft) [1, 2]. It can also be defined as a process to gather evidence about a specified object without physically touching it. The information gathered through this process provides several types of details about the object. The information is collected in the form of images and later interpreted by experts [3].

To capture and observe the object during the process of remote sensing, different types of sensors are used to perform this, capturing the image using an aircraft or satellite to ascertain the object [4]. Two main types of sensing technologies, active or passive, are used with remote sensing. Most of the information captured by active and passive sensors are comprised of digital images [2].

Under active-sensing technology, signals are emitted by the satellite or aircraft with which to examine the object and zones in question, which the sensor then senses and observes the radiation which is returned from the object to the sensor [5] The time taken by the emitted signal to return back to the sensor is calculated and, based on that location, the speed and the direction of the object is determined. The use of a satellite (or possibly aircraft) covers a large area of the Earth's surface and provides us with comprehensive information, compared with the information which can be captured by on-ground sensors [4]. Such a satellite-based sensor helps us to understand geographical issues from a spatial perspective.

In passive-sensing technology, the reflection of light from an identified object is measured and captured by the sensor [5]. The source of light, in most cases, is the sun. The image captured by a camera or infrared device is an example of passive sensing [6] The cameras installed on a satellite for passive sensing can provide enhance images of large objects with some additional information, such as the monitoring of melting ice in Antarctica [1].

Remote-sensing technology can have specific uses, being used for the scanning and monitoring of large areas on the ground, such as forests, for the monitoring of clouds to achieve accurate weather forecasts, to monitor volcanic activity, to observe storms, achieve the scanning of cities, the quantification of different crops on farms, the sighting and mapping of landscape of the marine base, etc. [7].

Remote sensing is a rapidly growing area in which much research is being carried out worldwide. To analyze the growth and development of remote-sensing research, the bibliometric technique is the most powerful tool. It is a technique based on the evaluation of the metrics of books, articles, or any other publications on this technology, by carrying out statistical analysis of the data generated. This strategy measures several characteristics, such as providing information on the most productive authors, institutions and countries, while also classifying collaboration among national and international networks. Furthermore, the bibliometric approach also achieves the mapping of multidisciplinary fields This technique is used in numerous scientific fields to analyze the ongoing trends of research [8–14].

18.2 METHODOLOGY

The well-known database, the Web of Science (WoS), was searched over the period 1989–2019 to extract the publications on 'remote sensing' in India. The query used was SU=(Remote Sensing) Refined by: Countries/Regions: (India) And [excluding] Publication Years: (2020) And Document Types: (Article Or Letter Or Proceedings Paper Or Review Or Note) And Timespan: All years. Indexes: SCI-Expanded, SSCI, A&HCI. The query was executed the same day in order to avoid the possible biasing of the updating of the database. In total, 3282 documents were retrieved, which were further analysed to extract the growth of research in remote sensing, the top sources, popular keywords, country collaborations, etc.

18.3 RESULTS AND DISCUSSION

In total, there were 3282 documents in the field of remote sensing, consisting of 3096 articles, 118 proceedings papers, 15 letters, 22 notes and 31 reviews. Table 18.1 represents the growth of the research in 'remote sensing' in India. To

TABLE 18.1
Growth of Publications in Remote Sensing

Year	No. of Publication	(%)	Total Citations
1989–1994	175	5.33	3483
1995–1999	191	5.82	3487
2000–2004	199	6.06	4761
2005–2009	634	19.32	11847
2010–2014	918	27.97	10978
2015–2019	1165	35.50	5311

compare the growth, 5-year spans of the 30 years of the study were considered and it was observed that there was a continuous increase in the research output. Of the 3282 documents, 63.47 % of the publication were from 2010–2019. In the context of citations, the highest number of citations were received for the period 2005–2009, with 11,847 citations being reported for 634 publications.

There were 50 sources (such as journals) in which the 3282 publications were published. Table 18.2 presents the top ten sources, along with the respective total number of publications, citations, impact factors and publication year start. The journal ranked first in terms of publications is the International Journal of Remote Sensing, with 937 publications and 15,344 citations. The Journal of the Indian Society of Remote Sensing' is in second position, with 791 publications and 3870 citations, with Geocarto International, with 205 publications and 1206 citations, in third position. Also in the top 10, three journals are from the IEEE group, namely IEEE Transactions on Geoscience and Remote Sensing, IEEE Geoscience and Remote Sensing Letters and IEEE Journal of Selected Topics in Applied Earth Observations and Remote Sensing, with 168, 145 and 136 publications, respectively. With reference to the h- and m-indices, the top two journals are International Journal of Remote Sensing and IEEE Transactions on Geoscience and Remote Sensing, with h-indices of 49 and 31, respectively, and m-indices of 93 and 55, respectively. The journal with the highest impact factor (IF) is IEEE Transactions on Geoscience and Remote Sensing, with a score of 5.630, followed by the International Journal of Applied Earth Observation and Geoinformation, with 4.846. Among the top 10 remote-sensing journals, the newest journal, started in 2010, is in third position, namely Geocarto International, with an IF of 2.365. Unsurprisingly, the oldest journal, *International Journal of Remote Sensing*, received the highest number of citations, 15344, but there are considerable numbers of citations for the journal, *International Journal of Applied Earth Observation and Geoinformation*, at 10th position, with 2483 citations for 97 publications, and in second position on the basis of the IF.

The five years' percent growth of the top journals (in terms of publications per year) is presented in Figure 18.1. The International Journal of Remote Sensing has showed continuous growth from 1989 until 2011, at which point there was

TABLE 18.2
Top-Ranked Journals of Remote Sensing Selected by Indian Scholars

Rank	Name of Journal	No. of Publications	Total Citations	IF	*PY_start
1	International Journal of Remote Sensing	937	15344	2.493	1989
2	Journal of the Indian Society of Remote Sensing	791	3870	0.869	2004
3	Geocarto International	205	1206	2.365	2010
4	IEEE Transactions on Geoscience and Remote Sensing	168	3608	5.630	1990
5	IEEE Geoscience and Remote Sensing Letters	145	1288	3.534	2004
6	IEEE Journal of Selected Topics in Applied Earth Observations and Remote Sensing	136	1316	3.392	2009
7	Marine Geodesy	114	587	0.962	1989
8	Radio Science	108	1328	1.658	1989
9	Journal of Applied Remote Sensing	99	407	1.344	2007
10	International Journal Of Applied Earth Observation and Geoinformation	97	2483	4.846	2007
	Other (40)	579	10913		

IF=Impact Factor as per JCR 2018, *PY_Start=publication year start

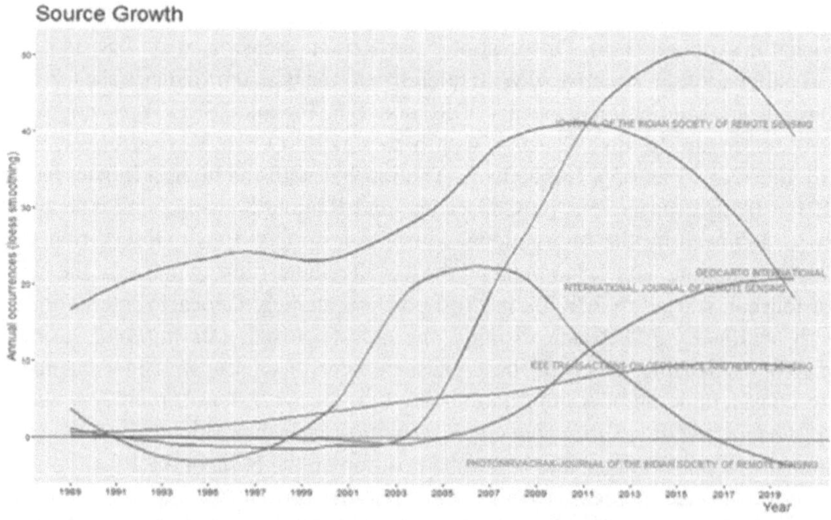

FIGURE 18.1 Growth of top remote sensing journals.

TABLE 18.3
Top-Cited Remote Sensing Papers, along with Citation Rate per Year

Title of the Paper	Total Citations	Total Citations per Year
Singh A, 1989, Int J Remote Sens	1742	54.44
Pal M, 2005, Int J Remote Sens	646	40.38
Xiao Xm, 2006, Remote Sens Environ	371	24.73
Pal M, 2010, IEEE T Geosci Remote	366	33.27
Nagendra H, 2001, Int J Remote Sens	327	16.35
Berthier E, 2007, Remote Sens Environ	269	19.21
Rao Pb, 1995, Radio Sci	238	9.15
Foody Gm, 2006, Remote Sens Environ	235	15.67
Rodriguez-Galiano Vf, 2012, Remote Sens Environ	232	25.78
Saraf Ak, 1998, Int J Remote Sens	227	9.87
Jat Mk, 2008, Int J Appl Earth Obs	219	16.85
Thenkabail Ps, 2009, Int J Remote Sens	203	16.92
Atkinson Pm, 2012, Remote Sens Environ	201	22.33
Mayaux P, 2006, IEEE T Geosci Remote	189	12.60
Kasetkasem T, 2005, Remote Sens Environ	188	11.75
Saha Ak, 2002, Int J Remote Sens	183	9.63
Bandyopadhyay S, 2007, IEEE T Geosci Remote	169	12.07
Maulik U, 2003, IEEE T Geosci Remote	168	9.33
Krishnamurthy J, 1996, Int J Remote Sens	168	6.72
Foody Gm, 2006, Remote Sens Environ-A	166	11.07

a decrease in publication numbers. The journal in second position is Journal of the Indian Society of Remote Sensing, which started publication in 2004, following which no articles were published for four years, but, since 2009, there has been continuous growth in the number of articles. The journal in third place is Geocarto International, which started publication in 2010, since when the number of articles has increased in a haphazard, zig-zag way, though maintaining its popularity with 205 publications over the study period. IEEE Transactions on Geoscience and Remote Sensing started publishing in 1990 but published very few articles, ranging from 0 to 16 every year, with a total number of publication of 168. The IEEE Geoscience and Remote Sensing Letters is in fifth place, starting publication in 2004.

Table 18.3 presents the top cited papers along with the total number of citations and the total number of citations per year. The top five cited papers are: A Singh (1989), Int J Remote Sens, M Pal (2005), Int J Remote Sens, Xm Xiao (2006), Remote Sens Environ, M Pal (2010), IEEE T Geosci Remote, M Pal (2010), IEEE T Geosci Remote, and H Nagendra (2001), Int J Remote Sens, with 1742, 646, 371, 366 and 327 citations, respectively. With respect to total number of citations per year, the top five papers are: A Singh (1989), Int J Remote Sens, 'A Singh A (1989), Int J Remote Sens, M Pal (2010), IEEE T Geosci Remote, Vf Rodriguez-Galiano (2012), Remote

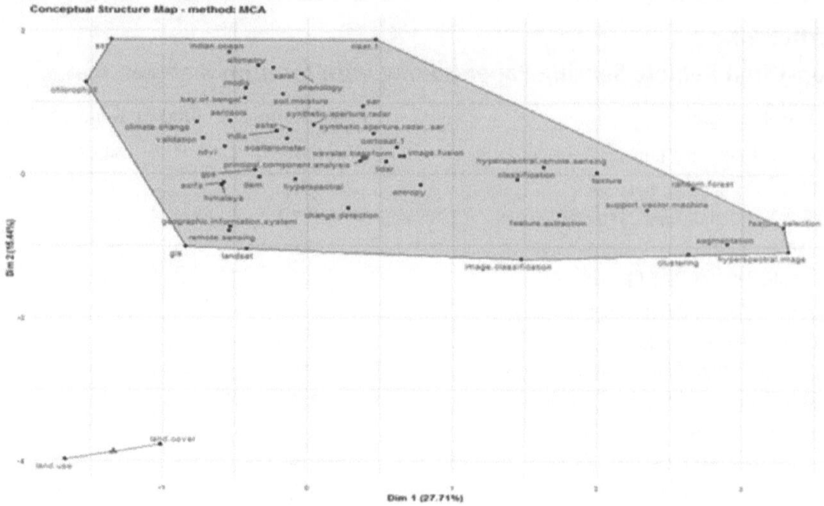

FIGURE 18.2 Conceptual networks of author keywords.

Sens Environ and 'Xm Xiao (2006), Remote Sens Environ', with 54.44, 40.38, 33.27, 25.78 and 24.73 citations per year, respectively.

The keywords of a paper give the essence of the research. Figure 18.2 gives the map of the top keywords assigned by the authors. Research on remote sensing revolves around geographic information system (GIS), Himalayas, GPS, Landsat, principal component analysis (PCA), wavelet transform, climate change and so on. The keywords clearly reflect that the research in remote sensing field is interdisciplinary in nature, as it covers the terminology of geography, electronics, computers, etc.

India has carried out collaboration on remote sensing with 73 countries. The highest level of collaboration is with the United States, with 211 collaborative documents, followed by the United Kingdom, France, Germany and the Netherlands, with 66, 63, 61 and 49 contributions. Italy and Japan have 46 contributions each, Australia, Canada, China and Thailand have 35, 31, 27 and 17 contributions, respectively, with Brazil and Spain having 16 contributions each, followed by Sweden, Norway and Vietnam with 15, 13 and 11 contributions. Greece and Korea have ten papers each, along with India. Malaysia, Sri Lanka and Taiwan have nine papers each, with the Philippines having eight papers. There are a further eight countries with seven articles, six countries with four articles, five countries with two articles, four countries with five articles, three countries with three articles, 11 countries with two articles and 22 countries with one article each (Figure 18.3).

Figure 18.4 demonstrates the mapping of the top-ranked organizations, authors and journals. The author RS Dwvedi, of the National Remote Sensing Agency, published the paper with the greatest number of citations in the International Journal of Remote Sensing. The top Indian organizations (in decreasing order) for publications are the National Remote Sensing Agency, the Indian Institute of

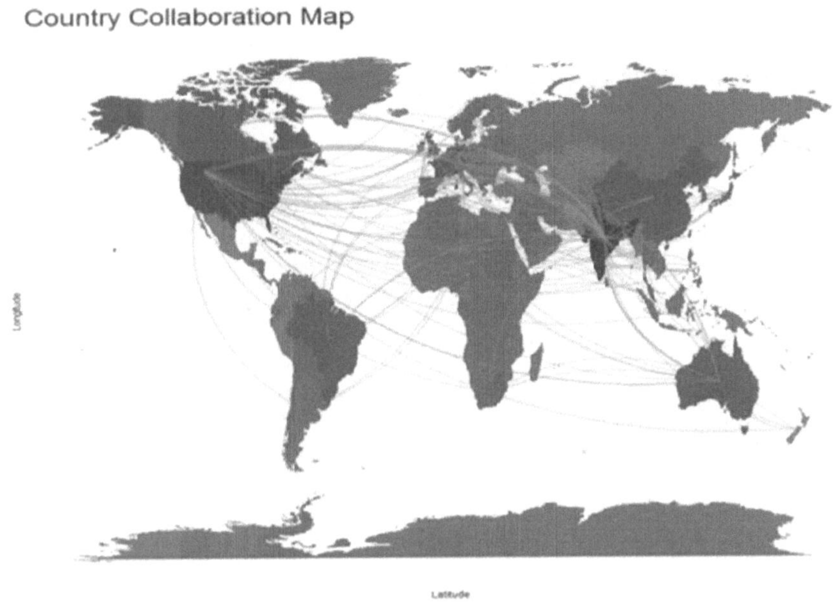

FIGURE 18.3 Country collaboration map.

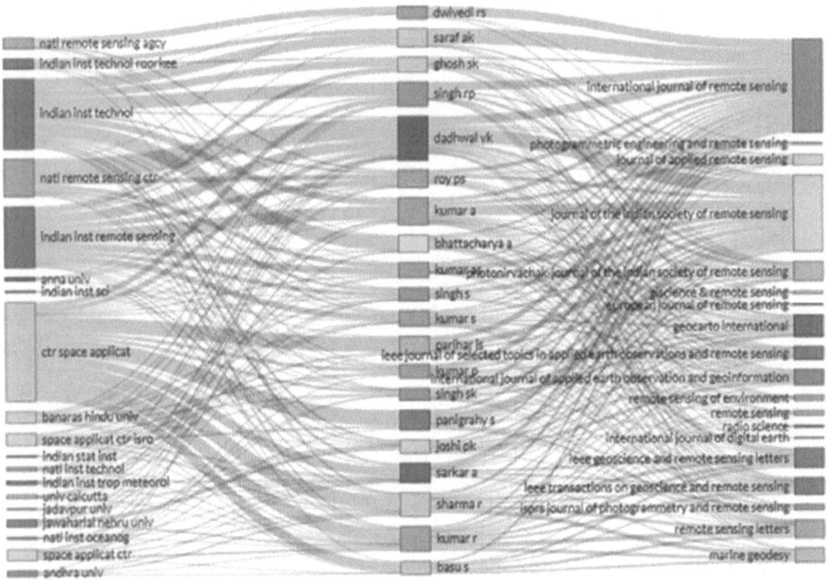

FIGURE 18.4 Collaborations with and contributions of top-ranked organisations, authors and journals.

Technology, Roorkee, the Indian Institute of Remote Sensing, Anna University, the Centre of Space Application, the Banaras Hindu University, and the Indian Space Research Organization. The highest-ranked authors (in terms of citations) affiliated to these organizations are RS Dwidedi, AK Saraf, SK Ghosh, RP Singh, VK Dadhwal, PS Roy, A Kumar and S Bhattacharya. The most-popular journals (in decreasing order in terms of citations) are the International Journal of Remote Sensing, Photogrammetric Engineering and Remote Sensing, the Journal of Applied Remote Sensing, and the Journal of the Indian Society of Remote Sensing.

18.4 CONCLUSION

Remote sensing is the science of obtaining information about objects or areas from a distance, typically from aircraft or satellites. It has its roots in many subjects. In the present work, the research trend of remote sensing in India has been analyzed by extracting the data from WoS for the period of 1989–2019. In total, there were 3282 documents, including 3096 research papers. These research documents were contributed by 5553 authors. There were only 180 single-authored documents, which clearly reflects the on-going collaborative nature of remote-sensing research. The top two journals for publication of research on remote sensing are the International Journal of Remote Sensing and the Journal of The Indian Society of Remote Sensing. The top three keywords of the remote-sensing field are geographic information system, Himalaya and GPS. India exhibits high levels of collaborative research, with 73 other countries in total. The collaboration index is 1.75. The greatest degree of collaboration is with the USA, in the form of 211 research papers.

REFERENCES

1. Ran, Lingyan, Zhang, Yanning, Wei, Wei, & Zhang, Qilin 2017. A hyperspectral image classification framework with Spatial Pixel Pair features. *Sensors*, 17(10): 421 doi:10.3390/s17102421. PMC 5677443. PMID 29065535.
2. Schowengerdt, Robert A. 2007. *Remote sensing: Models and methods for image processing* (3rd ed.). Academic Press. p. 2. ISBN 978-0-12-369407-2.
3. Guo, Huadong et al. 2013. Spatiotemporal analysis of urban environment based on the vegetation–impervious surface-soil model. *Journal of Applied Remote Sensing*, 8(1), 084597. https://doi.org/10.1117/1.JRS.8.084597
4. Schott, John Robert 2007. *Remote sensing: The image chain approach* (2nd ed.). Oxford University Press. p. 1. ISBN 978-0-19-517817-3.
5. Mills, J.P. et al. 1997. Photogrammetry from archived digital imagery for seal monitoring. *The Photogrammetric Record*, 15(89): 715–724. doi:10.1111/0031-868X.00080.
6. Twiss, S.D. et al. 2001. Topographic spatial characterisation of grey seal Halichoerus grypus breeding habitat at a sub-seal size spatial grain. *Ecography*, 24(3): 257–266. doi:10.1111/j.1600-0587.2001.tb00198.x.
7. Liu, Jian Guo, & Mason, Philippa J. 2009. *Essential Image Processing for GIS and Remote Sensing*. Wiley-Blackwell. p. 4. ISBN 978-0-470-51032-2.

8. Engler, Steven 2014. Bibliometrics and the study of religion\s1. *Religion*, 44(2): 193–219, doi:10.1080/0048721X.2014.893680
9. Mohan, Brij, Singh, H.P., & Singh, S. 2017. Education, research and practice in remote sensing and GIS: A study of Ph.D researches conducted in Indian Universities and Institutions from 1980 to 2016. *International Journal of Applied Business and Economic Research*, 15(21, Part 2): 509–519.
10. Chaman Sab M., P. Dharani Kumar, & B. S. Biradar. 2017. Remote sensing research in India: An analysis of publications output during 2011–2015. *International Journal of Library and Information Studies*, 7(4): 264–271.
11. Rajgoli, U., Mamdapur, N, & Ponniah, P. 2017. A bibliometric study of journal of the Indian society of remote sensing for the period 1973–2014. *Journal of Advances in Library and Information Science*, 1(6): 52–57.
12. Ram, Shri. 2015. A bibliometric assessment of liquorice (Glycyrrhiza glabra) research trends. *Annals of Library and Information Studies*, 62(1): 27–32.
13. Rana, Madan Sigh. 2010. *Scientometric Study of Wild Mammal Research in India: Authorship, Distribution and Research Trend.* LAP Lambert Acad. Publ.
14. Viana, Joao et al. 2017. Remote sensing in human health: A 10-year bibliometric analysis. *Remote Sensing*, 9(12): 1225; https://doi.org/10.3390/rs9121225

Epilogue

Future Research Agenda

Ripudaman Singh

In addition to offering innovative and exciting themes on applications of remote sensing and geospatial technologies, it is hoped that the present volume will inspire new researchers, with respect to approaches to be taken in novel and upcoming fields, domains and subjects with advanced applications of remote sensing. It is also anticipated that all the eighteen chapters presented in the book would serve as lighthouses to the young and enthusiastic researchers, in their respective fields of investigation, to apply remote sensing and geospatial technologies to create new knowledge and information, and to help address some of the major issues over the globe.

The present volume contains chapters on almost all major fields of remote-sensing applications, with the exception of the two broad groups of urban studies, and disaster management, models and planning. Because of the limited space in the current book, these topics will be addressed in another volume supplementary to this publication. Advances in satellite and information technologies have revolutionized the collection, production and analysis of geospatial data. In addition to the *geographers*, other groups of *scientists/technologists*, such as *agriculturists, anthropologists, archaeologists, botanists, civil engineers, computer scientists, forestry researchers, geologists, hydrologists, meteorologists, soil scientists, town planners,* etc. all are applying remote-sensing techniques to their respective researches and analyses.

As elaborated in the Prologue at the beginning of this book, the first airborne photographs were taken by *Gaspard-Felix Tournachon* (the French balloonist), who captured images of Paris in 1858 from a hot-air balloon, laying the foundations for aerial photography and remote sensing. Although such photographs were taken purely out of curiosity, to see '*how it looks like from above*', the information gathered was more useful than expected, and could be used for practical purposes. During World Wars I and II, for example, air photography was deployed and, with the development of satellite technologies during the Cold War era, improved sensors and global positioning system (GPS) further improved the precision of remote sensing and expanded its applications. From the early black-and-white photographs, taken with the aid of balloons, kites and pigeons; to the multispectral digital satellite imageries being taken by hundreds of satellites over different parts of the world every minute of the day, remote sensing has witnessed great transformations. With the *digital revolution*, remote sensing is improving every day, with greater resolution and wider analytical abilities for varied and diverse applications. Truly, these satellites are proving to be '*the human eyes in space*', and no place on Earth remains uncaptured, either in the day or at night, by these satellites and associated digital devices.

The use of these digital satellite imageries, taken through various wavelength regions of the electromagnetic spectrum, are providing cutting-edge information and analysis, and have a wide range of applications. Figure E.1 depicts the advances in remote-sensing technologies and their applications. It can be generalized that, in the early days of low-level remote-sensing technologies, black-and-white photographs were taken with the aid of balloons, kites or pigeons. With the development of aircraft, remote-sensing technologies moderated to panchromatic aerial photographs, usually limited to visible light. With further advances, satellites started capturing multispectral imageries in infrared, near-infrared and thermal infrared wavelengths during the 1960s, 1970s and 1980s, respectively. During the 1990s, radars and Lidars, using hyperspectral remote sensing with microwaves, further broadened the applications of remote sensing. With the commercialization of remote-sensing satellites and geospatial data products, the 21st century ushered in further advances in remote-sensing applications. At present, remote sensing is in the phase of advancing interactions with cloud computing, big data and artificial intelligence, and these interactions will strengthen and develop further in the future.

The crux of the future research agenda, with respect to re-envisioned remote-sensing applications, is illustrated in Figure E.2. With advances in computer technologies, the digital and information revolutions are opening up oceans (as well as skies) of digital knowledge and information, with enhanced capabilities for capturing, storing and analyzing boundless geospatial data. It is also expected that this increased availability of new information and technologies will generate new combinations and new applications. The future will see increased remote-sensing applications in combination with big data, cloud computing, machine learning and the Internet of Things. Open-source remote sensing is bound to open up new

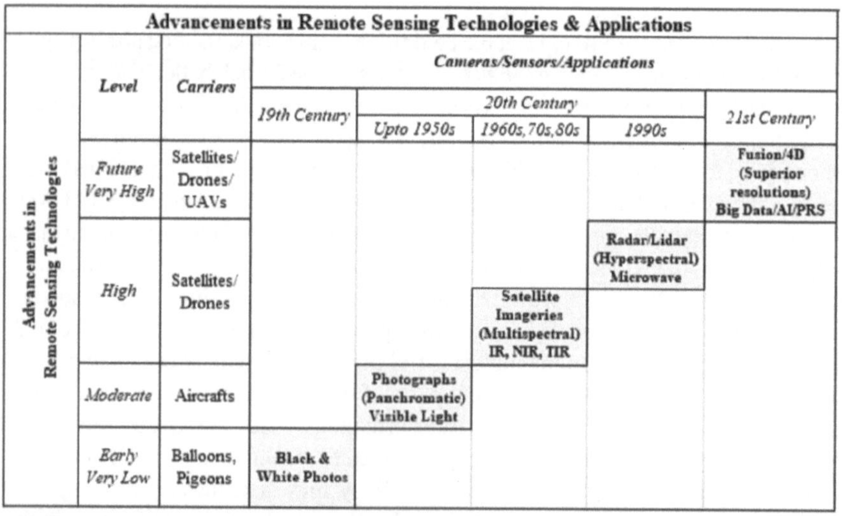

FIGURE E.1 Advancements in Remote Sensing Technologies & Applications.

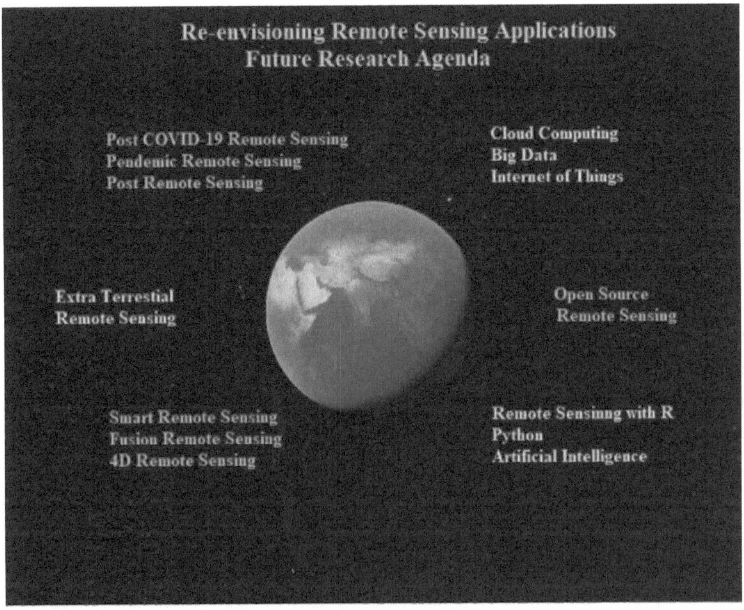

FIGURE E.2 Re-envisioning Remote Sensing Applications: Future Research Agenda.

vistas with the use of programming languages, such as R, Python and artificial intelligence (AI) and so on. Smart remote sensing is envisaged, with wider applications and interfaces with smart devices and technologies. Various civilian uses of GPS in automobiles and smartphones are also introducing smart GIS and smart remote- sensing applications through smart gadgets and devices. Fusion of remote sensing with other spatial/non-spatial data sciences is anticipated, to obtain high-quality spatio-temporal remote-sensing products and analysis. Remote sensing will re-create itself through 4D and higher-order technologies, as well as expanding its applications to extra-terrestrial domains.

Re-envisioned remote-sensing applications for the future research agenda include remote- sensing interfaces with cloud computing, big data, Internet of Things (IoT) and so on. As the availability of geospatial data increasing daily to the petabyte level, its applications, analyses and value to research are bound to expand in the coming years. Big data, cloud computing, data mining and machine learning in remote-sensing applications are emerging as new and upcoming fields (Dey et al., 2018; Sedona et al., 2019). Wang et al. (2020) analyzed remote-sensing cloud computing in detail, from the perspectives of cloud data acquisition to cloud storage and product analysis. These ever-growing geospatial data, and their applications to new disciplines, such as the social sciences, will allow, for example, the application of big data in remote sensing to fields such as disaster management, natural hazard monitoring, global climate change, smart city planning and other diverse domains. The interface of advanced technologies with remote sensing, such as with the IoT, will be useful in controlling numerous activities in smart learning environments, including ecological modelling (Mathews et al., 2019). Smart digital devices, like smart light-emitting diodes (LEDs), computing devices, smart mobile apps, tracking sensors and others in the learning environment can be useful in achieving smart learning. Integration of IoT with remote sensing can provide numerous prospects for teachers and learners to conduct research and training and to network knowledge in educational institutes, such as schools, colleges and universities. Such integrated remote-sensing technologies can digitally connect all the stakeholders, like academics, counsellors, students and teachers, in a single platform, and deliver appropriate monitoring and learning. With such big data applications, it is also being used from safety and security perspectives. As a consequence, remote-sensing applications to policing and traffic monitoring through big data and IoT will increase in the future.

The availability of open source and open access geospatial data are leading to the processes and applications of remote sensing in deep learning (Wegmann et al., 2016; Grizonnet et al., 2017; Cresson, 2020). Programming languages, such as R, Python and AI, are also being used for processing and classifying remote-sensing data and images, such as processing data in R (Ghosh and Hijmans 2019), using machine-learning algorithms (Kamusoko, 2019), and image analysis classification using Python for various applications (Morton, 2014). Wider applications of AI to remote-sensing data are going to generate actionable intelligence for businesses and organizations in the future as well. Smartphones and smart technologies are going to use larger geospatial data, and smart remote sensing is envisioned to have wider applications and interactions with smart devices and technologies. Various civilian uses of GPS in automobiles and smartphones are also achieving access to smart GIS and smart remote-sensing usages through smart gadgets and devices (Martinez et al., 2019). Now that smartphone apps, which can transform smart phones into remote-sensing devices, are available which can be used with drones and unmanned aerial vehicles (UAVs) for instant data capture. The fusion and integration of data have been envisioned as a promising means for obtaining high-quality geospatial and temporal datasets (Alparone et al., 2015). In the future, remote sensing will evolve, particularly through 4D technologies (Yang et al., 2016), and expanding its applications to extra-terrestrial domains as well. Remote sensing of extra-terrestrial domains is anticipated as more and more satellites are reaching the Moon, Mars and beyond. Such preliminary research, applying remote sensing to the search for water, minerals and related substances on other planets and moons, are in the pipeline and will expand further (Athanassas et al., 2018; Liu et al., 2020; Wu et al., 2020). The future applications of remote sensing are envisioned to reach a stage of '*post-remote sensing*', where traditional remote-sensing applications will be merging into the innovative arenas of very big data, extra-terrestrial domains as well as ultra-smartphone-based analysis and applications. On the one hand, it will be relying more on various data sciences and focusing on other planets, with side-by-side geospatial data would be readily analyzed and processed in next-generation ultra-smartphones and devices.

Eventually, it is anticipated that the current COVID-19 pandemic will also have its impacts on remote-sensing applications in the coming years. Additional remote-sensing applications to analyze the impacts of the pandemic are imminent. Global lockdowns, due to COVID-19, have improved environmental conditions, and air pollution levels have decreased drastically all over the world. This radical departure from the norm certainly opens up research avenues for comparing environmental patterns in the pre-, current and post-COVID-19 pandemic periods, using remote sensing. Comparison of such patterns and change detections from earlier, pre-COVID-19 and lockdown period satellite data will be a major focus of upcoming investigations. Preliminary observations by INSAT-3D confirm decreases of 28–40% in aerosol levels in the air across India from the pre-lockdown (15 to 24 March 2020) period to the lockdown (25 March to 5 April 2020) period (IIRS, 2020). Similar observations have also been made for European regions and the world as a whole (Wolters, 2020). The role of remote sensing *via* satellite monitoring is becoming vital in restricting the spread of corona infections, through identifying risk "hot zones", and facilitating fast responses to the pandemic situations world over. Earth observation (EO) data may shed new light on socio-economic impacts of the COVID-19 pandemic, but restrictions on sharing of such vast EO data with developing countries, like India, need to be re-assessed. The current pandemic may prove to be an eye-opener for improving such constraints in remote-sensing policies in developing countries (Ashok and Basu, 2020). This pandemic has increased the need for monitoring (GEO, 2020). Organizations with deployed remote-sensing technologies and wireless monitoring systems have access to critical infrastructures, which could be checked from anywhere and at any time throughout the world. The need for remote sensing and its applications have increased manifold in this pandemic situation (Zaragoza, 2020). Re-envisioning remote-sensing applications in such a pandemic scenario will look for EO to support response and recovery in various spheres, including health, socio-economic impacts, monitoring and surveillance, etc. It would be appropriate to complete '*Re-envisioning Remote-Sensing Applications: Perspectives from Developing Countries*' with the words of Sarah Parcak (2009), a world-renowned anthropologist and archaeologist, who applies remote-sensing techniques in discovering and identifying ancient archaeological sites of the Roman Empire and Egypt. For her, '*Google Earth is an incredible resource because, from hundreds of miles in space, we can zoom in, and we can find things. Everyone looks for their house first. That is the tip of the iceberg with remote sensing*'. Further, '*it is one of the methods of many to model past landscapes, to answer where we came from and to see far better where we are going*'.

REFERENCES

Alparone, L. et al. (2015). *Remote Sensing Image Fusion (Signal and Image Processing of Earth Observations).* CRC Press, Taylor & Francis Group: Florida.

Ashok, G.V. and Basu, P. (2020). COVID-19 Pandemic- Eye Opener for Better Remote Sensing Policies in India? *Geospatial World.* Retrieved from https://www.geospatialworld.net/blogs/covid-19-pandemic-eye-opener-for-better-remote-sensing-policies-in-india/

Athanassas, C.D. et al. (2018). Remote Sensing of Mars: Detection of Impact Craters on the Mars Global Surveyor DTM by Integrating Edge-and Region-Based Algorithms. *Earth Moon Planets*, 121: 59–72. https://doi.org/10.1007/s11038-018-9515-3

Cresson, B. (2020). *Deep Learning for Remote Sensing Images with Open Source Software*. CRC Press: Boca Raton.

Dey, N., Bhatt, C. and Ashour A.S. (2018). *Big Data for Remote Sensing: Visualization, Analysis and Interpretation: Digital Earth and Smart Earth*. Springer: Cham.

GEO (2020). *GEO Community Response to COVID-19*. Group on Earth Observations. Retrieved from https://earthobservations.org/covid19.php

Grizonnet, M. et al. (2017). Orfeo ToolBox: Open Source Processing of Remote Sensing Images. *Open Geospatial Data, Software and Standards*, 2(15): 1–8. doi:10.1186/s40965-017-0031-6

Ghosh, A. and Hijmans, R.J. (2019). Remote Sensing Image Analysis with R. *Rspatial.org*. Retrieved from https://rspatial.org/rs/rs.pdf

IIRS (2020). *Space Based Observation on Changes in Air Quality During COVID-19 Lockdown Period. IIRS Update*. Indian Institute of Remote Sensing: Dehradun. Retrieved from https://www.iirs.gov.in/iirs_slide_page?s=67

Kamusoko, C. (2019). *Remote Sensing Image Classification in R*. Springer Nature: Singapore.

Liu, D. et al. (2020). An Empirical Abundance of Nanophase Metallic Iron (npFe0) in Lunar Soils. *Remote Sensing*, 12(6), 1047. https://doi.org/10.3390/rs12061047

Martinez del Horno, M. et al. (2019). Calibration of Wi-Fi-based Indoor Tracking Systems for Android Based Smartphones. *Remote Sensing*, 11(9): 1072 https://doi.org/10.3390/rs11091072

Mathew, C., Mathukutty, R. and Madhanan, P. (2019). Controlling of Greenhouse Parameters based on IoT and Remote Sensing. *International Journal of Engineering Research & Technology (IJERT)*, 7(5): 1–3. Retrieved from https://www.ijert.org/research/controlling-of-greenhouse-parameters-based-on-iot-and-remote-sensing-IJERTCONV7IS05011.pdf

Morton, J.C. (2014). *Image Analysis, Classification and Change Detection in Remote Sensing with Algorithms for ENVI/IDL and Python*. CRC Press: Boca Raton.

Parcak, S.H. (2009). *Satellite Remote Sensing for Archaeology*. Routledge: New York.

Sedona, R. et al. (2019). Remote Sensing Big Data Classification with High Performance Distributed Deep Learning. *Remote Sensing*, 11(24): 3056. https://doi.org/10.3390/rs11243056

Wang, L., Yan, J. and Ma, Y. (2020). *Cloud Computing in Remote Sensing*. CRC Press, Taylor & Francis Group: Florida.

Wegmann, M., Leutner, B. and Dech, S. (2016). *Remote Sensing and GIS for Ecologists Using Open Source Software*. Pelagic Publishing: Exeter UK.

Wolters, E. (2020). *Air Quality before, during and After COVID-19 Lockdown*. Vito Remote Sensing. Retrieved from https://blog.vito.be/remotesensing/air-quality-monitoring-before-during-and-after-covid-19-lockdown

Wu, K. et al. (2020). Simulation Study of Moon-Base InSAR Observation for Solid Earth Tides. *Remote Sensing*, 12(1), 123. https://doi.org/10.3390/rs12010123

Yang, C.-H., Kenduiywo, B.K. and Soergel, U. (2016). 4D Change Detection Based on Persistent Scatterer Interferometry. Pattern Recognition in Remote Sensing (PRRS). *2016 9th IAPR Workshop on Pattern Recognition in Remote Sensing*. Cancun, Mexico. doi:10.1109/PRRS.2016.7867016

Zaragoza, A. (2020). Remote Sensing in times of COVID-19: Just What the Doctor Ordered. *Worldsensing*. Retrieved from https://blog.worldsensing.com/industrial-iot/remote-sensing-covid19/

Index